Building Industries at Sea:
'Blue Growth' and the
New Maritime Economy

RIVER PUBLISHERS SERIES IN RENEWABLE ENERGY

Series Editor

ERIC JOHNSON
Atlantic Consulting
Switzerland

Indexing: All books published in this series are submitted to the Web of Science Book Citation Index (BkCI), to CrossRef and to Google Scholar.

The "River Publishers Series in Renewable Energy" is a series of comprehensive academic and professional books which focus on theory and applications in renewable energy and sustainable energy solutions. The books serve as a multidisciplinary resource linking renewable energy with society, fulfilling the rapidly growing worldwide interest in energy solutions. All fields of renewable energy and their possible applications are addressed, not only from a technical point of view, but also from economic, social, political, and financial aspects.

Books published in the series include research monographs, edited volumes, handbooks and textbooks. They provide professionals, researchers, educators, and advanced students in the field with an invaluable insight into the latest research and developments.

For a list of other books in this series, visit www.riverpublishers.com

Building Industries at Sea:
'Blue Growth' and the
New Maritime Economy

Editors

Kate Johnson

Heriot Watt University
UK

Gordon Dalton

University College Cork
Ireland

Ian Masters

Swansea University
UK

LONDON AND NEW YORK

Published 2018 by River Publishers

River Publishers
Alsbjergvej 10, 9260 Gistrup, Denmark
www.riverpublishers.com

Distributed exclusively by Routledge

4 Park Square, Milton Park, Abingdon, Oxon OX14 4RN
605 Third Avenue, New York, NY 10158

First published in paperback 2024

Building Industries at Sea: 'Blue Growth' and the New Maritime Economy / by Kate Johnson, Gordon Dalton, Ian Masters.

Routledge is an imprint of the Taylor & Francis Group, an informa business

ISBN: 978-87-93609-26-6 (hbk)
ISBN: 978-87-7004-403-5 (pbk)
ISBN: 978-1-003-33743-0 (ebk)

DOI: 10.1201/9781003337430

Contents

Marcel J. C. Rozemeijer, Sander W. K. van den Burg,
Robbert Jak, Laura E. Lallier and Karel van Craenenbroeck

PART II: The Blue Economy Sectors

9 Tourism   **285**

*Dimitrios Pletsas, Sara Barrento, Ian Masters
and Jack Atkinson-Willes*

PART III: Planning by Sea Basin

10 Regulation and Planning in Sea Basins – NE Atlantic   313

Anne Marie O'Hagan

Kate Johnson, Gordon Dalton and Ian Masters

Preface

The purpose of this book is to publish a detailed analysis of prospective (Blue Growth) and established (Blue Economy) maritime business sectors. The contents of the book are based on deliverables and output material from the MARIBE (Marine Investment for the Blue Economy) project. This EU H2020 funded research project was completed in 2016. It identified the key technical and non-technical challenges facing maritime industries and placed them into the social and economic context of the coastal and ocean economy. Working with industry, MARIBE developed business plans and real projects for the combination of marine industry sectors into multi-use platforms (MUPs). MARIBE is the first extensive study to compare and contrast the traditional Blue Economy with the Blue Growth newcomers.

Throughout the world there is evidence of mounting interest in marine resources and expansion of maritime industries to create jobs and economic growth. Energy and food security are key priorities. Expanding populations, insecurity of traditional sources of supply and the effects of climate change add urgency to the need to address and overcome the challenges of working in the maritime environment. Five promising areas of activity for 'Blue Growth' have been identified at European Union policy level including Aquaculture; Renewable Energy (offshore wind, wave and tide); Seabed Mining; Blue Biotechnology; and Tourism. Work is well advanced to raise the technological and investment readiness levels (TRLs and IRLs) of these growth industries drawing on the experience of the traditional maritime industries such as Offshore Oil and Gas; Shipping; Fisheries and an already established tourist sector. An accord has to be struck between policy makers and regulators anxious to encourage research and business incentives into sustainable development; and developers, investors and businesses anxious to reduce the risks of such innovative investments and ensure profitability.

In this book, sector experts working to a common template explain each of these industries and their capacity to combine into multi-use platforms.

Factors essential to prospective business plans are identified – market, structure, lifecycle, employment, innovation and investment. The book goes on to describe progress with reformed regimes of maritime governance within which these industries must operate. The introduction of new planning and regulatory regimes in four Sea Basins – North East Atlantic, Baltic and North Seas, Mediterranean and Caribbean – are examined.

Acknowledgements

Firstly, and most importantly, the editors would like to thank the many authors who have contributed to this book. The process of drawing together such a wide range of expertise has been an enriching and informative experience for us all. Secondly, this book was made possible by the MARIBE project, which received funding from the European Union's Horizon 2020 research and innovation programme under grant agreement No. 652629. Listed below are all the contributors to this manuscript. Some of the authors have been funded by other sources and individual acknowledgments, where required, are given after the author list.

Introduction – The EDITORS: Gordon Dalton (UCC); Kate Johnson (HWU); Ian Masters (Swansea)

Section A. The Blue Growth Sectors

1. Aquaculture – Tamás Bardócz (ABT); Henrice Jansen (WUR); Junning Cai (FAO); José Aguilar-Manjarrez (FAO); Sara Barrento (Swansea); Shane A. Hunter (ABT); Marnix Poelman (WUR)
2. Blue Biotechnology – Jane Collins (eCOAST, KU Leuven); Arianna Broggiato (eCOAST, ABS-int); Thomas Vanagt (eCOAST, ABS-int)
3. Seabed Mining – Marcel J.C. Rozemeijer (WUR); Sander van den Burg (WUR); Robbert Jak (WUR), Laura E. Lallier (eCOAST, UGent), Karel van Craenenbroeck (GeoMarEx)
4. Wave and Tidal Energy – Gordon Dalton (UCC)
5. Offshore Wind Energy – Mike Blanch (BVGA); Clare Davies (BVGA); Alun Roberts (BVGA)

Section B. The Blue Economy Sectors

6. Fisheries – Kate Johnson (HWU)
7. Offshore Oil and Gas – Irati Legorburu (HWU); Kate Johnson (HWU); Sandy Kerr (HWU)
8. Shipping and Shipbuilding – Irati Legorburu (HWU); Kate Johnson (HWU); Sandy Kerr (HWU)

9. Tourism and Recreation – Dimitrios Pletsas (Swansea); Sara Barrento (Swansea); Ian Masters (Swansea); Jack Atkinson-Willes (Swansea)

Section C. Planning by Sea Basin

10. Atlantic – Anne Marie O'Hagan (UCC)
11. Baltic Sea and North Sea – Hester Whyte (FEC/UCC); Shona Paterson (FEC/UCC)
12. Mediterranean and Black Sea – Christine Röckmann (WUR); Tomás Vega Fernández (CNR & SZN); Carlo Pipitone (CNR)
13. Caribbean – Irati Legorburu (HWU); Kate Johnson (HWU); Sandy Kerr (HWU)

Section D. Combining Uses

14. Multi Use Platforms (MUPs) and Multi Use of Space (MUS) – Gordon Dalton (UCC)

Endwords

The EDITORS: Gordon Dalton (UCC); Kate Johnson (HWU); Ian Masters (Swansea)

Affiliations

ABS-int, Technologiepark 3, B-9052 Zwijnaarde, Belgium

ABT – AquaBioTech Group, Central Complex, Naggar Street, MST 1761 Mosta, Malta G.C.

BVGA – BVG Associates, The Blackthorn Centre, Purton Road, Cricklade, Swindon, SN6 6HY, UK

CNR – Consiglio Nazionale delle Ricerche, CNR-IAMC, Via Giovanni da Verrazzano 17, 91014 Castellammare del Golfo, Italy.

eCOAST – eCOAST, Esplanadestraat 1, 8400 Ostend, Belgium;

FAO – UN Food and Agriculture Organization

FEC – Future Earth Coasts, MaREI Centre, Environmental Research Institute, University College Cork, Ringaskiddy, Co. Cork, Ireland

GeoMarEx – Geological Marine Exploration – Luchterenstraat – 9031 Ghent, Belgium

HWU – Heriot-Watt University, Old Academy, Stromness, Scotland KW16 3AW

KU Leuven – Faculty of Pharmaceutical Sciences, Clinical Pharmacology and Pharmacotherapy, KU Leuven, O&N II Herestraat 49 - box 521, 3000 Leuven, Belgium

Swansea – Marine Energy Research Group, Energy and Environment Research Group, Zienkiewicz Centre for Computational Engineering,

College of Engineering, Swansea University, Bay Campus, Swansea, SA1 8EN, UK; Biosciences Department, Centre for Sustainable Aquatic Research, College of Science, Swansea University, SA2 8PP, UK

SZN – Stazione Zoologica Anton Dohrn, Villa Comunale, 80121 Naples, Italy

UCC – MaREI Centre, Environmental Research Institute, University College Cork, Ringaskiddy,
Co. Cork, Ireland

UGent – Maritime Institute, Faculty of Law, University of Ghent, Universiteitstraat 4, 9000 Ghent, Belgium

WUR – Wageningen Marine Research, Stichting Wagening Research, WUR, P.O Box 68 1970 AB Ijmuiden, The Netherlands; Wageningen Economic Research, Stichting Wagening Research, WUR, P.O Box 29703, 2502 LS Den Haag The Netherlands

Individual Funding Acknowledgements

In addition to EU funding for the MARIBE project (Grant reference number 652629), the following agencies are acknowledged:

Gordon Dalton, Hester Whyte
Additional support was given by the Centre for Marine and Renewable Energy Ireland (MaREI), funded by Science Foundation Ireland (12/RC/2302)

Ian Masters
Additional support was given by EPSRC projects EP/N509826/1, EP/P008682/1, EP/M014738/1. The authors acknowledge the financial support of the Welsh Assembly Government and Higher Education Funding Council for Wales through the Sêr Cymru National Research Network for Low Carbon, Energy and Environment.

Anne Marie O'Hagan
This material is based upon works supported by the Centre for Marine and Renewable Energy Ireland (MaREI), funded by Science Foundation Ireland (12/RC/2302).

Shona Paterson
This material is based upon works supported by Future Earth Coasts in collaboration with the Centre for Marine and Renewable Energy Ireland (MaREI), funded by Science Foundation Ireland (12/RC/2302)."

List of Contributors

Alun Roberts, *BVG Associates, UK*

Anne Marie O'Hagan, *MaREI Centre, ERI, University College Cork, Ireland*

Arianna Broggiato, *1. eCOAST, Esplanadestraat 1, 8400 Ostend, Belgium 2. ABS-int, Technologiepark 3, B-9052 Zwijnaarde, Belgium*

Carlo Pipitone, *Consiglio Nazionale delle Ricerche, Via Giovanni da Verrazzano 17, 91014 Castellammare del Golfo, Italy*

Clare Davies , *BVG Associates, UK*

Dimitrios Pletsas, *Swansea University, Wales*

Gordon Dalton, *University College Cork, Ireland*

Henrice Jansen, *Wageningen Marine Research, The Netherlands*

Hester Whyte, *Future Earth Coasts, MaREI Centre, ERI, University College Cork, Ireland*

Ian Masters, *Swansea University, Wales*

Irati Legorburu, *Heriot-Watt University, Scotland*

Jack Atkinson-Willes, *Swansea University, Wales*

Jane Collins, *1. eCOAST, Esplanadestraat 1, 8400 Ostend, Belgium 2. Faculty of Pharmaceutical Sciences, Clinical Pharmacology*

José Aguilar-Manjarrez, *Wageningen Marine Research, The Netherlands*

Junning Cai, *UN Food and Agricultural Organisation*

Karel van Craenenbroeck, *GeoMarEx – Geological Marine Exploration – Luchterenstraat – 9031 Ghent, Belgium*

Kate Johnson, *Heriot-Watt University, Scotland*

Laura E. Lallier, *eCOAST, Esplanadestraat 1, 8400 Ostend; Maritime Institute, Faculty of Law, University of Ghent, Universiteitstraat 4, 9000 Ghent, Belgium*

Marcel J. C. Rozemeijer, *WUR, Droevendaalsesteeg 4, 6708 PB Wageningen, the Netherlands*

Marnix Poelman, *Wageningen Marine Research, The Netherlands*

Mike Blanch, *BVG Associates, UK*

Robbert Jak, *WUR, Droevendaalsesteeg 4, 6708 PB Wageningen, the Netherlands*

Sander W. K. van den Burg, *1. WUR, Droevendaalsesteeg 4, 6708 PB Wageningen, the Netherlands*
2. eCOAST, Esplanadestraat 1, 8400 Ostend; Maritime Institute, Faculty of Law, University of Ghent, Universiteitstraat 4, 9000 Ghent, Belgium

Sandy A. Kerr, *Heriot-Watt University, Scotland*

Sara Barrento, *Swansea University, Wales*

Shane A. Hunter, *Aqua BioTech Group, Malta*

Shona Paterson, *1. Future Earth Coasts, MaREI Centre, ERI, University College Cork, Ireland*
2. Wageningen Marine Research, Stichting Wagening Research, WUR, P.O. Box 68 1970 AB IJmuiden, The Netherlands

Tamás Bardócz, *AquaBioTech Group, Malta*

Thomas Vanagt, *1. eCOAST, Esplanadestraat 1, 8400 Ostend, Belgium 2. ABS-int, Technologiepark 3, B-9052 Zwijnaarde, Belgium*

Tomás Vega Fernández, *1. Consiglio Nazionale delle Ricerche, Via Giovanni da Verrazzano 17, 91014 Castellammare del Golfo, Italy 2. Stazione Zoologica Anton Dohrn, Villa Comunale, 80121 Naples, Italy*

List of Figures

List of Tables

List of Abbreviations

ACS	Association of Caribbean States, ACS Group (industry)
BG	Blue Growth
BP	BP plc (industry)
CAPEX	Capital Expenditure
CFP	Common Fisheries Policy
CIESM	Mediterranean Science Commission
CSA	Communication and Support Action (H2020)
DG	Directorate General (EU)
DWT	Deadweight tonnage
EC	European Commission
EEZ	European Economic Zone
EMFF	European Maritime and Fisheries Fund
EPRS	European parliamentary research service
ERA	European Research Area (co-ordinated funding)
ESF	European Social Fund
EU	European Union
EUNETMAR	European Networking Group for Maritime Policy
EUROSTAT	EU statistics DG
EWEA	European Wind Energy Association
FAO	UN Food and Agriculture Organization
FIT	Feed in Tarrif
FPP	Floating Power Plant (Industry)
GDP	Gross Domestic Product
GES	Good Environmental Status
GVA	Gross Value Added
GW	GigaWatts
HELCOM	Helsinki Commission
ICZM	Integrated coastal zone management
IEA	International Energy Agency
IMO	International Maritime Organization

IMP	Integrated Maritime Policy
IRR	Internal Rate of Return
ISA	International Seabed Authority
JRC	Joint Research Centre
LCA	Life Cycle Analysis
LCOE	Levelised Cost of Energy
LNG	Liquified Natural Gas
LOSC	United Nations Convention on the Law of the Sea
MAP	Multi-annual management plans
MARE	EU Marine Directorate
MARIBE	Marine Investment for the Blue Economy
MPA	Marine Protected Area
MRE	Marine Renewable Energy
MSC	Marine Stewardship Council
MSFD	Marine Strategy Framework Directive
MSP	Marine Spatial Planning
MUP	Multi Use Platform
MUS	Multi Use of Space
MW, MWh	MegaWatt, MegaWattHour
NE	North East
NOC	National Oil Company
OCT	Overseas Countries and Territories
OECD	Organisation for Economic Co-operation and Development
OECS	Organisation of Eastern Caribbean States
OES	IEA Ocean Energy Systems
OPEX	Operational Expenditure
OR	Outermost Regions
OSPAR	Convention for the Protection of the Marine Environment of the North-East Atlantic
PRC	Policy Research Corporation
PTO	Power Take Off
RO, ROC	Renewables Obligation (UK), Renewables Obligation Certificate
RTDI	Research, Technological Development and Innovation
SEA	Strategic Environmental Assessment
SME	Small and Medium Enterprise
SMS	seafloor massive sulphide
SPC	Secretariat of the Pacific Community

TRL	Technology Readiness Level
TWh	TeraWatt hour
UN	United Nations
UNCLOS	United Nations Convention on the Law of the Sea
UNEP	UN Environment Programme
UNESCO	United Nations Educational, Scientific and Cultural Organization
UNWTO	World Tourism Organization
USGS	United States Geological Survey

Introduction

Ian Masters[1,*], Kate Johnson[2] and Gordon Dalton[3]

[1]Swansea University, Wales
[2]Heriot-Watt University, Scotland
[3]University College Cork, Ireland
*Corresponding Author

Background

For thousands of years the oceans have been highly prized and have provided us with efficient transport and a plentiful supply of food. Therefore, it seems obvious that our modern society should continue to use the oceans and maximize the benefits. There might be great treasures of valuable materials, new bio-compounds and endless energy. However society is reluctant to change and resistant to new ideas so it is often found that new uses are not being accepted as quickly as their advocates would like. This book aims to show that traditional uses of the sea can coexist alongside novel technology. Furthermore we present evidence that old and new ideas will complement each other and the whole will be greater than the sum of the parts.

The synthesis of established and growing industries at sea is covered by the overarching label of the Blue Economy. Traditionally this has been split into two subsectors, the marine and maritime sectors. The maritime sector is easier to define because it relates specifically to the transport of goods, including ports and shipping and the supply chain that enables these activities. The marine sector by contrast includes industrial and business activity at sea, and those activities required to enable them. In this book we define the blue economy in a different way, and consider nine sectors which are split according to their maturity. The established *blue economy* sectors are fisheries, offshore oil and gas, shipping and shipbuilding, tourism and recreation, while the new *blue growth* sectors are aquaculture, blue biotechnology, seabed mining, wave and tidal energy, offshore wind energy. Other sectors and sub-sectors do exist, but these nine give a broad overview of the whole.

More than 40% of the European population inhabits coastal areas, with approximately 5.4 million jobs and almost €500 billion of GVA a year resulting from activities directly related to the Blue Economy. The Blue Economy also largely supports aspects necessary for the welfare of the society such as trade and transport, food and health, energy and raw materials, labour and leisure, protection and environmental development. These activities not only have a direct impact on their own value chains, but they largely revert indirect benefits and very positively in other related economic sectors. In Europe, the existing blue economy has challenges to maintain itself as a healthy and profitable business sector. Taking each area in turn, the obvious issues are clear. Overfishing has reduced fish stocks in some areas to critical levels, reducing outputs. Offshore oil and gas is declining, due to reducing reserves and the global concerns over climate change. Shipping and ship building is a globalised sector and faces significant competition from countries with lower wage levels. Tourism makes up a significant part of the economy of many southern European countries and has been significantly hit by the global financial crisis, both within those countries' economies and with the reduced spending power of visitors. However, seeing that there are limits to growth for land based sectors, blue growth is seen as an opportunity for jobs and socio-economic development. Significant growth will not come from existing industries but from new industries that complement established revenue streams. For example, the building of the first offshore wind turbines in the North Sea was made possible by the skills and vessels used by the offshore oil and gas sector. The European Commission therefore includes Blue Growth as part of the Europe 2020 strategy (European Commission, 2014), which aims to make the most of Europe's seas potential in order to create long-term and sustainable socio-economic growth, while safeguarding the natural resources provided by the sea.

The purpose of this book is to publish the detailed analysis of each prospective and established maritime business sector. Sector experts working to a common template explain what these industries are, how they work, their prospects to create wealth and employment, and where they currently stand in terms of innovation, trends and their lifecycle. The book goes on to describe progress with the changing regulatory and planning regimes in the European Sea Basins including the Caribbean where there are significant European interests. The remainder of this introduction is concerned with the introduction of concepts that are common to many of the industrial sectors that are considered.

Lifecycle

Maritime industries have very different characteristics. Well-established activities, such as tourism or shipping employ a large number of people and create a high value added. Given the maturity of these sectors their main challenges relate to the development of new strategies, the adoption of sustainable practices or internationalisation. Sectors that are growing, such as aquaculture or offshore wind, have considerably increased their activity over the past years and show a good potential for development and employment generation. However, given their lack of consolidation in the markets, they still need policy support and various forms of investment. Finally sectors under development such as biotechnology, sea mining or certain renewable technologies (e.g., tidal, wave) will require significant investments in research, development and testing before society can take the advantage of their full commercial potential. Therefore, a good knowledge of the lifecycle stages and performance of these sectors provides a promising starting point for the creation of new business models. It also gives a complete image of the potential synergies and barriers that may arise between different economic sectors (technical, regulatory, social, environmental, etc.), which might be especially important for the development of combined and multi technology projects.

Innovation

The stereotypical view of entrepreneurs is the inventor who takes a brilliant idea and turns it into a multi-million-dollar industry. In reality, successful businesses rarely look like the original concept and key personnel will have changed as the required skills change. Having said that, start-up companies continue to be the most effective way to drive innovation forward. Large organisations may have the spare capital to be able to afford to innovate, but they are focused on generating revenue from existing products and services and are reluctant to take risks with shareholder funds. Therefore, a recognised business model is for a start-up company to be funded by venture capital until a point where a product is developed and risk is reduced. This then allows a larger company to buy out the start-up and grow it into a revenue generating product line. Some of the businesses and sectors described in later chapters are early in the innovation cycle and therefore an overview is given here of the process. In particular, blue biotechnology is characterised by start-up companies, while many ocean energy companies are at the final prototyping and commercial scale up.

Development

The early stages of innovation are focused on ideas. Costs are low in this stage as there is a small number of paid staff. The focus is on the business plan, proving the business concept at lab scale and protection by filing patents. Funding is provided by the founders, business angels and innovation grants. The value of the company is in the Intellectual Property. On the strength of the management team, value proposition and the IP, venture capital funding is secured. A full time management team is employed and technical work creates prototypes and small scale operations. Some businesses may start to generate modest revenues. Early patents are taken through the expensive international stages and new patents are filed. The value of the business is in the IP and technical know-how, which has less risk than earlier stages.

Transition

The business is now at the transition from testing to revenue. This requires significant funding for full scale prototypes or trials, and may take several funding rounds. A skilled development team will be in place so the business is spending cash on salaries. Patent protection costs will be continuing together with funding for labs, workshops, etc. This is the most difficult stage for most start-up companies and the point where many fail. Even though the business may have significant sunk costs in the value of patents and prototypes; the capital requirements to reach a profit making situation may be larger than venture capital funds are comfortable with and the risks may not be low enough for corporate investment. Therefore, the monthly cash burn may simply mean that the business fails by running out of money – 'the valley of death'. An alternative threat of competitors appears, because patents are published and prototypes are in the public domain, other businesses will enter the same market and could overtake the business.

Upscaling and Revenue

First commercial projects are delivered which may not be profitable but do generate revenue. Confidence in the business builds with each successive project and the structure changes from innovation to delivery. Revenues increase gradually. The company is bought out by a larger corporate, or a market floatation takes place and the original founders may exit the business at this stage. Many books have been written about the innovation process and how to avoid 'the valley of death' and that will not be repeated here. In the context of Blue Growth, the ideas of new technology are quite attractive, but this leads to the danger of 'technology push', where a great idea exists but unless there is a market willing to pay for it, it will ultimately fail. Much more successful is 'market pull' where there is a genuine gap in the market for the value proposition of the company. In some sectors, offshore wind for example, this market has been created through government intervention in the form of a subsidy scheme, gradually reducing as the sector grows in scale. The Blue Growth case studies within the EU funded MARIBE (Marine Investment for the Blue economy project (https://maribe.eu) discuss a number of successful companies that have found appropriate niche markets willing to purchase their product. The other distinguishing feature of the successful Blue Growth businesses is that they are continuously raising funds through new rounds of investments to aid transition and avoid the risks of early failure.

Policy and Regulation in Europe

Making space for the Blue Economy, encouraging growth while avoiding conflict with the environment and other activities, requires a significant effort in public policy and regulation for good governance. Marine space is a commons, there is no private ownership. The preservation of public rights, like the right to navigate and fish, and the introduction of quasi private rights, like the right of an offshore wind farm company to exclusively occupy space, is a matter for cooperation, negotiation and politics and, in the end, enforcement.

There is a global move towards multi-objective marine spatial planning (MSP) designed to balance the competing objectives of maritime industries with each other and jointly with the environment. Europe has taken a lead initially with the Integrated Maritime Policy (IMP) developed over the early years of the new century and promoted by the EU marine directorate (DG MARE). The DG MARE interest is essentially economic. It is responsible for the Common Fisheries Policy and promotes the 'Blue Growth Agenda' through a series of policy pronouncements and measures. The environmental pillar of the IMP is the Marine Strategy Framework Directive (MSFD) which came into effect in 2008 and requires Good Environmental Status (GES) in EU waters by 2020. The MSFD is actually the responsibility of the EU environmental directorate (DG ENV). The MSFD calls for a transboundary approach to conservation with monitoring and measurement of indicators supported by a network of Marine Protected Areas (MPAs). MPAs do not necessarily inhibit sustainable development. Multi-use of even protected areas is encouraged provided the issues are fully understood and provided for. Even though the MSFD regulates use of marine resources, it does not regulate the industrial activities or applications of marine resources (Van der Graaf et al., 2012). The Water Framework Directive (WFD) also applies to coastal waters up to one nautical mile from shore. It aims to control pollution and improve water quality in a holistic way to provide 'Good Status'. It is organised by river basin, and so in several cases the management plan for a river and adjacent coastal waters will encompass more than one country.

Policy and regulation affect the Blue Economy industries most directly in the consenting procedures which are adopted for the licensing of new developments. Developers have to show the environmental, social and economic impacts of their proposals and their adherence to the requirements of the law, and frequently a high degree of acceptance by affected stakeholders and communities. This is most commonly achieved in their Environmental Impact Statement (EIA) and through lengthy consultation and participatory procedures, both of which are also required by legislation.

Combining Sectors

By focusing on the potential for combinations of economic activities, multi – use of space (MUS) and multi-purpose platforms (MPP), the EU funded MARIBE project explored new business models and investment opportunities to encourage the further development of the Blue Economy. However, there is no doubt that the development of new maritime industries is accompanied by an increase of the human pressures in the environment and their associated challenges: conflicts between (Kadiri et al., 2012) traditional and emerging industries; environmental, social and economic sustainability; regulatory conflicts. The development of MUS/MPP facilities may offer a relevant solution for many of these conflicts. The search for synergies between different economic sectors may increase the economic performance of activities, promoting at the same time a more efficient use of infrastructure and logistical resources. Furthermore, the grouping and combination of activities, enables marine spatial planning to facilitate an efficient and environmentally sustainable management of maritime industries.

At first glance the idea behind MUS/MPPs is rather simple, two or more maritime industries sharing the same space and infrastructure in order to optimise the use of space and benefit from operational savings. However, they are novel concepts, and as such their development will require either the creation of new business models or the participation and investment from both public and private agents. In order to optimise these investments and make MUS/MPPs commercially viable, a good knowledge of the maritime industries involved becomes crucial. The knowledge of their operational methods, strengths, weaknesses and potential for growth is not only useful for management planning, but also it opens the door for the consolidation of emerging maritime sectors. The extensive coastal areas of the EU cover a large part of the global maritime zones. While continental EU borders with two oceans (Atlantic and Arctic) and four seas (Baltic, North, Mediterranean and Black Sea Sea), many of the EU outermost regions (ORs) are located in the Caribbean Sea and the Indian Ocean. These regions not only differ in environmental conditions, but also in their regulatory and socio-economic frameworks. Hence, the design of business models for MUS/MPPs must take into account the specific characteristics, needs and opportunities of each region.

This book is motivated by the opportunities presented by the Blue Economy. In the chapters which follow, the opportunities of new and old industries and their combinations are discussed. This material should be read in the context of the concepts outlined above. Where is this industry in terms of its lifecycle? What is the business opportunity and how does that opportunity survive the 'valley of death' from concept to reality? How does that business need to be structured to align with policy? Finally, and perhaps most importantly, how can that opportunity combine sectors for greater benefit?

References

EUROPEAN COMMISSION 2014. Innovation in the Blue Economy: realising the potential of our seas and oceans for jobs and growth.

KADIRI, M., AHMADIAN, R., BOCKELMANN-EVANS, B., RAUEN, W. & FALCONER, R. 2012. A review of the potential water quality impacts of tidal renewable energy systems. *Renewable and Sustainable Energy Reviews*, 16, 329–341.

VAN DER GRAAF, A. J., AINSLIE, M., ANDRÉ, M., BRENSING, K., DALEN, J., DEKELING, R., ROBINSON, S., TASKER, M., THOMSEN, F. & WERNER, S. 2012. European Marine Strategy Framework Directive-Good Environmental Status (MSFD GES): Report of the Technical Subgroup on Underwater noise and other forms of energy. *Brussels*.

PART I

The Blue Growth Sectors

1

Aquaculture

**Tamás Bardócz[1,*], Henrice Jansen[4], Junning Cai[2],
José Aguilar-Manjarrez[4], Sara Barrento[3],
Shane A. Hunter[1] and Marnix Poelman[4]**

[1]AquaBioTech Group, Malta
[2]UN Food and Agricultural Organisation
[3]Swansea University, Wales
[4]Wageningen Marine Research, The Netherlands
*Corresponding Author

1.1 Introduction

As the world population is growing and poverty is gradually being alleviated, the world is searching for new sources of protein in order to guarantee food security. Aquaculture has been identified as a sector with high potential for increased protein production without excessive burdens on the ecosystem. Predictions by OECD-FAO for fish and seafood production and trade (OECD/FAO 2016) indicate that future growth in seafood production will originate from aquaculture. Although 70% of the globe consists of water, aquaculture cannot be practised everywhere; it requires a unique set of natural, social and economic resources which must be used wisely if development of the sector is to be sustainable. In the EU and around the globe, the availability of areas suitable for aquaculture is becoming a major problem for the development and expansion of the sector. Care must be taken in the management of existing aquaculture facilities and the setting up of new production sites to ensure that there are appropriate environmental characteristics and that good water quality is maintained. Additionally, the consequences of social interactions and the appropriation of marine, coastal and inland resources must be well understood. In this chapter, we outline the status of this industry and discuss the key issues and opportunities. Firstly, we define what we mean by marine aquaculture.

Definitions

1. EU definition: 'aquaculture' means the rearing or cultivation of aquatic organisms using techniques designed to increase the production of the organisms in question beyond the natural capacity of the environment, where the organisms remain the property of a natural or legal person throughout the rearing and culture stage, up to and including harvesting; (REGULATION (EU) No 1380/2013)

2. FAO definition: Aquaculture is the farming of aquatic organisms including fish, molluscs, crustaceans and aquatic plants. Farming implies some sort of intervention in the rearing process to enhance production, such as regular stocking, feeding, protection from predators, etc. Farming also implies individual or corporate ownership of the stock being cultivated, the planning, development and operation of aquaculture systems, sites, facilities and practices, and the production and transport.[1]

3. According to FAO glossary of aquaculture, mariculture is cultivation, management and harvesting of marine organisms in the sea, in specially constructed rearing facilities e.g. cages, pens and long-lines. For the purpose of FAO statistics, mariculture refers to cultivation of the end product in seawater even though earlier stages in the life cycle of the concerned aquatic organisms may be cultured in brackish water or freshwater or captured from the wild. This term is interchangeable with marine aquaculture.

1.1.1 General Overview of the Sector

The Food and Agriculture Organization of the United Nations (FAO) produces regular authoritative statistical reports and publicly available databases (FishStatJ) on aquaculture sectors and subsectors, which include production volumes and values. This data is drawn on in this section, and refer the reader to their reports for further information (FAO, 2016). While world freshwater aquaculture and mariculture had similar growth rates over the past decade and each accounted for about half of the total aquaculture production, their species composition differs significantly. Freshwater aquaculture has been concentrated on finfish, while aquatic plants and shellfish (including Crustaceans and molluscs) were dominant in mariculture (Figure 1.1). For freshwater production, highest farm-gate value (production value calculated by using the on-farm, whole fish prices) is reported for fish production

---

[1]FAO Tech. Guidelines for Responsible Fisheries (5):40p. Rome, FAO.

(~80%) matching the largest production sector, while in mariculture the largest production sector (aquatic plants) only contribute marginally to the total farm-gate values. Freshwater carps, tilapia and catfish are globally the most important aquaculture species in terms of both volume and value. These are generally low-value fishes for domestic consumption, providing low-cost animal protein to ordinary consumers, but tilapia and some catfish species (e.g. Pangasius) have become increasingly popular global commodities. Marine shrimp and salmon are major commodities in international seafood trade, and are two high-valued species. Marine perch-like fishes (e.g. seabass, seabreams, groupers) do not belong to the top-10 of most important species in terms of volume but are among the top-10 in terms of value.

Aquaculture production is now fully comparable to capture fisheries landings when measured by volume of output on global scale. World aquaculture production of fish accounted for 44.1 percent of total production (including for non-food uses) from capture fisheries and aquaculture in 2014, up from 42.1 percent in 2012 and 31.1 percent in 2004 (FAO, 2016) From 2014 the total aquaculture production surpassed capture fisheries for human consumption and it is expected that by 2025 aquaculture production will be larger in volume than the capture fisheries. (OECD/FAO 2016.). According to the latest OECD – FAO forecasts (OECD/FAO 2016) expanding aquaculture production will remain amongst the fastest growing food sectors with a 3% annual growth rate, which is however significantly lower than the annual growth rate of 5.6% experienced in the previous decade. This slowdown in expansion will mainly be due to restrictions caused by environmental impacts of production and competition from other users of water and coastal spaces (World Bank. 2013). For example, aquaculture farming along coasts, lakes or rivers can conflict with urban development or tourism. This can create problems related to water quality and scarcity and push aquaculture expansion into less optimal production locations and consequently increasing costs, this is therefore encouraging the industry to seek innovative technologies and partnerships to maintain existing production costs.

Asian countries will remain the main producers with a share of 89% of total production in 2025, but Aquaculture will also show an impressive increase in developed countries, growing 26% during the same period. In Africa, the capacity building activities of the last decade and local policies promoting aquaculture also will raise the recent 1.7 million tonnes to 2.2 million tonnes.

The product groups listed in Table 1.1 are cultured by using various technologies, influenced by the environment and determining the social,

Table 1.1   Volume of main product groups in the various culture environments in 2015 (brackish water production is included in the marine environment.)

Product (in Thousands of Tons)	Freshwater Aquaculture	Mariculture/Marine Aquaculture	Aquaculture Total
Finfish	44,108	7,800	51,907
Crustacean	2,857	4,495	7,351
Molluscs	284	16,148	16,432
Aquatic plants	90	29,273	29,363
Other aquatic animals and products	523	427	950
Total	47,861	58,143	106,004

Data from: © FAO – FishStatJ 2017 March.

economic and environmental sustainability of the production. The majority of freshwater fish are carp produced in Asia (37.5 million tons) in pond based systems, thus ensuring the local protein supply for underdeveloped regions. Asian countries produce the majority of mariculture farmed species which are mainly extractive species such as molluscs (e.g. mussels and oysters) and aquatic plants (e.g. seaweeds). Products from marine aquaculture also have an important role in the food supply and application of aquaculture technologies to different (new) species in marine areas have a potential to supplement the global shortage in capture fisheries.

1.1.2 Marine Aquaculture as a Blue Growth Sector

In this book, the focus is on the mariculture technologies having a potential for combination with other Blue Growth industries. According to the distance from the coastline and characteristics of marine aquaculture activities; coastal, off the coast and offshore mariculture (or marine aquaculture) subsectors can be distinguished, where coastal and off the coast can be considered also as nearshore (Table 1.2):

1. **Coastal and off the coast marine fish culture**: Fish farming activities less than 3 km from the shore using various technologies also including flow-through and recirculation systems but mostly apply the open floating cage net technology. In Europe Atlantic salmon, Sea bream and Sea bass are the fish species produced in the largest quantity in marine cage aquaculture systems.

2. **Coastal and off the coast farming of molluscs and crustaceans**: Crustacean production is mostly inland or on shore pond based farming and because of the required technology, there are only very limited

opportunities to move the production to offshore farms (lobster cultures). Mussels and oysters are produced in large volume (Figure 1.1) using various techniques and molluscs cultures are considered as the most promising candidates for aquaculture on offshore energy platforms (Wever et al., 2015).

3. **Coastal and off the coast production of aquatic plants** (macro and micro): While off coast micro algae production is still in experimental stage, the production of seaweed is a well-established off the coast technology having a potential to be moved further offshore and combined with other offshore activities.

4. **Coastal and off the coast Integrated Multi-Trophic Aquaculture systems (IMTA)**: The basic concept of IMTA is the farming of several species at different trophic levels, that is, species that occupy distinct positions in a food chain. This allows one species' uneaten feed and wastes, nutrients and by-products to be recaptured and converted into fertilizer, feed and energy for the other crops (Chopin, 2012). As an example we can combine the cultivation of fed species (finfish or shrimp) with inorganic extractive species (seaweeds or aquatic plants) and organic extractive species (oysters, mussels and other invertebrates).

5. **Offshore mariculture or offshore marine aquaculture**: Adopting the FAO definition, offshore mariculture is classified as existing or potential activities where the distance of the production unit is more than 2 km from the coast. These are within continental shelf zones and possibly

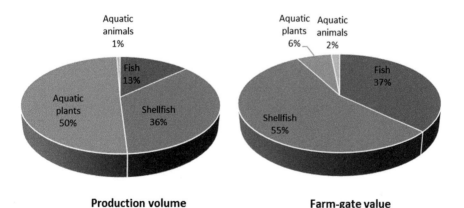

Figure 1.1 Relative Production volume (left) and Farm-gate value (right) by species type in 2014 for the Marine culture environment at a global level (data from FAO FishSTATJ).

Table 1.2　General criteria for defining coastal, off-the-coast and offshore mariculture. 1 Hs = significant wave height, a standard oceanographic term, approximately equal to the average of the highest one-third of the waves

Parameters	Coastal Mariculture	Off the Coast Mariculture	Offshore Mariculture
Location/ hydrogra- phy	<500 m from the coast <10 m depth at low tide within sight usually sheltered	500 m to 3 km from the coast 10–50 m depth at low tide often within sight somewhat sheltered	>2 km generally within continental shelf zones, possibly open ocean >50 m depth
Environment	Hs1 usually <1 m short-period winds localized coastal currents possibly strong tidal streams	Hs <3–4 m localized coastal currents some tidal streams	Hs 5 m or more, regularly 2–3 m oceanic swells variable wind periods possibly less localized current effect
Access	100% accessible landing possible at all times	>90% accessible on at least once daily basis landing usually possible	usually >80% accessible landing may be possible, periodic, e.g. every 3–10 days
Operation	manual involvement, feeding, monitoring and more	some automated operations, e.g. feeding, monitoring and more	remote operations, automated feeding, distance monitoring, system function
Exposure	sheltered	partly exposed (e.g. >90° exposed)	exposed (e.g. >180°)

Source: Lovatelli et al., 2013.

open ocean areas. The economic interest of offshore mariculture is today primarily related to finfish (Lovatelli et al. 2013), but from a technological point of view, seaweed and molluscs production have good opportunities for offshore farming. Offshore finfish farming has a specific technology using submersible floating cages and automatized feeding system paired with remote monitoring.

1.2　Sector Industry Structure and Lifecycle

The concept of business lifecycle (sometimes referred to as product lifecycle) is well established in economics. Influenced by Darwinian theories, Alfred Marshal considered how industries and firms were not in a steady state and appeared to evolve over time (Kerr and Johnson 2015.) As aquaculture moves offshore there will be important lessons to be learnt from existing installations and companies. Studying existing businesses provides the opportunity to see how barriers to growth were overcome in the past and learn lessons for future Blue Growth. This lifecycle analysis is especially useful to define the lifecycle stage of different aquaculture subsectors which helps to identify the benefits of each subsectors when they are combined with other Blue Growth industries. Life cycle description of the subsectors are summarised in the Table 1.3 according to the characteristics of life stages by Kerr and Johnson and the production and economic data of the subsectors. Based on the description of Table 1.3 the life cycle stage of the most relevant aquaculture subsectors are identified in Table 1.4.

Investigating these results from a business development point of view, it can be seen that aquaculture subsectors in different life stages could benefit from combinations with other Blue Growth sectors in various ways. Mature subsectors like salmon and sea bass, sea bream production are considerably limited by the available marine space. Investments in the combination of fish farming with other industries using off the coast and offshore areas could support the mature aquaculture subsectors to get licenses and increase their production.

Subsectors in the growth stage are in the process of increasing their capacity and reduce production costs. Mussel production in certain areas as well as organic fish production could benefit significantly from investments in combined coastal and off the coast platforms.

Offshore fish farming has only just started and can thus be classified as development or embryonic stage in the selected regions. Combination with other BG industries that are already in the mature and growth stage could facilitate the technology transfer of offshore technologies to aquaculture. There is also potential for the more mature industries to facilitate the investments in offshore aquaculture sectors.

1.3 Market

1.3.1 Products and Trade Flows in the World

The expansion of seafood consumption and thereby aquaculture production has dramatically changed the major seafood trade pathways (FAO 2016). Salmon and trout products have increased market share over a number of years and now represent the largest single commodity by value in the fish

Table 1.3 Analysis of business life cycle stages (Kerr and Johnson 2015) of aquaculture sub-sectors

Sub-sector	Demand/Products	Technology/Manufacturing	Trade/Competition	Key Success Factors	Finance/Investment
Coastal and off the coast marine fish culture	Stable mass market for salmon, sea bream and sea bass, customer knowledge is high. Branding phase emerging markets for species and products.	Well diffused technical knowhow, available marine space is one of the main limitation. Overcapacity in the Mediterranean. New EU regulation on organic aquaculture, innovations for new species.	Production shifts to less developed countries. Price competition; customers focus. New entries with new technologies for niche markets.	Cost efficiency achieved mainly through scale and driving down input costs. Technology innovations are still important. New production areas have to be opened.	Bank finance and institutional investments are common. R&D grants for new species and technologies.
Coastal and off the coast farming of molluscs and crustaceans	Stable market for molluscs (mussels, oysters). Emerging markets for abalone. Crustaceans market is large but highly dependent on global economy (Asia).	Hatchery is still a limiting factor. Diseases and natural events Structures need to be better prepared for storms. Production of new crustacean species is difficult.	The market for both crustaceans and molluscs can suffer major shifts depending on extreme weather events, food safety regulations of individual countries.	New communication streams are being used to inform consumers about the advantages of eating extractive species. Sustainability of the production methods is getting more important.	In Europe, the main investments are made mainly by existing producers in marketing, and combination of farming and tourism. New investors from BG industries are needed.

Coastal and off the coast production of aquatic plants (macro and micro algae/seaweed)	Stable, but growing market for different species of sea weeds. Product specification is oriented to added value products (compounds).	Technology in place is usually relatively low tech., and based on manual labour. Off the coast technology for micro algae production is in the experimental phase.	Seaweed trade and market in Asia is huge, but very little in Europe. Micro algae has an emerging market as raw material for food, health, chemical and biofuel products.	Technology innovations are needed to guarantee high crop quality and cost-effective production and processing. Product cost price is the main current bottleneck.
				R&D grants and EU or governmental funding in Europe.
Coastal and off the coast Integrated Multi-Trophic Aquaculture systems	No specific demand, as this is a production system and not a particular product.	Technology is available but needs to be developed and adapted to different environments and market trends.	High value markets, niche markets concerned with sustainability. Difficult to compete with low price products.	Communication channels, marketing tools and education are key factors to develop IMTA. Market diversification.
				Dependent on subsidies (EMFF aqua-environmental measures) and joint ventures between companies producing complementary products (e.g. fish and seaweed).
Offshore mariculture	Well known products with high demand are the main target species (e.g. Atlantic salmon). No specific market yet.	Technology is available, but innovations through technology transfer are needed to reduce the costs and solve some problems.	Only a few producer countries and companies. Competition with the coastal and off the coast production.	Main driving force is the easier licensing. R&D work to reduce OPEX and CAPEX costs.
				Only investments in large capacities can be economically feasible. The high CAPEX costs requires investors. Bank finance, share issue. Institutional investors. Corporate partners, merger.

Table 1.4 Life cycle stage of the sub-sectors

Sub Sector	Life Cycle Stage	Justification of the Development Stage (Including Regional Variations)
Coastal and off the coast marine fish culture	Growth and Mature stage	Salmon, sea bass, sea bream in the MATURE stage. Organic aquaculture and new species in the GROWTH stage.
Coastal and off the coast farming of molluscs and crustaceans	Life cycle stage depends on the species and region. Embryonic to Mature stage	Production conditions in moderate climates are investigated for biomass production optimisation. R&D on the most suitable technical approaches is done. To support this action, prototypes and pilot sites have been installed at certain areas (Norway, Portugal, Netherlands, Germany, Ireland, etc). Coastal molluscs culture is in the Mature stage as well as the crustacean (shrimp) culture in Asia. While lobsters farming in Europe is at a – development/embryonic stage. Bottom cultivation of blue mussels in the Netherlands is a sector in stage in growth stage, while suspended cultivation of spat collectors is embryonic stage.
Off shore production of aquatic plants (macro algae)	Embryonic to growth stage	Seaweed, macro algae production in Asia, worldwide and coastal micro algae production is in a GROWTH stage. Others in Development or Embryonic stage. Production conditions in moderate climates are investigated for biomass production optimisation. R&D on the most suitable technical approaches is done. To support this action, prototypes and pilot sites have been installed at certain areas (Norway, Portugal, Netherlands, Germany, Ireland, etc).

Coastal production of aquatic plants (macro algae)	Embryonic to growth stage	Production in moderate climates is currently in its first commercial stage. This is generally following previous wild harvesting activities. Production in Asia and tropical conditions is in its expansion stage. Development of sea weed culture areal is still increasing.
Coastal and off the coast production of aquatic plants (micro algae)	Development to Embryonic stage	Microalgae cultures under offshore conditions are in R&D stage, some prototypes have been installed. The developments are currently inhibited by productivity and thus economics of the production methods. Further development needed to optimise technologies.
Coastal and off the coast Integrated Multi-Trophic Aquaculture systems	Development/Embryonic stage	Mostly still in the pilot scale in Europe, only a few farms use the technology.
Offshore mariculture	Development/Embryonic stage	Companies in the Caribbean (for cobia) and in the Atlantic region (Atlantic salmon) use the offshore technologies. However, these businesses are in the embryonic stage already, there is high need for new technical solutions. Offshore fish farming in the Mediterranean and in the Baltic region is still in the development stage focusing on the research and pilot testing.

trade (16.6% of the world trade in 2013), this is clear evidence of the impact of aquaculture on the market. The majority of salmonids are exported from a limited number of countries which have a suitable natural environment for salmon farming (Norway, Chile, Canada), mostly to high-income countries although China also accounted for 4.4 percent of world salmon import in 2011. Shrimps and prawns are high value seafood commodities and are mostly exported from developing countries to developed countries. Trade data indicates that most shrimp import in EU comprises warm-water species which originate mostly from aquaculture. International trade of molluscs (excluding cephalopods) is more disperse among countries although China holds a strong position in both import and export of molluscs.

1.3.2 Market Trends, Prices and a View of Future Demand in the EU

Although aquaculture in the European Union is very diverse with production spread across more than 100 species categories, a limited number of species dominate. In 2014, 1,275,902 tonnes total aquaculture production comprised (EUMOFA online query): 35.7% Mytilus mussels (455,079 tonnes), 14.8% Atlantic salmon (189,476 tonnes), 13.6% other salmonids (mainly rainbow trout, 185,663 tonnes), 7.1% oysters (91,460 tonnes), 6.2% carp (79,994 tonnes), 11.7% Sea bass and Sea bream (149,317 tonnes together). Although the reported harvest from freshwater appears to be small relative to harvest from seawater and brackish water, it must be recognised that Atlantic salmon (and other salmonids harvested from seawater) are initially reared in freshwater and freshwater species are also reared in brackish and seawater (large trout).

Five Member States dominate EU-28 aquaculture (Figure 1.3), accounting for 75% of production (Spain: 266,594 tonnes; United Kingdom: 205,594 tonnes; France: 205,107 tonnes; Italy: ca. 160,000 tonnes; Greece: 108,852 tonnes). The relative importance of the different aquaculture sectors varies between Member States, e.g.:

- Molluscs dominate production (>60% of national tonnage) in Spain, France, Netherlands and Ireland;
- Atlantic salmon and other salmonids (mainly rainbow trout) dominate in the UK, Denmark, Finland, Sweden, Slovakia, Slovenia and Estonia;
- Marine finfish (including seabass and seabream) dominate in Greece, Malta and Cyprus;
- Freshwater finfish (including carp) dominate in Germany, Poland, Czech Republic, Hungary, Romania, Lithuania and Latvia.

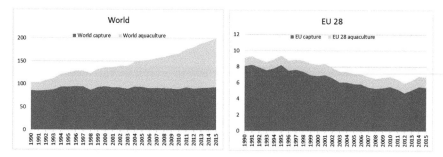

Figure 1.2 Development of Aquaculture production (in million tonnes) in the world (left) and European Union (right) indicating that the EU does not follow the high growth rates as displayed in other parts of the world (data from FishStatJ 2017 March).

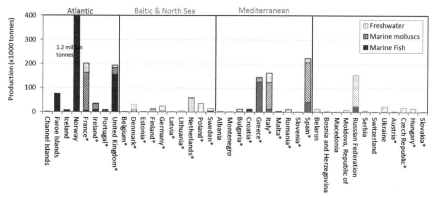

Figure 1.3 Marine aquaculture production in the Europe in 2015 specified for each Maribe basin and divided per production type (Marine finfish, Marine shellfish, Freshwater). *indicates if member state is part of the EU-28. (Data from: FishStat FAO).

The aquaculture production in the EU (Figure 1.2) has stagnated for many years in terms of the total production volume where the increases in salmonid (Atlantic salmon, large trout) and mussel production have been cancelled out by reductions in production of eels and other freshwater fish.

According to the predictions of aquaculture development in the period 2016–2025 (OECD/FAO 2016), developing countries will consolidate their position as lead aquaculture producers, with a share of almost 95% of global aquaculture production. The latest trends and forecast studies also justify the 28% increase until 2025 in Europe, however the majority of this growth still expected from the salmon aquaculture of Norway. Numerous studies indicate a considerable increase of aquaculture production in the European Union.

Lane et al. (2014) projected a faster growth rate of aquaculture in the European Union than the OECD/FAO forecast for Europe estimating a total increase of 772,000 tonnes (+56%) in volume from 2010 to 2030 with a corresponding value increase of 2.7 billion euros and requiring an additional 395,000 tonnes of feeds.

The latest production trend estimations for the European Union can be calculated from the data of the Multiannual National Aquaculture Plans prepared by each member states as a strategy for the use of the European Maritime and Fisheries Fund. The common structure of the plans required an estimation of the impact of EU funds on the production and the estimated projection for EU aquaculture volume in 2020 is expected to increase by about 300,000 tonnes (25%) to a total of more than 1.5 million tonnes.

The European Aquaculture Technology and Innovation Platform (EATIP) provides a forecast for the whole of Europe in a vision document (EATiP 2012) predicting that by the year 2030, European aquaculture will provide annually 4.5 million tons of sustainable food products (recent production is 3 million tonnes), worth 14 billion euros, and supporting more than 150,000 direct jobs.

1.3.3 Market Trends, Prices and Supply & Demand Gaps

The European Union is a major consumption market of seafood products in the world with 13,8 million tonnes representing EUR 49,3 billion in 2015 (EUMOFA 2016).

It is the largest importer of seafood products, absorbing 24% of total world exchanges in value. Seafood consumption per capita in the EU seems to have reached a plateau after a decade of dynamic growth. The consumption per capita in 2015 was 25,5 kg which is 1 kg more than it was in 2013. The increase in the consumption was more significant for farmed products (+6%) than for fisheries products (+2,7%), but the consumption in the EU market is still dominated by products originating from fishing activities (75% of total). Tuna, cod and salmon are the main species consumed in the EU in volume. Shrimps are the first imported species in value ahead of salmon, tuna and cod. Seafood consumption varies a lot from one Member State to the other. Northern Member States are more focused on processed fish (frozen, smoked) while Southern Member States still favour fresh products and devote a larger part of household expenditures to fish. Central and Eastern European countries are below the EU average but register increase in consumption (EUMOFA 2017). The data of the detailed EU wide consumer survey showed, that the per capita consumption trend (in kg) is slightly negative, while the per capita real expenditure (in Purchasing Power Parities which was calculated by multiplying the EU28 expenditure in real terms in EUR with the volume indices of real expenditure per capita (for fish)) trend is basically flat. (EUMOFA 2017). This widening gap between the two indicators supposes an increase in the consumption of high value products (high quality fresh products, processed products). Of course, these average figures hide the different situation in member states and while for example, the average unit value increased in the Central and Eastern European countries, in the most Mediterranean countries this value decreased.

In 2015 Atlantic salmon was the most consumed aquaculture product in the EU reaching 2.09 kg/capita/year consumption showing 9% and 38% increase from 2013 and 2005, respectively. The main producer of this species in Europe is Norway, selling the half of its yearly 1.3 million tons production (FEAP Production Report 2007–2015) to Europe. The second largest producer is the UK with 186 thousand tons mainly from Scotland. The main production area for this species is the Atlantic Ocean where all production countries continuously increase their production and looking for innovative technologies to support this raise.

In the Baltic region fish production in the marine environment is less developed and the main produced species is the large trout (>1.2 kg). The biggest producers of large trout using the marine cage technology are Denmark (10,500 t), Sweden (9436 t) and Finland (12,500 t) and according to their EMFF Operational Programs all countries want to increase its aquaculture production.

The fish production in the Mediterranean is dominated by the sea bass and sea bream production mainly in coastal and off the coast cages. The main producer countries are Greece (93,000 t) and Turkey (142,000 t) competing with each other for the leading producer position and for the markets (FEAP Production Report 2007–2015).

EU self-sufficiency for seafood (i.e. the production relative to its internal consumption) increased from 44,5% to 47,5% during 2013–2014, but to keep up with the rise of the internal demand needed an increase of catches and production as well. While the EU covers fully its needs for small pelagics (and even produces surpluses) it is increasingly and highly dependent on external sourcing for groundfish, salmonids and tuna. In terms of the aquaculture products the self-sufficiency is much lower and only 10% of the total EU seafood consumption (13 million tonnes) is currently come from EU aquaculture (1.3 million tonnes). These statistics suggests that demand is greater than supply and there is great potential to expand aquaculture production in the EU to meet the demand, improve food security and improve the economy.

1.4 Working Environment

1.4.1 Economic Indicators for the Aquaculture Sector in the EU

In the EU, aquaculture production is an important economic activity in many coastal and inland regions (COM 2012a), often providing employment in marginal and remote areas. The sustainable development of European aquaculture has been identified as a priority under reforms of the Common Fisheries Policies (CFP) to strengthen long term food security (EU 2013). These regulations require actions to improve the competitiveness of the sector, whilst ensuring its long term environmental, economic and social sustainability. Aquaculture has thereby been identified as one of five value chains that can deliver sustainable growth and jobs within the blue economy (COM 2012b).

Reliable data on key economic indicators are difficult to obtain for the aquaculture sector in all the 28 Member States, but the latest report of Scientific, Technical and Economic Committee for Fisheries (STECF 2016) provide a good overview. While the countries participated in the report indicated more than 12,000 enterprises, the study estimated that the total number of companies with aquaculture as their main activity in the EU-28 is between 14 and 15 thousand. In 2014 the majority (90%) of the companies were micro-enterprises (with less than 10 employees) and tend to be family owned. Micro-enterprises are usually small scale rather than large companies using capital intensive methods. The number of enterprises with more than 10 employees has increased from 1040 in 2012 to 1230 in 2014. The reported data displays an employment of about 69 700 people in 2014, but the study estimates that EU-28 aquaculture sector directly employs around 80,000 people. The EU aquaculture sector has an important component of part-time work which is due to the importance of the shellfish sector that has a significant percentage of part-time and seasonal work. Women accounted for the 24% of the EU aquaculture sector employments, but only 19% when measured in FTE. There is a lot of variability within the salaries paid in each country and subsector, varying from 3,300 Euros per year in Bulgaria to 72,100 Euros per year in Denmark.

EU aquaculture sector provided about €1 596 billion in Gross Value Added (GVA) in 2014. This is an increase of 14% from the €1 294 billion reported in 2012. EBIT (Earnings Before Interest and Taxes or Operating

Profit) data from 19 countries (excluding Poland) show that the EU aquaculture sector was more profitable in 2014 with a reported total EBIT of €402 million, which is an increase of 24% from the €324 million reported in 2012.

1.4.2 Driving Forces and Limitations of Aquaculture Sector

The majority of global aquaculture production is concentrated in developing countries, in particular in Asia, while aquaculture development in more developed countries and especially in the European Union is relatively stagnant. This partly due to a range of governance challenges, regulatory framework and the scarcity of suitable locations. The main constrains of aquaculture development in the EU-28 countries are often listed as the followings (Lane et al. 2014):

- Fierce and often unequal competition with third countries that brings market prices down. Fish farmers association in the EU says that the strict regulation often creates a sloped playing field for third countries having for example less stringent environmental or food security regulation.
- High labour and capital costs and administrative burdens slow down investments in the sector.
- Lack of understanding of the spatial needs and infrastructure for the industry among the planning authorities.

The annual growth rate of the world aquaculture in the next decade is expected to be 2.5% (FAO/OECD), which is significantly lower than the growth rate of 5.6% p.a. experienced in the previous decade. Driving forces of aquaculture growth on a global level are (Guillen & Motova 2013, Lane et al. 2014):

- Overfished and decreasing wild fish stocks, while the demand for fish is growing
- Aquaculture is more efficient in terms of freshwater use and energy than other animal production sectors.
- The availability of marine space for aquaculture is larger than availability of agricultural land
- Technology development makes aquaculture more and more profitable.

Limitations of aquaculture growth on a global level:

- Dependency on and availability of sustainable fish meal sources
- Direct environmental interactions: pollutions, predators, diseases, algal blooms

- Poor husbandry practices: use of antibiotics, antifungal, herbicides …
- Consumers attitudes and trends
- Deterioration of the quality of water bodies suitable for aquaculture

1.4.3 Regulatory Framework of Marine Aquaculture in the European Union

Aquaculture is an integral part of the reformed Common Fisheries Policy (CFP) (REGULATION (EU) No 1380/2013). The basic regulations define aquaculture as an important economic and food supply industry and encourage the development of the sector. Aquaculture has thereby been identified as one of five value chains that can deliver sustainable growth and jobs within the blue economy (COM 2012b). The Commission published Strategic Guidelines for the Sustainable Development of EU aquaculture (COM 2013a) which highlighted four priority areas to unlock the potential of the sector: i) simplification of administrative procedures, ii) co-ordinated spatial planning, iii) competitiveness and, iv) a level playing field. Using these guidelines, Member States (MS) has developed or are now developing multiannual national plans for the development of sustainable aquaculture.

One of the main tool to achieve the goals of the CFP is the European Maritime and Fisheries Fund (EMFF) which is one of the five European Structural and Investment (ESI) Funds which complement each other and seek to promote a growth and job based recovery in Europe. The EMFF regulation (REGULATION (EU) No 508/2014) lay down the principal rules how this fund is used to co-finance projects, along with national funding. Each country is allocated a share of the total 5.7 billion Euro Fund budget, based on the size of its fishing, aquaculture and processing industry. Member states then draws up an operational programme (OP), saying how it intends to spend the money. Once the Commission approves this programme, it is up to the national authorities to decide which projects will be funded. Recently, Member States are submitting their OPs to the commission and preparing their national legislation and system for the distribution of the fund. The regulations allow aquaculture investments to be supported with a maximum funding rate of 50% of the total investment.

The development of sustainable aquaculture is dependent on clean, healthy and productive marine and fresh waters. A prerequisite for sustainable aquaculture activities is compliance with the relevant EU Legislation. The Water Framework Directive (WFD) (Directive 2000/60/EC) and the Marine Strategy Framework Directive (MSFD) (Directive 2008/56/EC) aim to protect and enhance aquatic environments and ensure that the uses to which they are put are sustainable in the long term. All mariculture activities in the Member States has to be carried out in line with the common regulation of MSFD to minimise the risk of the introduction of non-indigenous species (NIS), to keep under limits the amount of discharged nutrients, organic matter, contaminants including pesticides and litter. New aquaculture technologies also have to reduce the disturbance to wildlife, and the possibility for escape of farmed fish. However, the magnitude of these impacts from aquaculture in comparison with impacts from other sources (e.g. agricultural runoff) are not known, aquaculture, alongside all other sectors, will need to reduce impacts in order to reach Good Environmental Status (GES) under MSFD. The role of the MSFD is becoming increasingly important to ensure that aquaculture activities provide long-term environmental sustainability.

1.5 Innovation

1.5.1 Innovation Trends in Coastal and Off the Coast Marine Aquaculture Subsectors

To identify the innovation trends in the aquaculture the mapping of the research needs of the sector can provide a good indication of the main directions. EFARO, the European Fisheries and Aquaculture Research Organisations, identified the following "game changer" technical and scientific topics enabling a European aquaculture vision to happen (EFARO 2017): 1) Develop sustainable fish feeds based on aquaculture ingredients, 2) Diversification of activities, 3) Breeding: Development of breeding programmes for the production of robust animals, 4) Seaweed Production and Value Chains: Innovation and optimization of seaweed products and processes, 5) Develop research on aquaculture productions associations and the integration of aquaculture productions with other productions or service productions, 6) Bivalve production: Innovation and optimization of shellfish products.

Marine and freshwater aquaculture is already an efficient user of land and freshwater while also has a lower carbon footprint than the production of terrestrial animals, but due to its dependence on plant products such as soy in the feed, aquaculture still uses a considerable amount of these limited resources. Therefore, exploring new or alternative feed resources and production sites along with the necessary technology and delivery systems are urgently needed.

EFARO members also launched the Cooperation in Fisheries, Aquaculture and Seafood Processing (COFASP ERA-NET) to develop and strengthen the coordination of national and regional research programmes. The COFASP Strategic Research Agenda identified the following topics to be of importance for future development of aquaculture as a whole: 1) Market demand (species that can be cost effectively produced), 2) Organic aquaculture (lower the production costs relative to conventional methods), 3) Technology development (Recirculation facilities & multi-trophic aquaculture), 4) Species enhancement (Aquatic animal health and welfare, and Breeding Programmes) (COFASP 2016). The ERA-NET COFASP collects and analyses a list of projects on aquaculture, fisheries and seafood processing funded at European/national level and maintain a database available at http://www.projectsdatabase.cofasp.eu. Recently a total of 1203 aquaculture research projects can be found in the database with €816 million total funding. From these projects 28 have activities to develop offshore aquaculture. Many of the projects funded focussed technological development, but environmental impact studies are also listed. The recent and planned future calls within the H2020 program have a significant focus on multi-use possibilities to make better use of marine space and resources.

1.5.2 Recent Technology and Expected New Technologies in Offshore Mariculture, Opportunities and Challenges

Sturrock et al. (2008) identified offshore aquaculture and IMTA as emerging technologies supporting the European aquaculture development. The current development of mariculture of species such as Atlantic salmon, Sea bream and European seabass and experimental/pilot farming of other species such as cobia (*Rachycentron canandum*) and amberjacks (*Seriola spp.*) provides excellent and promising technological advances for moving marine aquaculture farther offshore. However, the economic viability of offshore mariculture is a major challenge and better technologies still need to be developed. There are also concerns about the availability of capital for investments in research and development (R&D) and for the development of commercial farms. Moreover, there is no clear candidate species of finfish available that has proved both economic and physiological feasibility for offshore production and, while species of shellfish and aquatic plants are better identified, the economic viability of their production is still questionable. A transition from coastal to off-the-coast and offshore mariculture will demand the development of new or suitably adapted technologies throughout the value chain, with obvious scientific challenges. This is what is needed if global seafood supply is to be increased in a way that minimizes impacts on benthic and pelagic ecosystems as demanded by society. One of the main driving force of aquaculture innovations recently is the so called "green licensing" system of the Norwegian salmon industry. Norwegian governments have publicly declared that further growth is impossible until the problems of sea lice, escapes and pollution have been solved or, at least, considerably reduced. Since the salmon farming companies have shown that they have the capacity and market access for an increase of production, the government opened up new salmon farming licenses subject to strict environmental criteria mainly on sea lice, escape risk and other environmental factors known as "Green Licenses" (Hersoug 2015). The interest is still very high for the valuable available licenses and the requested new technological solutions encouraged other marine and offshore industries like shipbuilding and offshore oil and gas to bring in their experience to the aquaculture industry and team up with aquaculture companies. Of course, the large salmon farming companies like SalMar AS (www.salmar.no) and Marine Harvest AS (marineharvest.com) also invested in innovation and developed their own technological solutions to fulfil the requirements of green licenses. SalMar is more focused on the offshore production and invested in Ocean Farm-1 full scale pilot facility, a 110m wide cage system under a floating platform also hosting the feed barge, control room and maintenance facilities.

1.6 Investment

Investments in aquaculture stem from the sector itself, from private investment funds or investors and from public sources. The investment environment of the EU aquaculture sector can be characterized as follows.

- The Future Expectations Indicator (FEI) indicates whether the industry in a sector is investing more than the depreciation of their current assets. According to a recent research (STECF 2016) FEI in 2014 for 19 EU countries for the whole aquaculture sector (freshwater and marine) was negative at 5.8% while net investments in marine aquaculture increased by 16% from 2013–2014.
- The figures of fast growth of the sector in certain areas and increasing needs for aquaculture products attracts numerous private investors. While we will not recommend specific funds, some examples of

funds that specifically invest in aquaculture include: Oceanis Partners, A-Spark Good Ventures, Watershed Capital Group, Fish 2.0.

Public investments are mostly linked to the European Maritime and Fisheries Fund (EMFF), which is the EU financial instrument to support Common Fisheries Policy (CFP) implementation. The Commission is keen to use the opportunities presented by EMFF to boost aquaculture growth. It therefore requires Member States to produce Multiannual National Plans (MANPs) outlining how each member state intend to foster growth in the aquaculture industry. Each country is allocated a share of the total 5.7 billion Euro Fund budget, based on the size of its fishing, aquaculture and processing industry. The MANPs will provide information on how each member state will allocate the funds to stimulate sustainable aquaculture (Figure 1.4), including a prediction of the expected growth of the sector. Under priority 2 of the national operational programme the following objectives can be funded:

- support for technological development, innovation and knowledge transfer;
- the enhancement of the competitiveness and viability of aquaculture enterprises, including the improvement of safety and working conditions, in particular of SMEs;
- the protection and restoration of aquatic biodiversity and the enhancement of ecosystems related to aquaculture and the promotion of resource-efficient aquaculture;

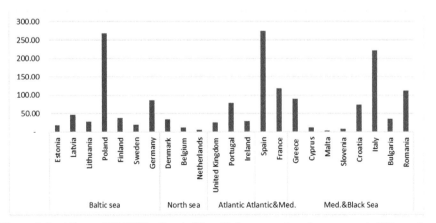

Figure 1.4   Allocated EMFF funding (million €) for aquaculture investments in EU member states with marine aquaculture production for the 2014–2020 period.

Source: Operational Programs of listed countries.

- the promotion of aquaculture having a high level of environmental protection, and the promotion of animal health and welfare and of public health and safety;
- professional training, skills and lifelong learning.

The sum of funding budgets between 2014–2020 consisting of national and EU contributions for promoting environmentally sustainable, resource efficient, innovative, competitive and knowledge based aquaculture is €1.7 billion allowing at least €3.4 billion supported investments in European Aquaculture (http://ec.europa.eu/fisheries/cfp/emff/index_en.htm). Good access to information on the economics of offshore mariculture can help would-be investors and coastal States in developing economically feasible technologies for offshore mariculture. Member government actions are also needed to create conditions for increased investment in mariculture and to allocate funds for R&D, including funding demonstration and pre-commercial projects for a variety of species. Governments should also encourage international cooperation and technology transfer among stakeholders.

1.7 Uncertainties and Concluding Remarks

Marine aquaculture is a well-developed industry in the Atlantic and Mediterranean regions while it is under development in Baltic, North Sea and Black sea regions. The sector is dominated by the fish production sector in terms of the value, but this subsector also needs higher investment and operating costs. In terms of environmental interactions, seaweed and mollusc aquaculture is considered to have a positive impact on the marine environment.

The most limiting uncertainties of the marine aquaculture sector are:

- Lack of available marine space for aquaculture production and different licensing strategy of coastal countries.
- Availability and price fluctuation of fish meal which is still an important ingredient of the fish feed.
- Even though numerous promising offshore aquaculture technologies have been developed in the last years, none of them was tested yet for commercial scale production.

In spite of these uncertainties, the recent technology and socio-economic developments in the sector can provide good opportunities for the investment in offshore aquaculture in Europe. The main driving force of offshore production is to release the pressure from coastal areas which often have special nature values and can't provide more space for the expected growth of production. However offshore development is between the development and embryonic stage of business lifecycle, there are strong interests from government, institutions and commercial sectors to explore and develop offshore potential. These developments require large investments that can only be realized by large industries or with the support of external investment (private or governmental subsidies). The recent known attempts for offshore aquaculture development shows that the high market value of Atlantic salmon and increasing demand for salmon products has encouraged the existing large producers and other large companies having experience in other marine industries (shipping, oil) to invest in offshore salmon production projects.

Even though cultivation of shellfish or seaweeds may be better adapted to offshore conditions, the market value does not (yet) guarantee a profitable business case, while their minimal environmental impact on coastal areas does not force these technologies to move offshore. Seaweed and shellfish production can have an important role in offshore aquaculture by reducing the nutrient discharge of fish production in IMTA systems.

It is expected that a rapid growth in offshore aquaculture production will be triggered when the feasibility of large scale production will be demonstrated with the profitable business cases (economy of scale) of the recent projects. This growth stage will also require a next level in technical engineering of aquaculture (structures, remote sensing, safety at sea are important issues) which will facilitate the combination of aquaculture with other, mature, offshore industries.

The recent state and future opportunities for combination of aquaculture with other maritime industries in different marine basins can be summarised as follows:

Basin	Summary	Opportunities and Justification
Atlantic	The Salmon industry in Norway and Scotland (UK) is in expansion looking for marine space for new production sites and new technologies.	The companies are motivated to find partners and share the marine space with other industries. Salmon aquaculture is in the mature stage and ready for feasible combinations.
Baltic/North Sea	Mussel and crustacean culture is relatively more important and considerable amount of national research was done to combine their production with offshore wind energy.	More than 1 billion € investment in aquaculture is planned in the region (according to the adopted OPs). There is also a high need for Blue Energy investments providing good base for combinations.

| Mediterranean and Black sea | Sea bass and sea bream industry is very well developed and production of new species is also emerging.   Mussel production in the Black sea region has a growing interest. | High interest to invest in combined offshore platforms in the region.   To reduce the environmental impact of fish production there is opportunity to establish IMTA systems. |

References

Aquaculture glossary: http://www.fao.org/faoterm/collection/aquaculture/en/

Aguilar-Manjarrez, J., Soto, D. and Brummett, R. (2017). Aquaculture zoning, site selection and area management under the ecosystem approach to aquaculture. Full document. Report ACS113536. Rome, FAO, and World Bank Group, Washington, DC. pp. 395. Available at: http://www.fao.org/3/a-i6992e.pdf

CEFAS (2014). Background information for sustainable aquaculture development, addressing environmental protection in particular: SUSAQ (Part 1), Cefas contract report <C6078>

Chopin, T., Cooper, J. A., Reid, G., Cross, S., Moore, C., (2012) Open-water integrated multi-trophic aquaculture: environmental biomitigation and economic diversification of fed aquaculture by extractive aquaculture. Reviews in Aquaculture, 4(4), 209–220. doi:10.1111/j.1753-5131.2012.01074.x

COFASP (2016) Strategic Research Agenda For Fisheries, Aquaculture and Seafood Processing – www.cofasp.eu

EFARO 2017 A vision on the future of European Aquaculture, IJmuiden, The Netherlands, May 2017 www.efaro.eu

EUMOFA (2016) The EU Fish Market, 2016 edition, European Commission – European Market Observatory for Fisheries and Aquaculture Products 2016.

EUMOFA (2017) EU Consumer Habits Regarding Fishery and Aquaculture Products – Final Report, European Commission – European Market Observatory for Fisheries and Aquaculture Products

European Commission (2014). Facts and figures on the Common Fisheries Policy – Basic statistical data – 2014 Edition, Luxembourg: Publications Office of the European Union 2014 – 44 p. – 14.8 × 21 cm ISBN 978-92-79-34192-2

FAO & World Bank. (2015). Aquaculture zoning, site selection and area management under the ecosystem approach to aquaculture. Policy brief.

Rome, FAO. Available at: http://www.fao.org/documents/card/en/c/ 4c777b3a-6afc-4475-bfc2-a51646471b0d/

FAO. 2016. The State of World Fisheries and Aquaculture 2016.; Contributing to food security and nutrition for all. Rome. 200 pp.

World Bank. 2013. Fish to 2030: prospects for fisheries and aquaculture. Agriculture and environmental services discussion paper; no. 3. Washington DC; World Bank Group. http://documents.worldbank.org/curate d/en/458631468152376668/Fish-to-2030-prospects-for-fisheries-and-aq uaculture

OECD/FAO (2016), "Fish and Seafood", in OECD-FAO Agricultural Outlook 2016–2025, OECD Publishing, Paris. DOI: http://dx.doi.org/10.178 7/agr_outlook-2016-12-en

Helen Sturrock, Richard Newton, Susan Paffrath, John Bostock, James Muir, James Young, Anton Immink & Malcolm Dickson. Ilias Papatryfon (editor) (2008) Prospective Analysis of the Aquaculture Sector in the EU. PART 2: Characterisation of Emerging Aquaculture, Systems European Commission, Joint Research Centre, EUR Number: 23409 EN/2

Hersoug, B. (2015) The greening of Norwegian salmon production Maritime Studies (2015) 14:16 DOI 10.1186/s40152-015-0034-9

Kapetsky, J. M., Aguilar-Manjarrez, J., Jenness, J., (2013). A global assessment of potential for offshore mariculture development from a spatial perspective. FAO Fisheries and Aquaculture Technical Paper. No. 549. Rome, FAO. 181 pp. (also available at http://www.fao.org/ docrep/017/i3100e/i3100e00.htm).

Kerr, S., Johnson, K., (2015) Identifying and Describing Business Lifecycle Stages, MARIBE Internal publication Version 1.1, Briefing paper prepared by ICIT Heriot-Watt University.

Lovatelli, A., Aguilar-Manjarrez, J., Soto, D., (Eds.) (2013). Expanding mariculture farther offshore – Technical, environmental, spatial and governance challenges. FAO Technical Workshop. 22–25 March 2010. Orbetello, Italy. FAO Fisheries and Aquaculture Proceedings No. 24. Rome, FAO. 73 pp. Includes a CD–ROM containing the full document (314 pp.). (also available at http://www.fao.org/ docrep/018/i3092e/i3092e00.htm).

Lane, A., Hough, C., Bostock, J., (2014). The long-term economic and ecologic impact of larger sustainable aquaculture, Study for the European Parliament's Committee on Fisheries, European Union, 2014.

Meaden, G. J., Aguilar-Manjarrez, J., Corner, R. A., Ó Hagan, A. M., and Cardia, F. (2016). Marine spatial planning for enhanced fisheries and aquaculture sustainability its application in the Near East. FAO Fisheries and Aquaculture Technical Paper No. 604. Rome, FAO. Available at: http://www.fao.org/3/a-i6043e.pdf

Scientific, Technical and Economic Committee for Fisheries (STECF) 2016 – Economic Report of the EU Aquaculture Sector (EWG-16-12); Publications Office of the European Union, Luxembourg;

Soto, D., (2009). Integrated mariculture: a global review. FAO Fisheries and Aquaculture Technical Paper. No. 529. Rome, FAO. 2009. 183 pp. (also available at http://www.fao.org/docrep/012/i1092e/i1092e00.htm).

Wever, L., Krause, G., Buck, B. H., (2015) Lessons from stakeholder dialogues on marine aquaculture in offshore windfarms: perceived potentials, constraints and research gaps. Marine Policy 51:251–259.

2

Blue Biotechnology

Jane Collins[1,2,*], Arianna Broggiato[1,3] and Thomas Vanagt[1,3]

[1]eCOAST, Esplanadestraat 1, 8400 Ostend, Belgium
[2]Faculty of Pharmaceutical Sciences, Clinical Pharmacology
and Pharmacotherapy, KU Leuven, O&N II Herestraat 49 - box 521,
3000 Leuven, Belgium
[3]ABS-int, Technologiepark 3, B-9052 Zwijnaarde, Belgium
*Corresponding Author

2.1 Introduction

Biologists categorise living things into 36 divisions (technically phyla) and members of 34 of these divisions are found in the marine environment. In fact, the marine realm represents 70% of the biosphere. Life forms are estimated to have appeared at the bottom of the world's ocean approximately 3.6 billion years ago, compared to only several hundred million years ago for terrestrial life. Due to the ancient history and diversity of life forms encompassed, the oceans are considered a unique reservoir for a wide variety of potentially useful molecules (Arrieta et al., 2010). However, until recently, marine molecules remained largely unexploited due to difficulties associated with accessing them. Our ability to access remote parts of the ocean has greatly improved over the last century, and particularly in recent decades, as a result of advances in oceanographic technology. The technology used to screen molecules of interest has also improved over the last few decades. Recent estimates show an exponential increase in the use of marine molecules or sequences of nucleic acids extracted from marine organisms in a variety of biotechnological fields. Industries involved encompass a broad range of applications including human health, agriculture, aquaculture, food, cosmetics and bioremediation (Arrieta et al., 2010; Blunt et al., 2011; Leal et al., 2012; Marine Board, 2010). Marine molecules have also been used to develop pharmaceutical drugs such as anti-cancer medication, as well as treatments against HIV and Alzheimer disease which have already been commercialised (Molinski, 2009). The market for such biotechnologies appears to be vast and has been expanding consistently over the past few decades. The market value of a number of commercialised products had already surpassed several billion USD per annum by the year 2010 (Leary, 2009).

2.1.1 Definition of Blue Biotechnology and Marine Biotechnology

Biotechnology is broadly defined by the Organization for Economic Co-operation and Development (OECD, 2005) in the following way:

- **OECD statistical single definition of biotechnology**: The application of science and technology to living organisms, as well as parts, products and models thereof, to alter living or non-living materials for the production of knowledge, goods and services (OECD, 2016).
- **OECD list-based definition for biotechnology**: The following list of biotechnology techniques functions as an interpretive guide in using the single definition. The content of the list-based definition is indicative rather than exhaustive and is expected to change over time as data collection and biotechnology activities evolve (OECD, 2016).
  - **DNA/RNA**: Genomics, pharmacogenomics, gene probes, genetic engineering, DNA/RNA sequencing/synthesis/amplification, gene expression profiling, and use of antisense technology.
  - **Proteins and other molecules**: Sequencing/synthesis/engineering of proteins and peptides (including large molecule hormones); improved delivery methods for large-molecule drugs; proteomics, protein isolation and purification, signalling, identification of cell receptors.
  - **Cell and tissue culture and engineering**: Cell/tissue culture, tissue engineering (including tissue scaffolds and biomedical engineering), cellular fusion, vaccine/immune stimulants, embryo manipulation, marker-assisted breeding technologies.
  - **Process biotechnology techniques**: Fermentation using bioreactors, bio-refining, bioprocessing, bioleaching, biopulping, biobleaching, biodesulphurisation, bioremediation, biosensing, biofiltration and phytoremediation, molecular aquaculture.
  - **Gene and RNA vectors**: Gene therapy, viral vectors.

- **Bioinformatics**: Construction of databases on genomes, protein sequences; modelling complex biological processes, including systems biology.
- **Nanobiotechnology**: Applies the tools and processes of nano/ microfabrication to build devices for studying biosystems and applications in drug delivery, diagnostics, etc.

This very clearly shows what is involved in biotechnology in general. There is however no single, official definition of blue biotechnology or marine biotechnology. Blue biotechnology is generally considered the use of marine bioresources as the source of biotechnological applications (Figure 2.1). In other words, marine resources and marine organisms are used to develop products or services for biotechnological gain (ECORYS, 2014). In contrast, marine biotechnology also includes the application of biotechnology developed using any resource (marine, terrestrial, freshwater or a combination) to the marine environment, and human activities therein.

Workshops and questionnaires were conducted in 2013 and 2014 in order to reach an agreement on a common understanding of the term marine

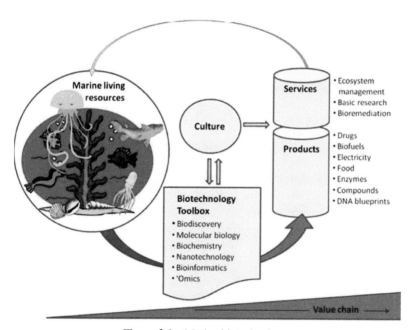

Figure 2.1 Marine biotechnology.

Source: OECD, 2013.

biotechnology. During workshop discussions, the European Commission (EC) highlighted the importance of consensus regarding marine biotechnology's definition for the development of new initiatives and policy options. It became apparent over time that adaptation of the existing OECD definition for biotechnology (single and list-based parts) could be the most straight forward way to reach an overall consensus and definition (OECD, 2016).

International definitions of **(marine) biotechnology**

- **Marine Board definition of marine biotechnology**: Marine biotechnology encompasses those efforts that involve marine bio-resources, as either the source or the target of biotechnology applications (Marine Board, 2010).
- **Mediterranean Science Commission (CIESM) definition for marine biotechnology**: Marine biotechnology is a category of products and/or tools relating to marine bio-resources, as either the source or target of their application. It provides goods and services for innovative industries and/or society as a whole (Not published, presented at an OECD workshop in 2012).

By adding reference to marine organisms or the use of biotechnology in the marine environment, the OECD's broad definition of biotechnology can be applied to define marine biotechnology. The OECD's list-based definition is particularly useful in this regard due to the fact that it can be adapted by adding specific technologies or elements of marine biotechnology (OECD, 2016). The definition for marine biotechnology can therefore be seen to approximate the OECD list-based definition for biotechnology.

The blue biotechnology sector is unique amongst biotechnology sectors in terms of the way that it is defined. For example, whereas red (medical, health and pharmaceutical), green (agricultural), yellow (environmental) and white (industrial) biotechnologies are delineated on the basis of the processes they entail or the markets they serve, blue biotechnology is the only biotechnology sub-sector to be defined by its source material, i.e. marine resources (see Table 2.1) (Kafarski, 2012). Therefore, the characterising feature of blue biotechnology is the first part of the development pipeline: from sampling to discovery and bioprospecting, to research and development (R&D) and initial product development (Figure 2.2). Blue biotechnology has the potential to contribute to a variety of other biotechnology and industry areas. As such, blue biotechnology is not a clear-cut sector. There are important overlaps associated with products of blue biotechnology feeding into other sectors of different colour, such as energy (marine algal biofuels), pharmaceuticals

Table 2.1 Biotechnology sub-sectors, associated colours and basis for delineation (Kafarski, 2012)

Biotechnology Sub-sector	Colour	Basis for Delineation
'Marine' or 'Blue'	Blue	Source of biomaterial
Medical and pharmaceutical	Red	Processes or markets
Agricultural	Green	Processes or markets
Nutritional	Yellow	Processes or markets
Industrial	White	Processes or markets
Environmental protection	Grey	Processes or markets
Management of deserts and arid regions	Brown	Processes or markets
Bioinformatics, computer science and chip technology	Gold	Processes or markets
Law, ethical and philosophical issues	Violet	Processes or markets
Bioterrorism and biological weapons	Dark	Processes or markets

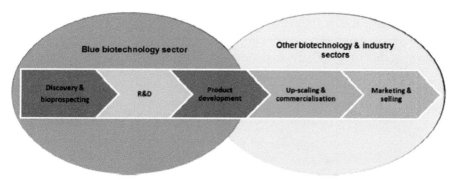

Figure 2.2 Visual representation of the blue biotechnology sector in Europe.
Source: ECORYS, 2014.

(novel antibacterials), cosmetics, aquaculture, food and nutrition, environmental protection and depollution (ECORYS, 2014; OECD, 2013; Marine Board, 2010). Subsequent stages or processes within the value chain become part of the wider biotechnology industry; these are separated from the marine component and should no longer be considered part of the blue biotechnology sector per se, but rather as part of any of the other classical biotechnology sub-sectors (ECORYS, 2014).

It is possible that definitions will change over time and that the distinction between 'blue' and 'marine' biotechnology may disappear. However, within this study, we strictly define blue biotechnology as requiring bio-material sourced from the oceans and define marine biotechnology more broadly as either involving sources from or applications in the marine environment.

2.1.2 Generic Value Chain of Blue Biotechnology

An alternative method for defining the blue biotechnology sector is through analysis of current marine biotechnology stakeholders. Building on the value chain approach, the position of stakeholders within the chain and the variety of activities conducted may then be considered (i.e. R&D, production, services and marketing) (see Figure 2.3) (ECORYS, 2014).

Key components of the generic value chain of blue biotechnology are listed in Figure 2.2 and include sectors such as discovery and bioprospecting. However, steps 2–5 in Figure 2.2 may not always be unique to blue biotechnology:

1. Discovery and bioprospecting: This initial phase of the value chain involves investigating environments and collecting living organisms. Extracts are made from organisms and genes may then be isolated to identify active gene products. Preliminary de-replication may take place at this time, as well as the establishment of preliminary evidence

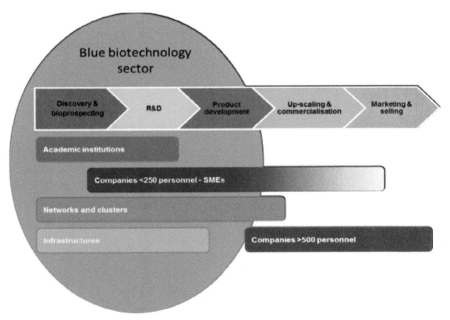

Figure 2.3   The value chain stakeholder composition in the marine biotechnology process. The sector is defined as in Figure 2.2.

Source: ECORYS, 2014.

for activity in lab-bench tests. This stage involves establishing the uniqueness and proprietary position of a particular environment.

2. Research and development: Extracts are taken from organisms during the R&D phase so that molecular components can be identified. Other activities which fall under R&D include: isolation of specific genes and gene products plus identification of their nature; de-replication of molecules and gene sequences/products; molecular characterisation of active molecules; structural identification; confirmation of proprietary position; synthetic strategies; and validation of preliminary bioactivity in further tests.

3. Product development: This step may involve the development of sustainable production strategies, chemical synthesis, gene isolation and the transfer to an industrially-useful organism with effective expression. Other potential activities include a demonstration of scale-up, stabilisation of the production process, preliminary demonstration of cost-efficiency and Life Cycle Analyses. Sufficient material is required during product development to confirm and extend the activity profile and to justify scale-up.

4. Up-scaling and commercialisation: Target organisms or molecules are produced economically and at an industrial-scale during this part of the blue biotechnology value chain. Other aspects include validated and stabilised extraction, purification and derivation processes for target molecules and materials. Positive production economics will also be considered.

5. Marketing and selling: This final step is based on the end-products of the value chain process. End-products may include pharmaceuticals, enzymes, hydrocolloids, nutraceuticals, cosmetic ingredients, biomimetic materials etc.

The value chain appears to become sub-sector specific at the stage of product development. Prior to that (i.e. discovery/bioprospecting, R&D and some aspects of product development) the value chain is normally common to all blue biotechnology applications and is a pre-requisite to the application of blue biotechnology in any given industry. The product development phase is often extensive and specific to the biotechnology or industrial sub-sector for which an application is intended. However, once a product has reached the stage of up-scaling and commercialisation, the 'blue' component diminishes and stakeholders are no longer limited to marine biotechnology, but are part of other biotechnology or industry sectors (ECORYS, 2014).

A number of risks are involved in bioprospecting. Firstly, too many novel organisms and molecules will be found, creating a bottleneck in screening, selecting and identifying desirable bioactivity. Another possible issue is the fact that organisms containing novel molecules may not be culturable in the lab. Even if organisms are culturable, the production of valuable molecules may vary between each batch that is grown. Other risks include the potential that molecules may be too complex for chemical synthesis, some genes may be isolatable but unable to express or transfer to a common industrial system, and successful production of target materials may not be replicable when culture is scaled-up. The associated risks are cumulative and may limit industry end-users' ability to see the opportunities present in blue biotechnology. Small and medium sized enterprises (SMEs), whether acting as facilitators or validators, need to be able to address this issue in order to attract end-user investment (ECORYS, 2014).

2.2 Market

2.2.1 Market Trends

The market associated with application of marine resource biotechnology has grown consistently over recent decades. For a number of commercialised products, the market exceeded several billion USD per annum by the year 2010, with a compound annual growth rate of 4–5% (or 10–12% under less conservative assumptions) (Leary et al., 2009). However, due to the absence of a universally accepted definition for the sector, it is difficult to evaluate its scope, structure and socio-economic performance (ECORYS, 2014). Global Industry Analysts, a market research agency, publishes reports on the approximate value of the blue biotechnology sector and estimate that the sector will reach USD 4.8 billion (EUR 3.5 billion) by the year 2020 (Global Industry Analysts, 2015). A study conducted by ECORYS (2014[1]) calculated that blue biotechnology currently contributes approximately 2%–5% of the total biotechnology industry. This suggests that in 2012 the European blue biotechnology sector may have been between EUR 302 million and 754 million (in terms of revenue). Healthcare biotechnology is considered the

---

[1]ECORYS calculation based on triangulation of ratio of Marine biotechnology compared to the whole biotechnology industry in terms of revenue using table Ernst & Young: Biotechnology Industry report 2013.

biggest and most rapidly growing end-use sector for marine biotechnology (Global Industry Analysts, 2011).

Potential applications of biotechnology in marine environments may include the following:

PUFAs

The discovery of polyunsaturated fatty acids (PUFAs, such as Omega-3 and Omega-6) and their importance for human health has long been established. The extraction of PUFAs mainly from fish has enabled its mainstream use in everyday life. Fish accumulate PUFAs through consumption of algae, and now that extraction of PUFAs directly from algae is possible, efficiency of extraction has increased (Medina et al., 1998). Application of PUFA-related knowledge to the aquaculture industry has for instance shown that PUFA-rich algae also benefits the growth and survival of shellfish (Reis Batista et al., 2013). Applying this knowledge to feedstock may in turn enhance future production of aquaculture and also result in aquaculture products with elevated PUFA concentrations.

Microbiomes

Possible applications in marine pest control include techniques to assess the composition and dynamics of microbiomes. The term microbiome originates from gene sequencing technology in microbiology and refers to an entire microbial population within a specific environmental niche. Microbiomes in different environments have been shown to change in population diversity and density as a function of changes in environmental conditions (for example: change in gut-microbiome in function of dietary shifts). Characterising microbiomes and their dynamics in and around ships (i.e. tanks, outer surfaces, bilges, etc.) can lead to new monitoring systems to check the emergence of environment-damaging organisms on board, and may also lead to advances in bioremediation to degrade organic pollutants in ballast water (Briand, 2011). The same technique can be used to assess fish health and response in rearing in aquaculture.

Coatings

Coatings with anti-fouling or anticorrosive properties are currently being developed and tested (Eduok et al., 2015). Analysis of an anti-fouling bio-coating containing encapsulated bacteria from a Saudi hot-spring has been found to inhibit corrosion. This biocoating may have potential applications for ship hull protection and protection of off-shore installations.

2.3 Sector Industry Structure and Lifecycle – Sub-sectors and Segments

2.3.1 Present and Future Centres of Activity

International Level

To date, blue biotechnology has mostly been confined to the European Union (EU), North America and Far East Asia. Countries that have been highly active in the field of marine biotechnology include: USA, Brazil, Canada, China, Japan, Republic of Korea and Australia (Lloyd-Evans, 2013). Thailand, India, Chile, Argentina, Mexico and South Africa have also displayed increasing interest in marine biotechnology research. The United States has established itself as the leader in marine algal fuels and Asia has taken a leading role in the field of bioinformatics.

India has been heavily pursuing the development of a biotechnology sector, and to this end has been providing financial incentives, venture capital and associated infrastructure. DNA sequencing costs in India and other regions in Asia are generally low and may entice European companies to outsource their operations to these Asian countries. This could potentially weaken Europe's ability to advance their own bioinformatics sector (ECORYS, 2014).

European Sea Basins

Blue biotechnology is analysed as follows:

2.3.2 Atlantic Sea Basin

2.3.2.1 Assessment

The Coordination and Support Action (CSA) study "Marine Biotechnology RTDI in Europe – Inventory of strategic documents and activities" (2012) underlined that in the Atlantic, marine biotechnology already contributes to almost all other industry sectors (e.g. healthcare, environmental bioremediation, cosmetics and food). Many parts of the marine environment are still poorly understood. Therefore, marine resources have so far been largely unexploited and there appears to be significant potential for the discovery of new enzymes, biomaterials, biopolymers, and other associated products such as bio-pharmaceuticals and nutraceuticals. These products could potentially meet the needs for innovation required by industry to remain competitive in global markets.

The Atlantic area plays host to many Centres of Excellence in science, technology and innovation, has a solid reputation in the field of engineering, a stable political and governance system and a number of knowledge-based SMEs. This represents an exclusive opportunity for collaboration to improve the existing resource base and create new knowledge-based and internationally-traded goods and services that will improve the quality of life for local populations (Calewaert et al., 2012).

The following research issues have been identified by Calewaert et al. (2012) as of high importance for the Atlantic Sea Basin:

1. Molecular biology investigation in life science. Genomic and meta-genomic analysis of systematically sampled marine organisms, including microorganisms (i.e. bacteria, viruses, archaea, pico- and microplankton), algae and invertebrates;
2. Cultivation of marine organisms and cell lines. Development of technologies to isolate and culture previously uncultivated microorganisms. Developing culture methods for vertebrate and invertebrate cell lines for the production of active compounds;
3. Bio-mass production. Development and application of new and effective systems, including bio-engineering, bioreactors and cultivation systems, for the production, use and transformation of biomass from marine organisms. The production systems and organisms are optimized to target specific applications (e.g. biorefinery and aquaculture);
4. Marine model organisms. Identify and prioritise new organisms of marine origin to increase life science knowledge and provide new opportunities for biotechnological exploitation;
5. Production of biofuel from marine algae.

2.3.2.2 Main initiatives

Many infrastructures and initiatives related to marine biotechnology R&D are already present in countries of the Atlantic Sea basin area. However, there are as yet no major capacities organised at the regional level. An Atlantic macro-regional strategy is currently under development which may help to create a wider framework for regional collaboration. This strategy could also assist with addressing common goals associated with science and technology as well as targets linked to marine biotechnology.

Regional funding is mostly provided by the European Regional Development Fund (ERDF)[2], as well as through various other interregional co-operation programmes that aim to encourage collaboration between different regions within the EU (Calewaert et al., 2012).

---

[2]Based on the Seas-ERA (www.seas-era.eu) Atlantic Sea Basin Strategic Research Agenda (SRA).

2.3.2.3 Way forward

The European Atlantic is in a good position to take full advantage of marine biotechnology potential. With an established maritime heritage, extensive marine territories covering a wide variety of marine habitats (including the deep ocean) and renowned capability in the field of marine sciences, the European Atlantic Sea Basin area has plentiful opportunity to develop and exploit marine biological resources (Calewaert et al., 2012).

An EU Strategy for the Atlantic Region (EUSA) was launched in 2011 and represents one of the main science-policy developments currently implemented in the area. The aim of the EUSA is to provide a strategic framework and action plan to enable improved cooperation at the Union level. This will be achieved by improving the coordination of actions across a number of policy areas (Calewaert et al., 2012). Science, R&D and the management of research infrastructures are aspects of policy which stand to gain from improved regional coherence with the added bonus of potentially promoting technology transfer and innovation. This strategy is likely to significantly influence and benefit regional marine biotechnology activities (Calewaert et al., 2012).

2.3.3 Baltic Sea Basin

2.3.3.1 Assessment

The "Study on Blue Growth, Maritime Policy and the EU Strategy for the Baltic Sea Region" conducted by the EC (2013), identified the potential for Blue Growth in each of the EU Member States (MS) of the Baltic Sea Region (BSR) and at sea basin level. The study revealed that the blue biotechnology industry in the BSR is still nascent and very much focused on R&D. Blue biotechnology still has limited economic performance (it doesn't rank among the largest or fastest growing maritime economic activities (MEAs) in any MS in terms of gross value added (GVA) and employment size) and plays only a small role in the development plans of the region. Data related to GVA and employment MEA is not available for the period 2008–2010 (this is mostly because the data is non-existent but also because data is too limited to be quantified or not captured by statistics). Only Germany could be said to have highly developed biotechnology in the region. While competence centres and private companies working on blue biotechnology topics can be found in all countries around the Baltic Sea, Germany and in particular the State of Schleswig-Holstein is considered as the leader in this field and was selected as the benchmark case for blue biotechnology within the Baltic Sea Blue

Growth study. Denmark has also made strides to foster this sector, setting a strategic direction for the nation's blue biotechnology industry. In addition to Germany, Poland also ranks this sector among the maritime economic activities with most future potential in the years to come.

2.3.3.2 Main initiatives

Initiation of the SUBMARINER (Sustainable Uses of Baltic Marine Resources) Project represents the start of strengthening institutional set-ups for transnational blue biotechnology cooperation within the Baltic Sea area.

Another important initiative is ScanBalt® fmba (or ScanBalt). ScanBalt is an organisation for the Baltic Sea or Nordic-Baltic Region's Health and Bio Economy community. ScanBalt is a non-profit member association and functions as a service provider for members and also promotes the development of the ScanBalt BioRegion as a globally competitive macro-region and innovation market (Calewaert et al., 2012).

2.3.3.3 Way forward

The Baltic Sea Region has a long-standing custom of pursuing transnational cooperative programmes, which is an essential requirement for converting blue biotechnology research into commercially successful products and applications. However, at present, blue biotechnology plays only a minor role in Member State development strategies. This sector could be supported at sea-basin level by establishing joint research initiatives and by bridging the gap between basic and applied blue biotechnology research. The development of suitable funding structures, research networks and clusters would also be helpful (EC, 2014). By creating a targeted research strategy for marine biotechnology in the BSR, regional differences could be turned into advantages. For example, joint ventures between laboratories in the Eastern Baltic and sophisticated pharmaceutical industries in the West would provide mutual benefits (Calewaert et al., 2012).

2.3.4 Mediterranean Sea Basin

2.3.4.1 Assessment

The capacity and potential for marine biotechnology in Mediterranean counties is currently being mapped and some profiles can already be viewed on the Mediterranean Science Commission (CIESM) website. Mapping results are expected to raise awareness of this field of R&D as well as encourage the development of new enterprises both within and beyond the Mediterranean Sea Basin region (Calewaert and McDonough, 2013).

2.3.4.2 Main initiatives

No comprehensive regional strategy focusing specifically on marine biotechnology R&D yet exists within the Mediterranean Sea Basin. However, as stated by Calewaert et al. (2012), general marine science topics in this area may be studied by organisations such as CIESM or through projects like the SEAS-ERA scheme. Since the Mediterranean is regarded as one of the world's most important locations in terms of marine biodiversity (contributing between 4% and 18% of the World's marine species) the SEAS-ERA Project has set the following research priorities in the field of blue biotechnology: Bioprospecting for Marine Drugs and Fine Chemicals; Technologies to Increase Sustainability of Aquaculture Production; Biofuels from Micro- and Macroalgae. The CIESM has a Committee for Marine Microbiology and Biotechnology and their specific research areas include ecology and biodiversity of marine prokaryotes (Archaea and Bacteria), viruses and hetero- and autotrophic protists (i.e. phytoplankton), microbial food web interactions and microbial pathogens. An additional research initiative coordinated by the CIESM, operating within their Marine Economics Research Program, involves marine genetic resources and has resulted in a study focused on the economic models of bioprospecting (ECORYS, 2014).

2.3.5 Caribbean Sea Basin

No comprehensive information is available for the Caribbean Sea basin with regards to marine biotechnology activity. However, the CSA report "Global landscape of Marine Biotechnology RTDI" (Lloyd-Evans, 2013) provides a comprehensive list of research centres in Mexico involved in marine biotechnology, and indicates that bioprospecting is a particular field of interest. It is likely that the Caribbean Sea basin will prove promising as an area for bioprospecting and sampling for European R&D.

2.3.6 Business Lifecycle Stage

2.3.6.1 Overview of sub-sectors

Blue biotechnology can contribute towards several other biotechnology sectors (Figure 2.4). Sectors chosen for review in the ECORYS (2014) study include health, cosmetics, food, energy, aquaculture, environmental services (such as environmental protection and depollution) and other industrial applications (see Table 2.2 for details of sectors). The proportion of marine biotechnology stakeholders associated with any of the other biotechnology

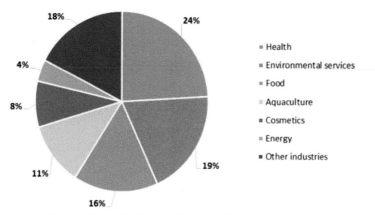

Figure 2.4 Distribution of stakeholders by sub-sector.

Source: ECORYS, 2014.

Table 2.2 Potential marine biotechnology products and services

Sub-sector	Potential Product Areas	Specific Product Areas
Health	Phrmaceuticals	Anti-cancer drugs, anti-viral drugs, novel antibiotics; wound healing; anti-inflammatory; immunomodulatory agents
	Biomaterials	Bioadhesives, wound dressings, dental biomaterials; alternative disinfectants (being more environmentally friendly and avoiding resistance development); medical polymers; dental biomaterials; coating for artificial bones that enhance biocompatibility; medical devices.
	Other	Tissues regeneration, 3D tissue culture
Cosmetics	Functional ingredients	UV-filter, after sun; viscosity control agents; surfactants; preservatives; liposomes, carrier systems for active ingredients; regulation of sebum;
	Raw materials	Micro and Macro-algae extracts; colourants, pigments; fragrances; hair-styling raw materials
Food	Functional foods	Prebiotics; omega 3 supplements;
	Nutraceuticals	Useful as antioxidants, anti-inflammatory; fat loss; reducing cholesterol; anti-HIV properties, antibiotic and mitogenic properties anti-tumour; iodine deficiency, goitre and myxoedema; anti-influenza; treatment of gastric ulcers;
	Food products and ingredients of marine origin	A stabiliser, suspending agents, bodying agents, makes a good jelly, prevents separation and cracking, suspending agent, foaming agent.

(Continued)

Table 2.2   Continued

Sub-sector	Potential Product Areas	Specific Product Areas
	Food packaging and conservation	Films and coatings with antimicrobial effects
Energy	Renewable energy processes (micro and macroalgae)	Microalgae; produce polysaccharides (sugars) and triacylglycerides (fats) that can be used for producing bioethanol and biodiesel. Macroalgae; large scale cultivation of macroalgae (seaweed) for the production of biofuel
	Microbial Enhanced Oil Recovery (MEOR)	Enhanced oil recovery and productive life oil reservoirs.
	Industrial additives	Anti-blur additives for textile printing, binding agent in welding rods, drilling fluid
Aquaculture	Seed	Surrogate broodstock technologies; transgenic approaches; developing culture species; selective breeding of existing cultured species for novel and disease resistant hybrids.
	Feed	Fish oils produced from algae; pigments in fish feed
	Disease Treatment	Diagnosis; treatment of disease; disease-resistant strains.
	Aquaculture systems	Treatment of re-circulated water.
Marine environmental health	Bioremediation	Biosurfactants (BS), bioemulsifiers (BE) induce emulsification, foaming, detergency, wetting dispersion, solubilisation of hydrophobic compounds and enhancing microbial growth enhancement; marine exopolysaccharides (EPs) induce emulsification.
	De-pollution	Removal of toxic elements including metals (lead, cadmium, zinc and metal ions); removal of dyes.
	Bio-sensing	Biomarkers and biosensors for soil sediment and water testing; to identify specific chemical compounds or particular physio-chemical conditions, presence of algal blooms, human health hazards.
	Antifouling	Reduce drag and fuel use for boat-going vessels without any negative environmental impacts.
	Bio-adhesives	Underwater industrial adhesives.
Other	Bio-refineries (separation of functional biomass components)	Biodiesel; feedstock for the chemistry industry; essential fatty acids, proteins and carbohydrates for food, feed for animals (replacement of feed with fishmeal) and production of proteins and chemical building blocks;

Source: ECORYS, 2014.

sectors can indicate the relative significance of marine biotechnology to these different fields. Stakeholders are commonly involved in more than one sector, indicating a variety of product portfolios. This also highlights the fact that academic groups routinely conduct a range of research activities associated with biological diversity rather than focusing on just one specific application field (i.e. one particular sector). Figure 2.4 indicates the distribution of stakeholders by sector. The key sectors in which marine biotechnology stakeholders participate are health (24%), environmental services (19%), food (16%) and other industrial applications (18%).

2.3.6.2 Sub-sector lifecycle stage

The lack of clear economic differentiation in blue biotechnology makes it difficult to find evidence for the stage of lifecycle that each associated sub-sector is in. Patents can be used as an indicator of sector development and, together with scientific publications, are a measure of output performance. Patents and publications can also be used to determine the potential strengths of a region, country or organisation with regards to this particular type of intellectual property protection (ECORYS, 2014). It should be recognised, however, that assessment of the patent situation does not always prove the economic potential of a specified sector. This is because other strategies for valorisation also exist. Patenting is regularly avoided due to the high associated costs and efforts, particularly when SMEs are involved. Therefore, this does not necessarily indicate a lack of commercialisation, but suggests a different approach. The majority of patents deal with compounds or genes rather than with particular production processes, leading to at least two possible consequences. Firstly, patents often concern more than one application field and thus, many patents belong to more than one sub-sector. For example, patents on "natural products" belong on average to three sub-sectors. By patenting the resources themselves, use of these compounds and genes in any process is more difficult for competing parties. Secondly, it indicates only the initial stages of product development. Nonetheless, since the costs of patenting are high, a patent can generally be interpreted to indicate high potential for commercialisation.

2.3.7 Trend Analysis of Patents

Trends in patenting-rate over time can indicate commercial profitability of patents in a subsector. According to ECORYS (2014), the number of patent publications has increased exponentially over the last 50 years, with a notable surge between years 2000 and 2010. Rates of increase were comparable across almost all sub-sectors. Analysis of trends up to the year 2020 suggests that the number of patents in most biotechnology sectors will stabilise whilst the cosmetics and energy sectors are likely to rise by a further 10–20% (ECORYS, 2014). In years 2006 and 2010 there was a decrease in the number of patent publications for nearly all fields linked to marine biotechnology. In 2011 and 2012 patenting increased again, but did not reach the levels observed in 2008 and 2009. These dates correspond with fluctuations in the global economy, suggesting that this sector is sensitive to larger economic factors. The majority of patents currently belong to the health sub-sector, indicating that this is likely to be the most financially interesting industry in the near future (see Figure 2.5) (ECORYS, 2014). At present, there is a lack of blue biotechnology products and services on the market, which corresponds with the fact that blue biotechnology is considered a 'young' field of biotechnology. Through observation of the patent categories, the health, cosmetics and food sectors appear to be the largest 'users' of blue biotechnology but their products have extensive trials and testing processes, extending the time taken to reach the market. Other associated subsectors are energy, aquaculture and marine environmental services. Collectively these subsectors are diverse and dynamic in nature, at different stages of development and have so far encountered different stages of growth.

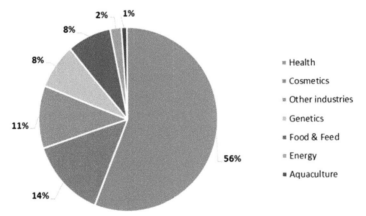

Figure 2.5   Distribution of patents across sub-sectors.

Source: ECORYS, 2014.

2.4 Working Environment

2.4.1 Employment and Skills Availability

Due to the broad nature of blue biotechnology, it is difficult to determine the economic value and employment that this sector creates. Furthermore, it is not possible to evaluate the working environment according to each Sea basin. Based on the stakeholder database developed by ECORYS (2014), total employment is currently thought to be between 11,500–40,000 people. These are usually high-end jobs and are the result of substantial public investment in education and training.

2.4.2 Revenues

Annual revenue for the European biotechnology industry is estimated to be approximately EUR 15 billion. Extrapolation from the entire EU bio-economic sector (using conservative estimates that blue biotechnology accounts for only 2–5% of the whole sector) suggests an annual turnover between EUR 302–754 million. Yearly growth rate of the EU blue biotechnology sector is in the region of 4–5%, slightly below that of biotechnology as a whole (6–8%) (ECORYS, 2014). In terms of end-use, healthcare biotechnology constitutes the largest and fastest growing end-use segment for blue biotechnology (Global Industry Analysts, 2015).

2.4.3 Stakeholders

An assessment conducted by ECORYS (2014) of a representative group of blue biotechnology stakeholders found that there are nine forms of stakeholder organisation. Academic institutions (universities or research institutes), SMEs and blue biotechnology network clusters are the main stakeholder categories. Large companies and infrastructure institutions were also found to be important stakeholders. The remainder were funding agencies, policy makers, medium companies (250–500 employees) and outreach professionals. Many stakeholders are involved in more than one industry sector, with the "other industries" sector as a common second field. This is particularly the case for SMEs that work in a number of product fields, e.g. developing processes for multiple purposes. Higher proportions of stakeholders are present in the health, environmental services and food sectors than in any of the other industry sectors.

ECORYS (2014) found that larger companies (more than 500 employees) do not typically specialise in or limit themselves to blue biotechnology.

Large corporations are typically broader in scope, work mostly within one particular biotechnology/industry sector and have links to blue biotechnology through specialised research centres. They play an important role in product up-scaling and commercialisation as well as in marketing.

2.4.4 Role of SMEs in Blue Biotechnology

SMEs are important actors in the blue biotechnology value chain as they bridge the gap between public sector R&D and commercialisation of products. Blue biotechnology SMEs are generally responsible for the initial product development stage of the value chain: identification, validation and de-risking of industrial opportunities related to marine bioresources (ECORYS, 2014). SMEs tend to be single-focus marine bioactives companies, operating at the high risk 'cash-burn' stage where screened products are converted into potential products for up-scaling and commercialisation. Due to the inherent risks associated with this phase, financing (often from venture capital) is unpredictable. SMEs can therefore be very vulnerable. A 17% fall in venture capital investment in SMEs was observed between 2008 and 2014, illustrating the unstable conditions that SMEs may have to deal with (ECORYS, 2014). This period corresponds to the global financial crisis, so is not unique to this sector. In addition, two SMEs focused on blue biotechnology experienced bankruptcies in 2013: AquaPharm[3] and BioAlvo[4].

The interface between SMEs and the downstream (large) corporations is emerging as one of the weakest links in the value chain (ECORYS, 2014). As noted by the Marine Board (Marine Board, 2010), most industrial contributions to marine biotechnology in Europe are generated through specialised SMEs, assuming most of the risks inherent in R&D and characterised by a rapid turn-over. Given the economic crisis in Europe and the consequent reductions in venture capital and public funding, there is a danger that the capacity of marine biotechnology SMEs to develop new technologies,

---

[3] Aquapharm Bio-discovery Ltd (founded 2000) was one of the first UK marine biotechnology companies dedicated to the discovery and commercialisation of novel compounds from the marine microbial biosphere, a relatively untapped renewable source of marine bio-diversity.

[4] Bioalvo, the Biotech for Natural Products, is a Portuguese start-up company that focused on fully integrated biotech solutions to maximise natural products market applications in areas as diverse as cosmetics, household products, nutraceuticals, pharmaceuticals or even industrial. It was ranked at the TOP 6 best companies in Europe's Most Innovative Biotech SME Award 2011.

processes and products may decline unless bigger companies are involved as investors.

The weak partnerships between researchers and industry has previously been underlined by the OECD (2013) in their report on blue biotechnology. According to this report, one big challenge is the timing of engagement between researchers and industry: '*Engagement with industry is often regarded as incidental to basic R&D or as post-research, downstream activity. This can leave R&D results stranded, either without a ready market or unable to reach the anticipated market for technical or feasibility reasons.*' Therefore, the OECD recommends an earlier collaboration with industry (within funded R&D projects) which would help to make sure that products of blue biotechnology research are appropriate for up-scaling and commercial production. However, this may also create concern in terms of divulging knowledge of downstream research, and therefore impede the development of research itself, due to confidentiality issues that the industry might want to push forward. EU rules on the management of Intellectual Property Rights (IPRs) in EU funded projects help in solving this issue by prescribing safeguards for confidentiality within the dissemination obligation.

2.4.5 Infrastructure and Clusters

As stated by ECORYS (2014), infrastructure institutions refer primarily to 'Marine Research Infrastructures (MRIs)[5] which support blue biotechnology activities and underpin the discovery and bioprospecting, R&D and to some extent product development stages in the value chain'. MRIs can be broken down into six clusters: research vessels and underwater vehicles; in situ data acquisition systems; satellites; experimental facilities for biology and ecosystem studies; marine data facilities; marine land-based facilities for engineering (for a comprehensive analysis refer to Annex 7 of ECORYS, 2014). Vessels and platforms required for prospecting and capturing marine resources can be extremely expensive to operate and these inherent costs must be properly understood when considering a blue biotechnology venture. Costs may be even higher if exploration takes place in deep water, particularly when extreme environments such as hydrothermal vents need to be sampled. Extreme marine environments are considered to have high potential for the

---

[5]Research infrastructures are facilities, resources and services used by the scientific community to conduct research and include libraries, databases, biological archives and collections (e.g. biobanks), large and small-scale research facilities (e.g. laboratories), research vessels, communication networks, and computing facilities.

discovery of innovative biological material, with specialised micro-faunal communities that have evolved to function under unusual temperatures, pressures and/or salinities. Therefore, the high costs of working in these areas must be anticipated (ECORYS, 2014).

Clusters and networks typically involve scientists, organisation of research activities and associated infrastructures. These groups can therefore be linked to the initial stages in the blue biotechnology value chain. For example:

- PôleMer France, consisting of the Pôle Mer Méditerranée and the Pôle Mer Bretagne, which has actively involved itself and its SME members in marine biotechnology projects;
- ScanBalt in northern Europe, which is working within the EU Strategy for the Baltic Sea and has established a flagship project SUBMARINER, sustainable uses of Baltic marine resources, with EU region support;
- The German industrial biotechnology cluster CLIB 2021 includes several marine-orientated SMEs amongst its members, including Bitop AG, C-LEcta GmbH, DIREVO Industrial Biotechnology GmbH, Evocatal GmbH and Swissaustral Biotech SA.

There are a number of initiatives and networks in Europe which specifically exist to coordinate marine research infrastructures and to facilitate access to them. For example, the Marine Biotechnology ERA-NET[6] is a consortium of national funding bodies to pool resources and undertake joint funding of transnational projects in the area of marine biotechnology.

2.4.6 Public Policy Regulatory Framework

2.4.6.1 International and regional legal frameworks

All activities undertaken in the marine environment are subject to international law of the sea, codified by the United Nations Convention on the Law of the Sea (UNCLOS) of 1982. However, this Convention does not refer to blue biotechnology, nor to marine genetic resources, as it pre-dates most of the

---

[6] The **vision of the Marine Biotechnology ERA-NET (ERA-MarineBiotech or ERA-MBT) project** is to support Europe's marine biotechnology community to participate in a lasting enterprise-driven network that adds value to marine biological resources in ways that nurture and sustain the lives of European citizens. The ERA-MarineBiotech is therefore designed to deliver **better coordination** of relevant national and regional Research, Technology, Development and Innovation (RTDI) programmes in Europe, **reducing fragmentation and duplication**, and paving the way for common programmes and cooperation in the provision and use of research infrastructures.

scientific discoveries that resulted in development of these sectors. Accessing marine genetic resources is also subject to the Convention on Biological Diversity (CBD) of 1992 and its Nagoya Protocol signed in 2010 and which entered into force in 2014.

United Nations Convention on the Law of the Sea

According to the law of the sea (1982), several obligations have to be fulfilled before and while undertaking marine scientific research, such as:

- Request to the coastal state for a permit to undertake marine scientific research in its Exclusive Economic Zone (EEZ) or continental shelf (Article 248 UNCLOS);
- Report and share with the coastal states the data, samples and research results (Article 249 UNCLOS);
- To cooperate on a global and regional level (Articles 242–244 UNCLOS);
- If marine scientific research of biological material or sampling of marine genetic resources is undertaken in Areas Beyond National Jurisdiction (ABNJ), access is free (so far) and needs to be conducted exclusively for peaceful purposes;
- Use of appropriate scientific methods;
- Shall not unjustifiably interfere with other legitimate uses of the sea;
- Shall be in compliance with all relevant regulations including adoption of necessary measures for protection of the marine environment (Part XII and XIII UNCLOS).

In 2015, a decision was made by member states of the United Nations to begin negotiations for an Implementing Agreement to UNCLOS with the aim of regulating biodiversity in ABNJ (UNGA resolution 69/292, 2015). This agreement will likely have implications for accessing and utilizing genetic resources derived from ABNJ, in terms of benefit-sharing (UNGA resolution 66/231, 2011).

Convention on Biological Diversity and Nagoya Protocol

Accessing marine genetic resources in maritime areas within national juris-diction is subject to the prior informed consent (PIC) of the provider country (in case the provider's legislation requires so); to the negotiation of mutually agreed terms (MAT) on utilisation of the accessed genetic resources and to the share of the benefits arising from such utilisation. Therefore, before sampling the seas in areas within national jurisdiction, it is crucial to verify whether the national legislation of that country prescribes any constraints in terms of access and benefit-sharing (ABS). This has an influence on and potentially raises the burden of every scientific expedition in the sea, which is usually undertaken with basic research purpose and which is at the basis of the pipeline of blue biotechnology (ECORYS, 2014).

The Nagoya Protocol has been implemented in the EU (Regulation (EU) No 511/2014, 2014). It does not regulate access (every EU Member State is free to regulate access to its own genetic resources), but it regulates users' compliance. Therefore, the Nagoya Protocol has a more significant impact on parts of the research pipeline following sampling and bioprospecting. Users are obliged to exercise due diligence in order to establish that genetic resources and associated knowledge have been accessed in accordance with applicable ABS legislation. In addition, benefits must be fairly and equitably shared upon mutually agreed terms, also in accordance with applicable legislation. Therefore, users shall transfer information on where the utilized genetic resources have been collected, when and under which legal circumstances (PIC-MAT and benefit-sharing). This regulation has only recently been implemented, so it is still too early to evaluate the impact it will have on blue biotechnology (ECORYS, 2014).

Beyond these international regulations, the research and product development steps of blue biotechnology have to comply with international, regional and national obligations on biosafety and any other relevant rules concerning biotechnology activities. However, these rules and obligations go beyond the scope of the present chapter as they are not unique to blue biotechnology, but instead apply to the whole biotechnology sector.

2.4.6.2 European policy framework

In common with all sectors of the Blue Economy, the primary strategic legal and policy framework is the Marine Strategy Framework Directive (MSFD). In addition, the Sea basins strategy elaborated by the EC has an influence on research activities in the field of blue biotechnology. A number of strategic documents have been published as a result of science, policy and research initiatives over the last decade. The EC has acknowledged the potential of blue biotechnology in Europe through its Communication on Blue Growth (COM/2012/494) and European Bioeconomy Strategy (COM/2012/60), both of which identify blue biotechnology as a sector that has the possibility to contribute to bioeconomy and to economic growth in general. Furthermore, EU research policy has been responsive to the growing awareness of the importance of blue biotechnology: the EU has funded key research on blue biotechnology in its Framework Programmes for Research FP6, FP7 and Horizon 2020. The EU's Horizon 2020 Strategy and support programme specifically addresses blue biotechnology and marine biomass as contributors to the economy of the future (COM/2012/494). However, no comprehensive and specific blue biotechnology policy yet exists in Europe, although Ireland, Denmark and Norway do have relevant national policies in place. Most countries support blue biotechnology R&D under a wider strategic umbrella, either within an overarching science and technology strategy, as part of a more general marine or biotechnology research plan or as a combination of both (Table 2.3) (ECORYS, 2014). Portugal, for example, does

Table 2.3 Overview of European countries with the level of focus and available mechanisms to support marine biotechnology activities, as identified by the CSA Marine Biotechnology project's preliminary landscape profiling exercise (Calewaert et al., 2012). (Adapted from: ECORYS, 2014).

Countries with a dedicated plan, programme or strong policy focus on marine biotech	Countries where marine biotech is supported via more wide-scope programmes and/or instruments (general science and technology plans, marine science plans and/or biotechnology plans/strategies)		
	Countries with considerable interest and/or activities in marine biotechnology research and development*	Countries with some interest and activities in marine biotechnology research and development*	Countries where there is only limited marine biotech focus and activities*
• Ireland	• Belgium***	• Croatia	• Austria**
• Denmark	• France	• Greece	• Bulgaria
• Norway	• Germany***	• Finland**	• Estonia**
	• The Netherlands	• Iceland	• Latvia**
	• Poland	• Romania	• Lithuania**
	• Portugal	• Slovenia	• Malta**
	• Italy**	• Turkey	• Switzerland**
	• Spain		• Ukraine**
	• Sweden		
	• UK		

Based on the information that could be collected within the scope of the CSA Marine Biotechnology;
**Countries for which no or only limited information could be collected within the scope of the CSA Marine Biotechnology;
***Countries with a federal structure with considerable activities in one or more specific coastal regions

not have a dedicated blue biotechnology strategy or plan, but a more generic marine strategy (National Strategy for the Sea) containing ample reference to the strategic importance of blue biotechnology research while currently, in practice, the R&D activities in this field still remain very fragmented. In a growing number of countries there is also significant focus on support for activities that stimulate what is called the "biobased economy", echoing largely the EC's strategy and action plan "Innovating for Sustainable Growth: a Bioeconomy for Europe" which was adopted in early 2012 (Calewaert et al., 2012). The report also underlined difficulties in gathering up-to-date information in the different countries.

The CSA Marine Biotechnology analysis (Calewaert et al., 2012) revealed that the national priorities identified include the following:

- Marine bioprospecting/biodiscovery (in particular for human health and new industrial compounds);
- Development of robust, biotechnology-based state of the art R&D tools and infrastructures tailored for blue biotechnology;
- Molecular aquaculture;
- Biomass production for bioenergy and fine chemicals;
- Marine environmental biotechnology applications and bio-sensors in the context of the European Marine Strategy Framework Directive (MSFD).

2.5 Innovation

2.5.1 State of Technology and Trends

Europe is active within the R&D stage of the blue biotechnology value chain and generates almost a third of the scientific publications in this field. However, a striking difference emerges when comparing scientific activity to trends in patent publication. Europe represents only 13% of patents filed in connection with new marine molecules, suggesting limited success in developing products from promising resources. In contrast, Japan and China appear far more active in patent publication than in scientific publication (ECORYS, 2014). Therefore, it seems that whilst Europe is strong in coordinating research activities in the early stages of the value chain, there may be a lack of coordination further along the chain between those conducting research or initial product development (mainly research institutes and SMEs) and investors (larger companies with the resources to up-scale and commercialise a product) and the industry within which the blue biotechnology application will be used.

2.6 Investment

Blue biotechnology is a new area of biotechnology that is considered rather 'invisible' by current key players. The sector is complex and from the outside there is little understanding of what exactly it is. As such, blue biotechnology is seen as fairly unattractive to investors and investment has so far been hard to come by (ECORYS, 2014).

In the context of blue biotechnology, research institutes and universities are fundamental to the discovery, bioprospecting and R&D phases, and are also central to research associated with the identification of new species and molecules from different marine environments. SMEs are similarly focused on the earlier stages of the value chain, concentrating efforts on identification, validation and de-risking of industrial opportunities linked to marine biological resources. This is because for SMEs these stages often represent a cost chain (in other words, the cash-burn stage prior to income-generation). Nevertheless, SMEs are commonly also the most active generators of innovation, with the generic business model based on a very diverse product portfolio, often comprising of non-marine in addition to marine related services. SMEs tend to be absent from industrial production of natural marine products, for reasons mostly linked to high capital expenditure. They will also not be involved in the commercial-scale or demonstration-scale levels of energy production from algae, again due to the associated high capital expenditure.

ECORYS (2014) found that financing is a major issue for SMEs involved in blue biotechnology. Typically, an investment company will have only one marine-orientated/ -involved company in its portfolio. Therefore, in the absence of easy access to investment, publicly funded research collaborations are usually part of a funding model and SMEs may work in collaboration with researchers at universities or institutes and also with larger industrial companies. Universities and research organisations are frequently involved in the stages from bioprospecting to identification and characterization, but may also be involved in industrial adaptation, often as part of contract funding by industry or publicly funded, industry-facing consortia. As a result of the cash-limitations associated with SMEs, plus the limited power they have to bring blue biotechnology products to market, they require downstream linkages to end-users to whom they can sell or license their innovations, products and processes or who may become their exits through trade sale, and to investors who can help them survive longer while they validate and de-risk their developments. The difficulty for SMEs in maintaining momentum through the value chain when blue biotechnology is being applied to biomedical and industrial applications has been recognised by CIESM. As an innovative policy initiative, the CIESM advocates linking SMEs with biotechnology associations, venture capitals, financing bodies and other stakeholders who can help them tackle financial challenges and constraints (Briand, 2011).

According to ECORYS (2014), there is at present no comprehensive European inventory of micro-, small- and medium-sized enterprises working in the field of blue biotechnology. A brief scan for this type of information returns more than 140 SMEs working on various aspects of the marine bioresource value chain.

A literature review and stakeholder discussions (conducted as part of the public consultation launched by the EC in November 2013 and also in various stakeholder workshops organised on blue biotechnology) indicated that the lack of coordination and collaboration between academia and industry at the EU level was the biggest barrier to the development of blue biotechnology, even though it was noted that some examples of productive partnership do exist, such as the open innovation approaches adopted both by Unilever and P&G. ECORYS (2014) also suggest that there may be a lack of collaboration between investors, SMEs and industry in relation to product development, up-scaling and commercialisation. Stakeholders identified the need for an interface between industry, research and policy because the approach to blue biotechnology research in Europe is still fragmented. The Marine Biotechnology ERA-NET does in fact aim to close this loophole and improve coordination between funding agencies. Important efforts have recently begun at the national, regional and European level to create clusters, initiatives and networks with the aim of providing a coherent framework for blue biotechnology activities. However, at present there are still too few platforms through which investors and SMEs can be brought together and in general the number of clusters remains small compared to the number of areas that could potentially use blue biotechnology to assist with regional development (ECORYS, 2014).

2.7 Uncertainties and Concluding Remarks

2.7.1 Bottlenecks and Way Forward

The EU blue biotechnology sector is not yet fulfilling its true potential. This is likely due to a number of barriers specific to the EU blue biotechnology sector (ECORYS, 2014):

- Difficulty in sampling the huge diversity of resources;
- Potential high cost of sampling some of these;
- The consequent preponderance of public funding for Research and Development;
- The complexity of property rights under marine governance mediated by UNCLOS;
- The lack of clarity on the mechanism for benefit sharing, particularly in marine systems with regards to the Nagoya Protocol;
- The uncertainty of the status of genetic resources in Areas Beyond National Jurisdiction;
- The dependence upon vulnerable SMEs and high risk investments to translate R&D results into a marketable product for commercialisation;
- Problems of economic data availability within a poorly defined sector, and
- Weak coordination between public research, SMEs and investors, due to a low number of clusters compared to other sectors.

Blue biotechnology still needs to deliver a huge amount of basic research, given that marine biotechnology is a relatively new area and considering the current low level of knowledge on marine biodiversity. It might be the case that incentives are needed for all key players to ensure that the whole innovation and development pipeline is established (OECD, 2013). ECORYS (2014) found that EU competitiveness in the field of blue biotechnology lies in support of R&D activities. The EU appears to be particularly strong in developing important infrastructure, financial support for companies involved in research and innovative new ways to access marine biological resources. The ability for researchers and companies to access new marine resources is crucial and may currently be limiting the European blue biotechnology sector. As competition between countries increases, it is thought that access to material (particularly from extreme environments) will become more difficult. Access will also be influenced by the development of legislation in coastal states concerning protection of genetic resources within their EEZs.

Several cross-cutting and interwoven barriers currently exist with regards to the development of the blue biotechnology sector. One of the most significant barriers is related to the fact that blue biotechnology has so far been sponsored and promoted mainly by policy bodies and rather ignored by "the sector" (i.e. large companies) which has all the means to make it a success. Other issues are associated with benefit sharing from the discovery of new marine biological resources, both on the high seas and between states.

The lack of clarity can cause legal uncertainty and risks to investment in terms of the source and traceability of material used in blue biotechnology products. These uncertainties also have implications for policy required to overcome barriers and to help the EU reach its full blue biotechnology potential.

References

Arrieta, J., Arnaud-Haond, S., Duarte, C. M., (2010). What lies underneath: Conserving the Ocean's Genetic Resources. Proceedings of the National Academy of Sciences 107(43): 18318–18324.

Blunt, J. W., Copp, B. R., Keyzers, R. A., Munro, M. H., Prinsep, M. R., (2013). Marine natural products. Natural Product Reports 30: 237–323.

Børresen, T., Boyen, C., Dobson, A., Höfle, M., Ianora, A., Jaspars, M., Kijjoa, A., Olafsen, J., Querellou, J., Rigos, G. and Wijffels, R. H., (2010). Marine Biotechnology: A New Vision and Strategy for Europe. Marine Board-ESF Position Paper, 15. www.marineboard.eu/file/45/

Briand F (Ed.) (2011). New Partnerships for Blue Biotechnology Development: innovative solutions from the sea. Proceedings of the CIESM International Workshop, Monaco, 11–12 Nov 2010. http://www.ciesm.org/WK_BIOTECH_REPORT_2010.pdf

Calewaert, J-B., Piniella, Á. M. and McDonough, N., (2012). Marine Biotechnology RTDI in Europe – Inventory of strategic documents and activities. Deliverable No. 3.5. Inventory report of marine biotechnology RTDI in Europe. Part of Task 3.1. Inventory of Marine Biotechnology RTDI Strategies, Programmes and Initiatives Report. Marine Board-ESF http://www.marinebiotech.eu/sites/marinebiotech.eu/files/public/library/CSA%20project%20reports/Marine%20Biotechnology%20RTDI%20in%20Europe%20Inventory%20of%20strategic%20documents%20and%20activities.pdf

Calewaert, J-B. and McDonough, N., (2013). Marine Biotechnology RTDI in Europe – Strategic Analysis. Deliverable No. 3.6 Report on strategic analysis of marine biotechnology RTDI in Europe. Part of Task 3.3. Preliminary Analysis of the European Marine Biotechnology RTDI Landscape. Marine Board-ESF http://www.marinebiotech.eu/sites/marinebiotech.eu/files/public/library/CSA%20project%20reports/Marine%20Biotechnology%20RTDI%20in%20Europe.pdf

Convention on Biological Diversity (CBD), Nairobi, (1992). In force 29 December 1993, 31 International Legal Materials 822.

Collaborative Working Group on Marine Biotechnology, (2009). Background and recommendations on future actions for integrated marine biotechnology R&D in Europe http://www.marinebiotech.eu/sites/marin ebiotech.eu/files/public/library/MBT%20publications/2009%20kbbenet %20report%20distributed.pdf

European Commission, (2012). Communication from the Commission to the European Parliament, the Council, the European Economic and Social Committee and the Committee of the Regions. Blue Growth: Opportunities for Marine and Maritime Sustainable Growth. http://ec.europa. eu/transparency/regdoc/rep/1/2012/EN/1-2012-494-EN-F1-1.Pdf

European Commission, (2012). Communication from the Commission to the European Parliament, the Council, the European Economic and Social Committee and the Committee of the Regions. Innovating for Sustainable Growth: A Bioeconomy for Europe http://ec.europa.eu/research/ bioeconomy/pdf/201202_innovating_sustainable_growth_en.pdf

European Commission, (2013). Study on Blue Growth, Maritime Policy and the EU Strategy for the Baltic Sea Region. https://webgate.ec.europa.eu/ maritimeforum/en/node/3550

European Commission, (2014). A Sustainable Blue Growth Agenda for the Baltic Sea Region. http://ec.europa.eu/maritimeaffairs/policy/sea_basins/ baltic_sea/index_en.htm

ECORYS, (2014). Study in support of Impact Assessment work on Blue Biotechnology, FWC MARE/2012706 – SC C1/2013/03 – 13 June 2014

Eduok, U., Suleiman, R., Gittens, J., Khaled, M., Smith, T. J., Akid, R., El Ali, B., Khalil, A., (2015) Anticorrosion/antifouling properties of bacterial spore-loaded sol–gel type coating for mild steel in saline marine condition: a case of thermophilic strain of Bacillus licheniformis. RSC Advances 5: 93818–93830. DOI: 10.1039/c5ra16494j

Global Industry Analysts Inc. (2011) "Marine Biotechnology: A Global Strategic Business Report"

Global Industry Analysts Inc. (2015) "Marine Biotechnology: A Global Strategic Business Report" http://www.strategyr.com/pressMCP-1612. asp

Hayes, M., Carney, B., Slater, J., Brück, W., (2008). Mining marine shellfish wastes for bioactive molecules: Chitin and chitosan – Part B: Applications. Biotechnology Journal 3: 878–889. DOI 10.1002/biot.200800027.

Heip, C. and McDonough, N., (2012). Marine biodiversity: a science roadmap for Europe. European Marine Board. http://www.marineboard. eu/images/publications/Marine%20Biodiversity-122.pdf

Kafarski, P., (2012). Rainbow code of biotechnology. CHEMIK nauka-technika-rynek, 1(66), pp. 811–816.

Leal M. C., Puga J., (2012). Trends in the Discovery of New Marine Natural Products from Invertebrates over the Last Two Decades – Where and What Are We Bioprospecting? PLoS ONE 7(1): e30580.

Leary, D., Vierros, M., Hamon, G., Arico, S., Monagle, C., (2009). Marine genetic resources: A review of scientific and commercial interest. Marine Policy 33(2): 183–194.

Lloyd-Evans, M., (2013). A Global Perspective: High-level analysis of key trends and developments in global marine biotechnology RTDI. BioBridge Ltd. Marine Biotechnology CSA – Task 3.2

Medina, A. R., Grima, E. M., Giménez, A. G. and González, M. I., (1998). Downstream processing of algal polyunsaturated fatty acids. *Biotechnology Advances*, *16*(3), pp. 517–580.

Molinski, T. F., Dalisay, D. S., Lievens, S. L., Saludes, J. P., (2009). Drug development from marine natural products. Nature Reviews Drug Discovery 8: 69–85.

Nagoya Protocol on Access to Genetic Resources and the Fair and Equitable Sharing of Benefits Arising from their Utilization to the Convention on Biological Diversity (the 'Nagoya Protocol'), Nagoya (2010). In force 12 October 2014. Convention on Biological Diversity.

OECD, (2005). A framework for biotechnology statistics. Paris.

OECD, (2012). OECD Global Forum on Biotechnology: Marine Biotechnology Enabling Solutions for Ocean Productivity and Sustainability. Workshop (Vancouver, Canada, 30–31 May 2012).

OECD, (2013). Marine Biotechnology: Enabling Solutions for Ocean Productivity and Sustainability, OECD Publishing. http://dx.doi.org/10.1787/9789264194243-en This project has received funding from the European Union's Horizon 2020 research and innovation programme under grant agreement No 652629.

OECD, (2016). Marine Biotechnology: Definitions, Infrastructures and Directions for Innovation. Working Party on Biotechnology, Nanotechnology and Converging Technologies. http://www.marinebiotech.eu/sites/marinebiotech.eu/files/public/DSTI_STP_BNCT_2016_10.pdf

Regulation (EU) No 511/2014 of the European Parliament and of the Council of 16 April 2014 on compliance measures for users from the Nagoya Protocol on Access to Genetic Resources and the Fair and Equitable Sharing of Benefits Arising from their Utilization in the Union Text with EEA relevance (OJ L 150, 20/05/2014, p. 59)

Reis Batista, I., Kamermans, P., Verdegem, M. C. J., Smaal, A. C., (2013) Growth and fatty acid composition of juvenile Cerastoderma edule (L.) fed live microalgae diets with different fatty acid profiles. Aquaculture Nutrition 20(2): 132–142. DOI: 10.1111/anu.12059 http://www.marinebiotech.eu/sites/marinebiotech.eu/files/public/library/CSA%20project%20reports/Marine%20Biotechnology%20RTDI%20in%20Europe.pdf

United Nations Convention on the Law of the Sea (UNCLOS), Montego Bay, (1982). In force: 16 November 1994, 1833 United Nations Treaty Series 396

United Nations General Assembly (UNGA) resolution 69/292, (06 July 2015). Development of an international legally-binding instrument under the United Nations Convention on the Law of the Sea on the conservation and sustainable use of marine biological diversity of areas beyond national jurisdiction. UN doc A/RES/69/292

United Nations General Assembly (UNGA) resolution 66/231, (24 December 2011). 'Oceans and the law of the sea.' UN docA/RES/66/231, paragraph 167.

Van der Graaf, A. J., Ainslie, M. A., André, M., Brensing, K., Dalen, J., Dekeling, R. P. A., Robinson, S., Tasker, M. L., Thomsen, F. and Werner, S., (2012). European Marine Strategy Framework Directive-Good Environmental Status (MSFD GES): Report of the Technical Subgroup on Underwater noise and other forms of energy. *Brussels*.

3

Seabed Mining

Marcel J. C. Rozemeijer[1],*, Sander W. K. van den Burg[1], Robbert Jak[1], Laura E. Lallier[2] and Karel van Craenenbroeck[3]

[1]WUR, Droevendaalsesteeg 4, 6708 PB Wageningen, the Netherlands
[2]eCOAST, Esplanadestraat 1, 8400 Ostend; Maritime Institute, Faculty of Law, University of Ghent, Universiteitstraat 4, 9000 Ghent, Belgium
[3]GeoMarEx – Geological Marine Exploration – Luchterenstraat – 9031 Ghent, Belgium
*Corresponding Author

3.1 Introduction

3.1.1 Challenges for Offshore Mining

The surface of the planet is approximately 71% of water spread over five oceans: the Arctic, Atlantic, Indian, Pacific and Southern ocean. In fact, it represents the largest habitat for life on Earth[1]. The deep ocean beyond the continental shelf is the most difficult to access but also very promising in available resources, like biodiversity and ores. These includes minerals like gold, silver, nickel, cobalt, Rare Earth Elements (REEs), phosphorytes and gas hydrates (Scott et al., 2008; SPC 2013a, b, c, d; SPC 2016; EPRS, 2015; Rogers et al., 2015; Petersen et al., 2016).

Ores are currently being mined in coastal waters. Sand and gravel are already in exploitation e.g. for use in coastal defence and use for infrastructural works such as roads and the production of concrete. Metal ore sands and precious stones such as tin in Indonesia, gold in Alaska, or diamonds in

---

[1]This diverse habitat is largely unknown. Biodiversity, general ecology, natural dynamics, responses to natural and human drivers are hardly studied.

Namibia are exploited as well (Cronan, 2000; Baker et al., 2017; Hannington et al., 2017).

Due to the nature of the deep ocean (the immense pressure, the hard to reach bottom, the lack of data and the offshore character), the exploration and especially the exploitation of the resources on the seabed pose immense technical and environmental challenges. The initial euphoria of the 1970s was generated by high prices combined with relatively easy access to minerals available at that time. Then a collapse in world metal prices and new land-based mines dampened interest in seabed mining. However, after decades 'on hold', there is renewed interest in the potential for commercial exploitation of marine minerals from the private sector and governments alike (SPC 2013a, b, c, d; Ecorys, 2014; Lange et al., 2014; Arezki et al., 2015; Rogers et al., 2015; Worldbank, 2016).

Deep seabed mining (as a sector) must therefore be considered a significant new and emerging use of the global ocean. It was included in the project of MARIBE as a form of Blue Growth. To completely understand its functioning and promote the development of seabed mining, this chapter aims to provide an extensive overview of the social and economic drivers that influence the performance of the industry (including industry lifecycle and structure, socio-economic impact and regulatory framework, among others). The purpose is that investors, governments, operators and other interested stakeholders generate insights for future developments.

Numerous reports already exist on the analysis of the metallurgic ores that are found subsea (e.g. SPC, 2013a, b, c, d; SPC 2016; Ecorys, 2014). This study therefore aims at adding to this discussion by comparing metallurgic ores with the other major deep seabed resources phosphorites and gas hydrates. Comparing the subsectors could yield additional insight and information.

The considered subsectors face more or less similar challenges and technical demands. The challenges for a viable offshore mining industry are to deliver products at competitive prices given a volatile market, high costs, low levels of development, and anticipated major environmental impacts. A major discussion point is to tackle an investment gap; is development of small-scale innovations by adapting existing vessels and gear sufficient or are thorough innovations needed?

3.1.2 Definitions and Demarcation

Major issues for mining of marine resources are depth and distance. The general rule of thumb is the further away the deeper. And the deeper one has to mine, the more complex the techniques. Some geological and practical definitions are introduced here as a general setting and to support the definitions used.

- Limits of conventional dredging: the depth of –150 m is the theoretical limit where the conventional dredging equipment like trailing suction hopper dredgers (TSHDs) can still be used without major accommodation ('business as usual'). In practice this depth appears to be –80 m. Below that –80 m, a degree of innovation of the equipment is needed or excessive amounts of energy need to be applied making the deeper dredging a new business case. From –80 m till –200 m adapted regular exploitation vessels can be used. E.g. in the diamond mining industry the type of underground determines whether conventional techniques (vertical mining with a rigid large diameter drill) or Remotely Operated Vehicles (ROVs, horizontal mining) are used (Scott et al., 2008[2]).
- An important limit is that of the continental shelf towards approx. –200 m depth (SPC, 2013d; Rogers et al., 2015). Beyond that the depth strongly increases from the continental slope to the abyssal plains at approx. 4000 m and deeper: the deep sea.
- Potential of river deposits: The sea-level fluctuations due to ice ages are normally till –130 m ($\pm$10 m; Liu et al., 2004; Cronan, 2000). In general riverine deposits are measured and exploited till that depth (Cronan, 2000). However, in southern Africa beach planes and riverine deposits (like sand, diamonds and ore sands) can be found till at least –500 m due to tectonic movements, lowering erosion ridges and former beach planes to deeper regions. Possibly similar tectonic movements can also be valid for Australia (Siesser & Dingle 1981; Gurney et al., 1991; Cronan, 2000).

A distinction can be made between nearshore mining and offshore mining. The words offshore and nearshore represent the distance component and illustrate the differences between the business cases we describe. Taking these limits and the aspects of depth and distance we define:

Nearshore mining, ranging from –0 till –200 m still on the continental shell as a measure for both distance and a markedly chance in geology (from plane to abyss). Typically riverine deposits can be found here.

Offshore mining starts from –200 m downwards. Till –500 m exploitations could still be profitable with adaptation of existing ships and technologies,

---

[2]https://www.debeersgroup.com/en/explore-de-beers/mining.html (d.d. 03-11-2015).

which implies low investment costs, and high exploitation costs with lower economic revenues (using e.g. TSHD with a flexible trailing head and an extended (and partially flexible) suction tube to dredge the nodules. Schulte, 2013[3]). From –500 m and deeper more adaptions seem to be required.

The seabed offers a variety of resources like i) polymetallic manganese nodules (nodules), ii) polymetallic seafloor massive sulphide (SMS) deposits, iii) polymetallic cobalt crusts (cobalt crusts), iv) phosphorites, polyphosphates and phosphate sands, v) gas hydrates, vi) metal ore sands, vii) sand and gravel, viii) precious stones ix) shells x) other chemicals (Baker et al., 2017). Offshore mining encompass an elaborate scale of potential resources, which differ from location to location. Some demarcation is necessary to limit the scope of this study, as given in the following sections. The sector needs to be a new developing business (Blue Growth), and not an established business (Blue Economy). To limit the vast field of ores the following resources are studied (defined as subsectors):

1. Nodules, SMS deposits and cobalt crusts because of their potential and the fact that they are part of a developing economy (SPC, 2013a, b, c, d; Ecorys 2014; Lange et al., 2014).
2. Phosphorites and polyphosphate sands are also an upcoming mineral and a developing economy (USGS, 2017 and e.g.[3]).
3. Gas hydrates are considered interesting because the reserves are estimated to exceed known petroleum reserves and governments are highly interested for geopolitical reasons (Lange et al., 2014).

3.2 Market – Investigating Market Trends

In this section market trends are described for the different subsectors. As dealt with in Section 3.3.2, the number of exploration licences issued by the International Seabed Authority (ISA)[4] or individual countries within their Exclusive Economic Zone (EEZ) is limited[5]. The number of licenses for exploitation is even scarcer, if any. Despite the low number of licenses, the claimed surfaces for exploration are rather large. Offshore mining demands high technological development and high capital expenditures (CAPEX) and operating expenses (OPEX) costs making it high-risk for commercial exploitation (see Section 3.3). On the other hand there is a general feeling that

---

[3]http://www.rockphosphate.co.nz/ (d.d. 16-07-2017).
[4]Responsible for the international area of the deep seabed (the Area).
[5]https://www.isa.org.jm/deep-seabed-minerals-contractors (d.d. 13-07-2017).

despite all the draw backs on the economic and commercial domain, offshore mining could be important in the future. Main pulling factors are cobalt and REEs supply, excessive environmental impacts of land mining (Section 3.4) and local needs and interests of countries lacking independent reserves or other means of income such as the Pacific States for metals (SPC, 2013d; Worldbank, 2016), Japan and Korea for gas (Lange et al., 2014).

Also having and demonstrating a leading position in dredging technology and abilities could be a driving force to be first (EPRS, 2015; Worldbank, 2016).

The challenge is to find those spots where concentrations and availabilities of ores are high enough to have commercially viable exploitation despite the low TRLs and resulting high costs of equipment and techniques. This results in a strong competition for suitable concessions.

Phosphorites can now be produced at normal market prices and are thereby in competition with the land-based producers (Schilling et al., 2013). When the distances are far between consumers and land-based operators, local nearshore production is especially interesting (Don Diego, 2015). Gas Hydrates are not yet commercially produced.

3.2.1 Market Trends, Product Demand, Prices

This section and its subsections give the general trends and interpretation of the different resources.

3.2.1.1 Metals

Table 3.1 presents an overview of metal resources and reserves on land for crusts and nodules and an example of SMS type deposits. Also the estimated yearly world production is given in absolute figures and as a percentage of the currently economically minable deposits today on land. The yearly production is ranging between 0.005 and ~6% of the currently economically minable deposits on land. The three bulk metals manganese, copper and nickel consume yearly ~3% of the reserves (meaning enough reserves for >30 years for most resources, not taking into account the sub-economic deposits on land (Table 3.1). In a global observation, the economic minable reserves are around 30 years for all metals. Thirty years is the normal financial horizon used by banks and other financial institutions. Mining companies will be reluctant to invest more in exploration beyond a 30-year stock/reserves (Arndt et al., 2017). Hence, based on Table 3.1 and Arndt et al. (2017), the economically minable reserves can be expected to be much larger than a 30-year stock.

Table 3.1 Metal resources and reserves on land and seabed for crusts and nodules (millions of tonnes) and an example of sulphide type deposits (data from Hein et al., 2013; Lange et al., 2014 unless stated differently). Also estimated amount of SMS deposits are given without Atlantis II and the estimates from Sulphide rich sediments Atlantis II separately

Location Elements	Cobalt Crusts in the Prime Crust Zone (PCZ)	Global Reserves on Land (Economically Minable Deposits Today)	Global Reserves and Resources on Land (Economically Minable as Well as Sub Economic Deposits)	Manganese Nodules in the Clarion-Clipperton Zone	Estimated Amount of SMS Deposits without Atlantis II[d]	Atlantis II Sulphide Rich Sediments[e]	Estimated Yearly Worldwide Production in 2016[f]	Estimated Yearly Worldwide Production in 2016 as a Percentage of the Economically Minable Deposits Today on Land
Manganese (Mn)	1714	630	5200	5992		3.8–4.3	16.0	2.5
Copper (Cu)	7.4	690	1000+	226	10	0.74–0.81	19.4	2.8
Titanium (Ti)	88	414	899	67				
Rare earth oxides	16	110	150	15				
Nickel (Ni)	32	78[f]	130[f]	274			2.3	0.3
Vanadium (V)	4.8	19[f]	38	9.4			0.076	0.04
Molybdenum (Mo)	3.5	10	19	12			0.0002	0.002
Lithium (Li)	0.02	13	14	2.8				
Cobalt (Co)	50	7[f]	13	44		0.0053	0.123	0.16

Tungsten (W)	0.67	3.1	6.3	1.3				1.5
Niobium (Nb)	0.4	4.3[f]	4.3[f]	0.46			0.064	3.7
Arsenic (As)	2.9	1	1.6	1.4			0.0365	
Thorium (Th)	0.09	1.2	1.2	0.32				
Bismuth (Bi)	0.32	0.3	0.7	0.18				
Yttrium (Y)	1.7	0.5	0.5	2			0.005–0.007	1–1.4
Platinum group	0.004	0.07	0.08	0.003				
Tellurium (Te)	0.45	0.025[f]	0.05	0.08			>0.000108	>0.4
Thallium (Tl)	1.2	0.0004	0.0007	4.2				
Gold (Au)	0.05–0.057[a,b,c,f]		0.1157[c]	0.000095	0.00102	0.000046	0.0031	5.4–6.2
Silver (Ag)	0.57[f]			0.0036	0.069	0.0065	0.027	4.7
Zinc (Zn)	220–230[b,f]		1900[b,f]	13	20[d]	3–3.8	0.012	0.005

[a] Estimate based on booked reserves in mining companies http://www.bullionmark.com.au/how-much-gold-is-there (d.d. 09-09-2015).
[b] USGS (2015).
[c] http://www.visualcapitalist.com/global-gold-mine-and-deposit-rankings-2013/ (d.d. 09-09-2015).
[d] Hammington et al. (2010, 2011): estimated total 600 millions of tonnes. In Hammington et al. (2011) copper and zinc are presented summed together 30 millions of tonnes. Based on Hammington et al. (2010) a simplified 1:2 Cu: ZN ratio is presented here.
[e] Bertram et al. (2011); Lange et al. (2014); Laurila et al. (2015).
[f] USGS (2017).

In general land-based mining is an inflexible economy. The investments in and cost structure of the mining infrastructure is so huge that they cannot flexibly react to market developments. This results in typical fluctuations between a state of oversupply and supply shortage. In Figure 3.1 an example is given of the price developments of the resources, showing largely stable prices with a peak in prices in the 1970's and another peak starting from approximately 2005 and going down after 2011. For the recent past, three trends can be distinguished:

1. An increase in demand of metals since the early 2000's due to economic development raising prices. Economic development of Brazil, Russia, India and China (the BRIC countries) has led to a higher demand. China is particularly consuming more and more metals. In addition, techno-logical development (smartphones etc.) has increased the demand for special metals like cobalt and rare earth elements (REEs) (Worldbank, 2012 and 2016; SPC, 2013d; Ecorys, 2014; Arezki et al., 2015).
2. The financial crises in 2008, which started with the bursting of the United States housing bubble in 2004–2006 (Tully, 2006; Worldbank, 2012) and that lead to both raising and lowering of prices.
3. A decrease in the quality of ores by the end of the 1990, early 2000's leading to higher prices (Worldbank, 2012; Mudd et al., 2013; SPC, 2013d).

3.2.1.2 Phosphorite

The same patterns in price development and demand that occur for metal ores occur for phosphorites as well. Prices remained stable around €30 to €50 per metric tonne from 2000 till 2007 (Figure 3.1). Around the time of the financial crises, prices rose sharply to almost €300 per metric tonne and then descended till a fluctuation plateau of €70 to €150 per metric tonne. In 2007–2008, world agriculture increased due to growing world population and associated food demand, leading to a strong rise in demand for phosphate-derived fertilizers. Currently economically minable deposits on land are 68,000 Million metric tons with a yearly consumption of 261 Million metric (0.4%). 74% of this reserve is located in Morocco and Western Sahara. Large phosphate resources have been identified on the continental shelves and on seamounts in the Atlantic Ocean and the Pacific Ocean. Total world resources of phosphate rock are more than 300 billion tons. There are no imminent shortages of phosphate rock (USGS, 2017).

Increasing concerns on both the supply market being dominated by a few suppliers (especially Morocco) which seems to become more extreme

A

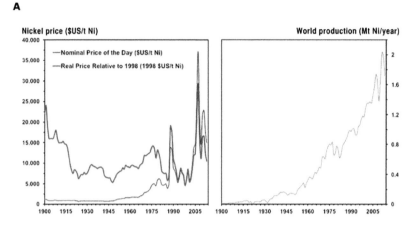

B

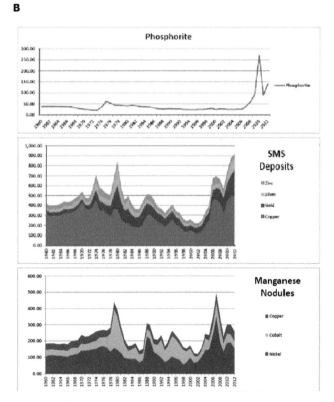

Figure 3.1 (A) Long term development of nickel https://theconversation.com/queensland-nickels-demise-yabulu-a-relic-refinery-53368, (d.d. 10-02-2017). (B) Price development of minerals, corrected for inflation, in Euro per ton till October 2012 (Schulte, 2012).

in the future and a need for phosphate rock with a lower cadmium content (de Ridder et al., 2012), urge for new supply source where offshore mining can offer options. Also the exploitation of local phosphorites can mean local employment and export potential for the region and even reduce the carbon footprint.

3.2.1.3 Natural gas

The price developments of gas depend on location (Figure 3.2) (BP, 2016). In the US, which is self-sufficient in gas-supplies, prices remained relatively stable (around $ 5 for a million British Thermal Units (mmBTU) except in the period of the financial crises. In Europe and Japan prices tend to be higher around $ 7–$ 10 for a mmBTU (probably reflecting the dependency of the import) (ECB, 2014).

The prices of natural gas depend on many factors, including macroeconomic growth rates and expected rates of resource recovery from natural gas wells. Natural gas prices, as with other commodity prices, are mainly driven by supply and demand fundamentals. However, natural gas prices may also be linked to the price of crude oil and/or petroleum products, especially in continental Europe. Higher rates of economic growth lead to increased consumption of natural gas, primarily due to gas usage in housing, commercial

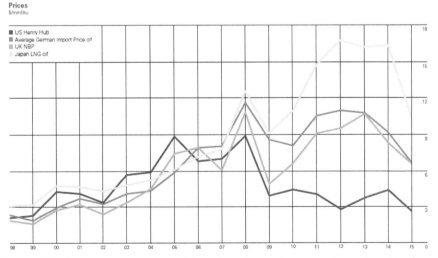

Figure 3.2 A long term overview of the price of natural gas (mostly methane) in $ per million British Thermal Units (mmBTU) for different regions in the world (BP, 2016).

floor space, and industrial output. Also an event like the earthquake in Japan leading to less nuclear energy and trust in nuclear energy can be noted in an international context. Weather conditions can also have a major impact on natural gas demand and supply. Cold temperatures in the winter increase the demand for space heating with natural gas.

3.2.1.4 A general model

De Ridder et al. (2012) developed mathematical models to explain the price development for phosphorites, which seem quite suitable for metals as well. In general ore production was insufficient, causing greater derived demand for ore. Meanwhile, supply tightened (ore degradation), with production and transport costs going up. This resulted in a higher price. Eventually, higher prices made more exploration and recycling activities economically feasible. It therefore became possible to restore supply. As demand remained stronger than before, new prices reached a slightly higher level than originally (de Ridder et al., 2012).

3.2.2 A View of Future Supply and Demand

3.2.2.1 Metals

The previous section described price developments and the drivers on supply and demand. It emphasized the demand by economic developments and the influence of price. Despite steadily increasing demand, the onshore deposits will in most cases continue to satisfy our growing appetite for metals and minerals (SPC, 2013d; Lange et al., 2014; Ecorys, 2014) (Figure 3.1, Table 3.1). Indeed, with an increasing political stability worldwide new land-based reserves are discovered in emerging market and developing economies (Table 3.1) (Figure 3.3, Arezki et al., 2015). Section 3.6 performs a sensitivity analysis for global prices and revenues. The analysis concludes that global metal prices are currently low, making offshore mining of metal ores unlikely in the short term. Metal prices will need to rise substantially before making offshore mining commercially viable.

On the long run the combination of increased absolute and relative demand combined with geopolitical issues can limit the availability of some metal resources. New technological developments demand more and more of special metals and REEs. A lot of these resources for new technology are situated in a few countries only, often with a political instable climate making it a geopolitical issue of availability. Geopolitical issues can make offshore mining an interesting option despite the high costs. Examples of components of geopolitical concern are the supply of cobalt (dominated by the Democratic

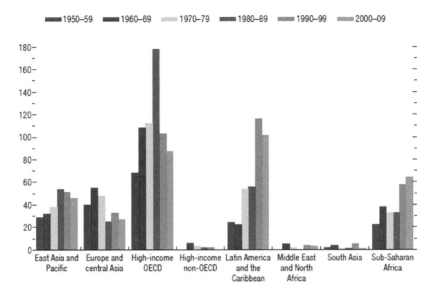

Source: MinEx Consulting.
Note: OECD = Organisation for Economic Co-operation and Development.

Figure 3.3　Number of metal deposit discoveries by region and decade (Arezki et al., 2015).

Republic of Congo), phosphorites (Morocco), as well as gas hydrates (Hein et al., 2013; de Ridder et al., 2012; Lange et al., 2014; USGS, 2015 and 2017). In addition, environmental concerns on land-based mining could turn the table towards seabed mining (Section 3.4). Currently, China is considered the only supplier of REEs. However, numerous large reserves have been discovered and are available in Australia (Mount Weld), Greenland (Kvanefjeld), Chili, Bolivia (Uyuni Salt Flats), and Afghanistan[6], as well as Brazil, India, Russia and Vietnam (USGS, 2017). With fluctuating market prices these mines open and close with profitability.

3.2.2.2 Phosphorites

With an increasing population, food production and phosphate demand will increase. In addition a need has arisen for phosphate with less calcium

_____

[6]http://geology.com/usgs/ree-geology/, http://www.australianrareearths.com, http://www.ggg.gl, https://www.masterresource.org/electric-vehicles/rare-earth-and-lithium-supplies-cloud-renewables/, http://www.popsci.com/science/article/2010-06/us-geologists-uncover-staggering-1-trillion-cache-unmined-mineral-resources-afghanistan, (all d.d. 27-03-2017).

concentrations. World stocks can easily meet the demand on phosphate; the calcium content is a different topic (USGS, 2017, De Ridder et al., 2012).

3.2.2.3 Gas hydrates

World consumption of gas is steadily increasing. Global proved natural gas reserves in 2015 were estimated to be 186.9 trillion cubic metres, sufficient to meet 52.8 years of current production (in most cases not taking shale gas into account). Proven gas reserves were dominated by the Middle East (43%). Also Russia holds large proven reserves (~17%). Other countries have substantial reserves. It appears that every ten years more proven reserves are determined (BP, 2016). Reserve estimates change from year to year as new discoveries are made, as existing fields are more thoroughly appraised, as existing reserves are produced, and as prices and technologies evolve. Sources also differ in actual estimates. It is estimated that there are about 900 trillion cubic metres of "unconventional" gas such as shale gas, of which 180 trillion cubic metres may be recoverable (another ~50 years).

Recent estimates of worldwide amounts of gas hydrate, which attempt to consider all of these aspects, are on the order of 5 to 15 times the land-based reserves (Lange et al., 2014).

3.2.2.4 Potential influence of offshore mining ores on global markets

The Ecorys study (2014) made some assumptions and calculations of the potential influence of offshore mining ores on global markets. As mentioned before, only a limited number of metals seem interesting and from those copper, gold and silver are the targets for SMS deposits and copper, cobalt and nickel for the crusts and nodules. The impact on the world market can only be estimated with assumptions since there is no production at this moment.

Taking the target metals: for **gold and silver** a production by offshore mining was estimated at ~3% and ~1% of the yearly terrestrial production respectively (USGS, 2015; Ecorys, 2014). These volumes are very small. In addition, metals like gold and silver are characterised by low production concentration and existing market exchanges, which however are only marginally influenced by physical demand and supply. Therefore offshore mining is not expected to have an influence on the price.

Currently, global annual production of **copper** is 18.7 million tonnes from different sources (USGS, 2015). Looking into an initial reachable estimated annual volume of 0.1 million tonnes of copper (~0.5%) from a typical offshore mining operation (Ecorys, 2014) it is unlikely to have a substantial impact on global prices. The same is valid for **nickel**.

In the case of **cobalt** (8 thousand tonnes, Ecorys, 2014) the impact on price may be more substantial as global annual production is around 112 thousand tonnes (USGS, 2015). An estimated annual output of $\sim$8% could have an impact on market prices and price fluctuation, particularly in view of cobalt's supply risk due to geopolitical reasons. Congo (Kinshasa), a potentially unstable country, has $\sim$50% of the world production. Any substantial new source will influence the market. The same line could be valid for the **REEs** as well (Ecorys, 2014; Worldbank, 2016).

Annually, 261 million tons of **phosphate** is produced. Namibia's offshore **phosphorite** mining aims at $\pm$10% market share of the traded phosphate market of 30 million tons a year[7]. Taking into account the potential impact of Don Diego (Mexico) and Chatham Rise (New Zealand) exploration, the seabed mining of phosphorites can have a substantial impact on world prices.

For **gas hydrates** it can be expected that once in full operation it will have a substantial impact on local and world prices (Lange et al., 2014).

3.3 Sector Industry Structure and Lifecycle

The polymetallic (manganese) nodules, cobalt crusts, SMS deposits, phosphorites and gas hydrates have different distributions over the world and in depth. Moreover, different techniques are required to harvest them based on depth and origin. This gives rise to distinct industries and sectors involved in the development of offshore mining.

3.3.1 Worldwide Offshore Resource Distribution

Resources for offshore mining are spread all over the world in both the deeper national waters and the international seas and oceans. The most interesting sites for exploration are not found in European waters. Below, information is presented on the presence of the considered deposits worldwide.

3.3.1.1 Nodules

Nodules occur widely on the vast, sediment-covered, plains of the abyssal ocean at depths of about 4,000 to 6,500 m (Hein et al., 2013; SPC 2013b). The greatest concentrations of metal-rich nodules occur in the Clarion-Clipperton Zone (CCZ), which extends from off the west coast of Mexico to as far west as Hawaii (map B, Figure 3.4). Nodules are also concentrated in the Peru

---

[7]http://www.namphos.com/project/sandpiper.html (d.d. 16-07-2017).

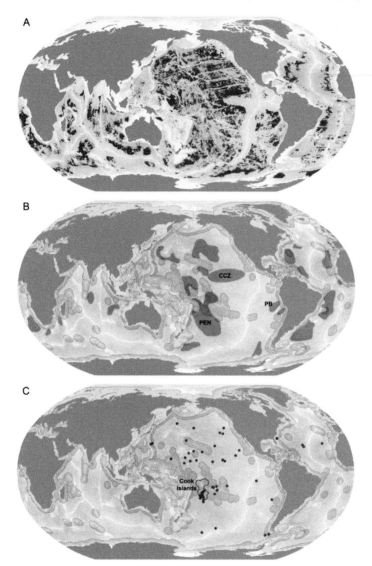

Figure 3.4 (A) Locations of areas within the abyssal plains that are important for manganese nodule formations based on seafloor classification, seafloor age (older than 10 My), sediment thickness (<1000 m), sedimentation rate (<1 cm/1000 years), and water depth (between 3000 and 6000 m). Note the lack of data below 70°S and above 80°N. (B) Areas with highest Mn-nodule potential based on seafloor morphology, age of crusts, and metal input. Light blue areas delineate the EEZ. Abbreviations: CCZ = Clarion-Clipperton Zone, PB = Peru Basin, PEN = Penrhyn Basin. (C) Location of manganese nodule samples in the ISA database with Co concentrations above 0.5 wt% (N = 211). Note the large number of Co-rich samples in the EEZ of the Cook Islands. (Petersen et al., 2016).

Basin, near the Cook Islands, and at abyssal depths in the Indian and Atlantic oceans (Hein et al., 2013 and 2015). In the CCZ, the manganese nodules lie on abyssal sediments covering an area of at least 9 million square kilometres (Figure 3.4).

No relevant concentrations of polymetallic (manganese) nodules have been found in basins within the scope of the MARIBE project (Atlantic, Baltic/North Sea, Mediterranean, and Caribbean). However, some spots with substantial amounts of nodules were discovered recently in the tropical Atlantic (north of French Guyana and west of Africa (Devey, 2015[8]). These findings await publications or reports that putting them into perspective. In addition in the Galicia Bank region (northwest Iberian margin, NE Atlantic), a complete suite of mineral deposit types was encountered including (1) phosphorite slabs and nodules, (2) Fe-Mn crusts and strata bound deposits, (3) Co-rich Mn nodules, and (4) Fe-rich nodules. The Galicia Bank nodules are exceptionally rich in cobalt (Gonzalez et al., 2016). Quantities for commercial exploitation need to be assessed.

3.3.1.2 SMS deposits

SMS deposits are rich in copper, iron, zinc, silver and gold. The total accumulation of sulphides is estimated to be on the order of 600 millions of tonnes (Hannington et al., 2010 and 2011). As compared to nodules and terrestrial reserves the amounts deposited in SMS are far less (Table 3.1, Figure 3.5), although the amount of precious metals is substantial. Gold and silver, together with copper, appear to be the commercially most interesting metals (Boschen et al., 2013, Ecorys, 2014).

Deposits are found at tectonic plate boundaries along the mid-ocean ridges, back-arc ridges and active volcanic arcs, typically at water depths of around 2,000 m for mid-ocean ridges (Figure 3.5). These deposits formed over thousands of years through hydrothermal activity, which is when metals precipitate from water discharged from the Earth's crust through hot springs at temperatures of up to 400°C. Because of the black plumes formed by the activity, these hydrothermal vents are often referred to as 'black smokers'.

SMS deposits can potentially be found in the Mediterranean, near the Azores (Marques & Scott 2011; Lange et al., 2014; Ortega, 2014) and in Norwegian waters at the Mid Atlantic Ridge. Future exploration is needed to get more indication of their values.

_____

[8]http://www.geomar.de/en/news/article/tiefseetiere-gesucht-manganknollen-gefunden/ (d.d. 15-07-2017).

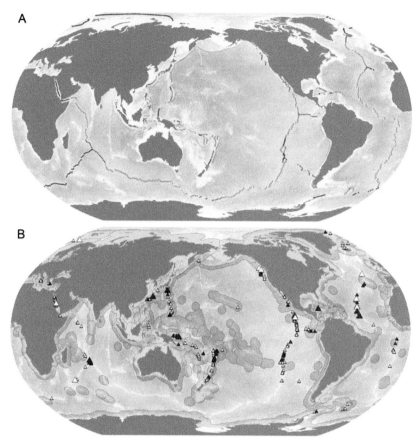

Figure 3.5 (A) Locations of mid-ocean ridges and back-arc spreading centres important for the formation of seafloor massive sulphides. Colours denote the spreading rate of
each segment. Dark blue = ultra-slow spreading <20 mm/yr); light blue = slow spreading
(20–40 mm/yr); green = intermediate spreading (40–60 mm/yr); orange = 1/4 fast spreading
(60–140 mm/yr); red = ultra-fast spreading (>140 mm/yr). (B) Location of high-temperature
seafloor hydrothermal systems and associated seafloor mineralization, where red colour
indicates occurrences with economically interesting metal concentrations (average grade of
the deposit is either 45 wt% Cu, 415 wt% Zn, or 45 ppm Au) and large symbols indicate
occurrences with size estimates above 1 million tonnes. Using these criteria, only a few
occurrences of economic interest have been identified. Note that geochemical analyses are
commonly only available for surface samples that are not representative for the entire occurrence. A quantitative resource assessment for seafloor massive sulphides is only available for
two occurrences (Solwara 1 and Solwara 12, both within the EEZ of Papua New Guinea).
Light blue areas delineate the EEZ. (Petersen et al., 2016).

3.3.1.3 Cobalt crusts

Cobalt crusts accumulate at water depths of between 400 and 7,000 m on the flanks and tops of seamounts. They are formed through precipitation of minerals from seawater. The crusts contain iron, manganese, nickel, cobalt, copper and various rare metals, including rare earth elements (Table 3.1). They vary in thickness from <1 to 260 mm and are generally thicker on older seamounts. Because cobalt crusts are firmly attached to the rocky substrate, they cannot simply be collected on the bottom like manganese nodules. They will have to be laboriously separated and removed from the underlying rocks. (Hein et al., 2013; Lange et al., 2014; Petersen et al., 2016).

Globally, it is estimated that there may be as many as 100,000 seamounts higher than 1,000 m, although relatively few of these will be prospective for cobalt crust extraction. As compared to the terrestrial reserves, cobalt crusts represent a substantial portion. The commercially most important metals seem to be copper, cobalt and nickel (Table 3.1) (Ecorys, 2014).

The world's oldest seamounts are found in the western Pacific. Accordingly, many metallic compounds were deposited here over a long period of time to form comparatively thick crusts. This area, around 3000 kilometres southwest of Japan, is called the Prime Crust Zone (PCZ) (Figure 3.6) (Hein et al., 2013; SPC, 2013b; Petersen et al., 2016).

For Europe some potentially commercially exploitable crusts can be found on seamounts near Madeira, the Canary and Azores islands, the Galicia Bank, Iberian margin and one sample from the western Mediterranean Sea (between –750 to –4,600 m). The resource potential of Fe-Mn crusts within and adjacent to the Portuguese EEZ is evaluated to be comparable to that of crusts in the central Pacific, indicating that these Atlantic deposits may be an important future resource (Muiños et al., 2013; Conceição et al., 2014; Gonzalez et al., 2016). The resources at the Galicia Bank, Iberian margin need to be evaluated in a commercial perspective. They are not as enriched in cobalt as the nodules from the Galicia Bank (Gonzalez et al., 2016; Hein et al., 2013).

3.3.1.4 Phosphorites

Phosphates are found in areas of oceanic upwelling and riverine deposits. They are most commonly formed off the western margin of continents and on plateaus (zones of upwelling, Figure 3.7). In this sense they are the result of marine and oceanographic processes and not (direct) land run off and deposits. Europe has some deposits at the continental shelf of Portugal

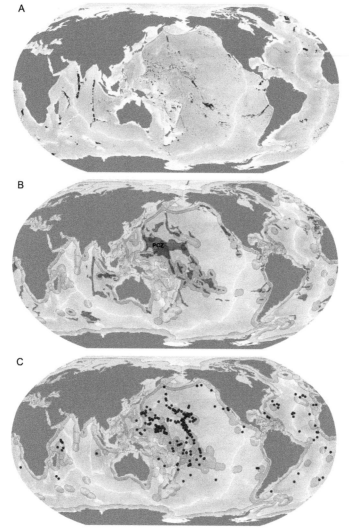

Figure 3.6 (A) Locations of seamounts, guyots, and oceanic plateaus that are important for the formation of ferromanganese crust based on seafloor classification, seafloor age (older than 10 My), sediment thickness (<500 m), sedimentation rate (<2 cm/1000 years), and water depth (peaks between 800 and 3000 m). Note the lack of data below 70°S and above 80°N. See text for details. (B) Area with highest ferromanganese crust potential based on morphology, age of the crust, and metal input. Light blue areas delineate the EEZ. Abbreviations: PCZ = Prime Crust Zone. (C) Location of ferromanganese crust samples from the ISA database with Co-concentrations above 0.5 wt% (N = 465). Note that most Co-rich ferromanganese crust samples lie in the western Pacific (Petersen et al., 2016).

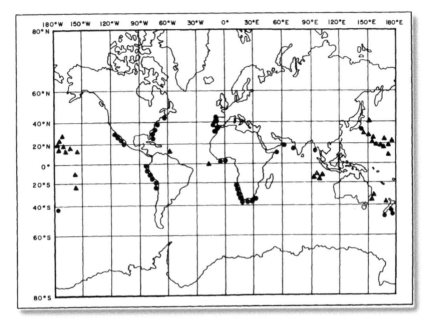

Figure 3.7   The presence of Phosphorites and other resources according to the status in 2008. Dots: continental shelves; triangles: seamounts.

Source: Thiel et al., 1998.

(measured $\sim$−400 m till $\sim$−2000 m Gaspar, 1982) and Spain (Galicia Bank, measured $\sim$−750 m till $\sim$−1900 m) (Gonzalez et al., 2016).

3.3.1.5 Gas hydrates

Methane is formed by the metabolisation and decomposition of dead biological material by anaerobic bacteria or by chemical decomposition by earth heat starting from −300 m to −3000 m. When gas molecules are trapped in a lattice of water molecules at temperatures above 0°C and pressures above one atmosphere, they can form a stable solid. These solids are gas hydrates which are trapped in the pore of the sediments (Boswell & Collett, 2011; Lange et al., 2014).

Methane hydrates develop in permafrost regions on land or beneath the seafloor. They are usually covered by a layer of sediments. Their formation under the seafloor requires an environment of sufficiently high pressure and low temperature. Thus, in the Arctic, methane hydrates can be found below water depths of around 300 metres, while in the tropics they can only occur below 600 metres. Most methane hydrate occurrences worldwide lie at water

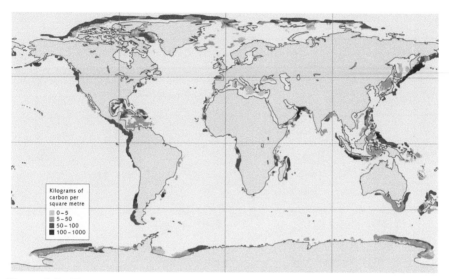

Figure 3.8 The occurrence of biogenic gas hydrates. Gas hydrate forms when methane and water combine at pressure and temperature conditions that are common in the marine sediments of continental margins and below about −200 m. The figure only shows biogenic gas hydrates. The amounts of thermogenic methane are not taken into account (Fig. from Lange et al., 2014).

depths between 500 and 3000 metres at the continental margins. According to current estimates the largest deposits are located off Peru and the Arabian Peninsula (Lange et al., 2014; Figure 3.8).

3.3.2 Centres of Offshore Activity

3.3.2.1 International areas

To date (20-07-2017), a number of contracts signed with the ISA for the *exploration* for mineral deposits are currently into force: 17 for polymetallic nodules, 6 for polymetallic sulphides, 4 for cobalt-rich crusts (Figure 3.9)[5]. Three States have notified the ISA of their prospecting activities (Fiji, Tuvalu, Samoa). There is no application or contract for *exploitation* of minerals as of yet in international areas.

3.3.2.2 National areas

Metallurgic deposits

In relation to metallurgic deposits, Nautilus Minerals Inc. holds a license for exploration and exploitation of SMS deposits at the Solwara site in

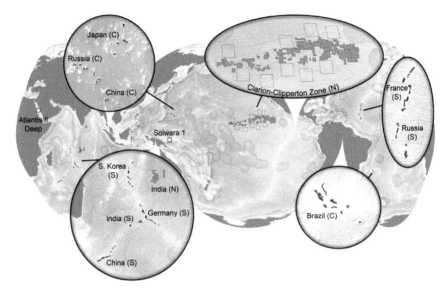

Figure 3.9   Locations of global exploration licenses for manganese nodules (N), Co-rich ferromanganese crusts (C) and seafloor massive sulfides (S) for licenses within "the Area", orange for licenses within EEZs. The locations of the only two seabed mining licenses (Atlantis II Deep in the Red Sea and Solwara 1 in Papua New Guinea) are indicated by the white squares. The location of the protected "Areas of Particular Environmental Interest" (size of 400 km by 400 km each) in the CCZ is provided as rectangles with a green outline (Petersen et al., 2016).

Papua New Guinea. For mining the Atlantis II Deep in the central Red Sea, positioned in the common EEZ of the Kingdom of Saudi Arabia and the Democratic Republic of the Sudan, the Diamond Fields Ltd. of Canada and Manafa of Saudi Arabia consortium has received a 30-year license for exploration and exploitation (Thiel et al., 2013; Petersen et al., 2016, Figure 3.9[9]).

Neptune Minerals[10], a company registered in the USA, is also conducting exploration for SMS since 2005. The company holds (or has held) prospecting and exploration licenses in Japan, Papua New Guinea, Solomon Islands, Vanuatu, Fiji, Tonga and New Zealand.

---

[9]http://www.diamondfields.com/s/AtlantisII.asp (d.d. 20-07-2017). Atlantis II Deep contains hot brines with metallurgic content. The upper 10 metres of sediment in the Atlantis II Deep at ∼2200 m, contains economically highly valuable metal deposits.

[10]http://www.neptuneminerals.com/ (d.d. 20-07-2017).

Phosphorites

There is no phosphorite nor gas hydrates exploration going on in international areas. Currently three regions are in various stages of exploitation: phosphate rich sands in Namibia (–180 m to –300 m, two companies), nodules in Chatham Rise (–250 to –450 m, New Zealand), and phosphate rich sands in the Don Diego deposit (–50 m to –90 m, offshore Baja California, Mexico). They are all currently on temporary hold due to environmental considerations. Environmental impact estimates are questioned by stakeholders fearing the impacts of large-scale exploitation (Sharma, 2017).

Offshore deposits located off Florida and Georgia in the south-eastern U.S. have been drilled, fairly well characterized and seem promising for exploitation (Scott et al., 2008).

Gas hydrates

Japan and South Korea are at the cutting edge of the exploration and exploitation of gas hydrates. In the coming years these two countries will carry out additional production tests on the seafloor. Significant efforts are also being undertaken in Taiwan, China, India, Vietnam and New Zealand to develop domestic gas hydrate reserves in the seafloor. A major technical barrier is the development of methods best suited for production. For this reason large amounts of money continue to be spent on research. To date, close to 1 billion US dollars have been invested in gas hydrate research worldwide. The first resource-grade gas hydrates in marine sands were discovered in the Nankai Trough area off Japan in 1999. In 2013, methane was produced there for the first time from a test well in the sea (Lange et al., 2014). This resource exploitation is still in an experimental phase.

3.3.3 Ownership

In general most of the exploitation of offshore metallurgic and gas resources is in the hands of governmental related companies[5]. Commercially exploitable, high grade phosphorites concessions seem more in the hands of private investors combined with national authorities, as further explained below.

3.3.3.1 Governmental companies

In most of the projects in international waters, the main contractors are governments (Korea, Russian Federation, India) or companies sponsored and funded directly or indirectly by governments through public funding. It is the case, for example, of KIOST (Korea), COMRA (China), JOGMEC102 and

DORD (both Japan) and the Federal Institute for Geosciences and Natural Resources (BGR, Germany)[5]. In the case of nodules, out of 16 contractors, nine are directly or indirectly government related, three operators and a science institute with potentially a strategic interest; only 3 private investors are involved.

Depending on the country, governmental institutes perform a more supporting task for a ministry (the final contract holder with ISA), or manages the contracts with ISA itself. The distinction is the relationship of the contractor with the governmental department, as well as the degree of (in)dependency.

3.3.3.2 Private companies

Private companies are encountered at two levels: operation and investment. Typically in metals most private companies provide services in the value chain (Figure 3.10). In the case of profitable phosphorites, private companies are investors as well.

The value chain of mining operations includes exploration and resource assessment, mining and extraction as well as processing (smelters) and distribution (Figure 3.10). The tendency for large aggregations is typical of more mature land-based mining rather than seabed mining (Ecorys, 2014). In offshore mining, smaller companies (as compared to the broader mining industry) can conduct exploration activities. However, specialised companies like Fugro and GSR (exploration) are bought by larger dredging firms like Boskalis and DEME, demonstrating vertical integration and the aggregation tendencies of maturing industries.

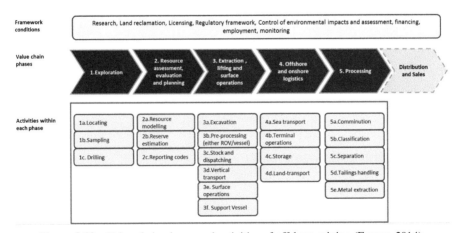

Figure 3.10   Value chain phases and activities of offshore mining (Ecorys, 2014).

The companies can be owned or are supported by investments of three groups of investors:

1. Large mining firms acting as investors (e.g Nautilus).
2. Large generalist investors (like Levi Levine and Namibia Marine Phosphate (NMP) for phosphorites).
3. Dredging companies and offshore construction companies.

3.3.4 Integration

Both vertical integration and horizontal integration takes place in the value chain of nearshore and offshore mining. Integration of different types of expertise also appear necessary to allow offshore mining to occur.

3.3.4.1 Vertical integration

A clear case of vertical integration is that of the Phosphorites mining companies in Namibia and Mexico. NMB and the Leviev group want to have their own refinery factory to increase the ore grades to commercially interesting grades (downstream) (Benkenstein, 2014). The Mexican Don Diego project also foresees a form of local, on site, processing of the ore to a more refined ore reduced in volume in order to reduce transport costs, e.g. a factory ship that refines the raw ores working next to a TSHD.

3.3.4.2 Horizontal integration

Horizontal integration is shown in the fact that dredgers offer their service to all kinds of marine resources: sand and gravel; phosphorites, metal ore sands etc. Exploration companies like Odyssey explore the oceans of the world locating valuable treasures and resources, archaeological sites and shipwrecks. Bosch Rexroth designs materials for both offshore mining and offshore oil and gas industry. Offshore knowledge, capacity and capability is highly valuable and adapted for new purposes. The dredgers have rather recently entered the offshore wind energy installation market. The key value here is general offshore knowledge (Rozemeijer et al., 2015).

3.3.4.3 Highly specialised operators

Because offshore mining is located in open seas, it is by definition a capital-intensive sector. All commercial activities on seas and oceans require high-end knowledge, extensive experience and large investments. Offshore and adapted nearshore mining represent an extremely demanding environment, which has to deal with both the very harsh conditions and remoteness of the open ocean and the extreme environment of the deep sea. In the role of operators, only established companies with a long history of operation can operate there, having developed a balanced view on investment, revenues, logistics, innovation etc. (Ecorys, 2014; Lange et al., 2014; EPRS, 2015). These companies operate in an international, global setting. Europe has some major players in the fields: renowned international dredgers and offshore-installation producers.

3.3.4.4 Buying in knowledge and reducing risks

More often than not, major companies buy in extra technology or local market knowledge of procedures with the local government and local stakeholders. For Don Diego, Boskalis is investing in Odyssey Marine Explorations and in a second Mexican company Dragamex, thereby getting access to knowledge on exploration techniques as well as the local governance procedures and stakeholders. Odyssey currently owns 54% of the outstanding shares of its subsidiary, Oceanica Resources S. de. R.L. (Oceanica). Oceanica itself owns Exploraciones Oceanicos, S. R.L. de CV, the Mexican operating company with the mining concession containing the Don Diego phosphate deposit. Next to buying in knowledge, it protects the mother company and implies minimal investments for maximum influence (staged 54% majority shareholding).

Similar combinations or networks of expertise also exist, e.g. around Chatham rise Phosphorites projects. Odyssey Marine Exploration has minority ownership stakes in Chatham Rock Phosphate Ltd. Once more, Boskalis is the operator for the Chatham rise concession. Odyssey Marine Exploration also has minority ownership stakes in Neptune Minerals. They are all companies controlling exclusive mineral licenses for areas believed to contain high-value ocean floor mineral deposits.

Both Boskalis and DEME bought in exploration knowledge with the smaller companies of respectively Fugro and GSR.

A network of interdependent investors and operators ensures the conservation of investment and essential knowledge.

3.4 Working Environment

In this section, attention is given to the various factors impacting the lifetime of a seabed mining project as well as the interactions of said project with the surrounding environment in its widest sense. This includes the governance and societal implications of project development, the employment aspects, the economical context and ecological concerns.

3.4.1 Employment and Skills

Although the typical ores extracted through offshore mining are in general not present within European waters, the interest of EU-based companies in the sector is of primary importance. The relevant experience in specific vessel design, construction and operations of extracting seafloor resources are mostly of European origin and Europe-based until today. Indeed, it is the European dredging and offshore construction industry – mainly concentrated in the Netherlands, Belgium and the UK, which is particularly involved in applying their knowledge and experience arising of nearshore dredging and mining around the globe (Rozemeijer et al., 2015).

The long-term employment opportunities that should arise from offshore mining are expected to be limited to a few hundred of high skilled positions per project, which is relatively low when compared to the sectors of land-based mining or recycling. This is explained by the need for technological tools rather than workforce on board mining vessels, requiring expertise from mainly crew, technicians, managers and other indirectly involved staff. However, when looking at the entire value chain, treatment and processing factories on land as well as commercial phases should open the door to a greater need for labour supply. Even though, the EU offshore industry has been qualified as marginal in terms of job creation by several studies (SRK Consulting, 2010; EPRS, 2015).

Despite the low impact on employment, this type of activities also has the potential to become an important driver for technological development and innovation (EPRS, 2015; Worldbank, 2016). Universities, public-private partnerships in R&D, and EU funding programs like H2020[11] play a consequent role in pushing and pulling this leading position in technology development, engineering, and adjacent fields such as environmental optimization, ecological impacts and sustainable governance.

Research on governance, policy and legal development is of particular interest to future mining projects. Indeed, the assessment of their impact on such projects, as well as the associated costs and liabilities, is rendered challenging by the status of the legislation which is still, to date, under development.

---

[11] https://ec.europa.eu/programmes/horizon2020/en/draft-work-programmes-2016-17 (d.d. 13-07-2017).

3.4.2 Rules and Regulations

The deep seabed spreads both over areas within national jurisdiction (EEZ, Continental shelf) and the Area. There are thus two different levels of regulatory framework depending on the specific location of mining activities:

1. International law: Part XI of the *United Nations Convention* on the *Law of the Sea* (LOSC, 1982), applicable to the Area and where the ISA is responsible to administer and regulate mining activities through the development of its Mining Code.
2. Domestic law: the legislation of the coastal State applicable to the seabed within its national jurisdiction.

3.4.2.1 International law

The Area and its mineral resources are reserved for the Common Heritage of Mankind, as provided in Part XI of the LOSC. The ISA is mandated by the LOSC to adopt rules and regulations to ensure that prospecting, exploration for and exploitation of minerals in the Area is conducted in accordance with the economic and environmental principles set forth in the LOSC. To this aim, the ISA has started drafting a Mining Code[12]. Components of this Mining Code on exploration have since then been adopted and implemented, but the exploitation phase remains to be regulated. Since 2015, the ISA has effectively begun the drafting process of exploitation regulations that will be incorporated into the Mining Code[13]. Their adoption is expected by 2018 or 2019. As it stands in its incompleteness, the current regime under which these resources are administered may be described briefly as follows:

- While scientific research is largely free of restrictions, prospecting may be conducted only after the ISA has received notification, accompanied with a written undertaking that the proposed prospector will comply with the LOSC and the ISA rules, regulations and procedures, and will accept verification of compliance by the ISA. This solely implies requirements on environmental and human safety considerations, and respect for other activities taking place in international areas.
- Exploration and exploitation may only be carried out under a contract with the ISA and are subject to its rules, regulations and procedures. Contracts may be issued to both public and private mining enterprises

---

[12] Available at (d.d. 13-07-2017): https://www.isa.org.jm/mining-code

[13] Updates on the drafting process are available at (d.d. 13-07-2017): https://www.isa.org.jm/legal-instruments/ongoing-development-regulations-exploitation-mineral-resources-area

provided that they are sponsored by a State Party to the LOSC (the Sponsoring State) and meet certain standards of technological and financial capacity. Although the contractual form allows for more flexibility than permitting or licensing, which is the traditional mean of authorization for land-based mining, most of the contract clauses are pre-set by the Mining Code.

The ISA has also emphasized provisions relating to environmental protection and safeguards (Benkenstein, 2014), although the requirements for the exploration phase are rather light. In March 2017, the Federal Institute for Geosciences and Natural Resources (BGR) and the German Environment Agency (UBA) held an expert workshop jointly organized with the ISA on environmental standards for seabed mining. In the current context where the ISA is still developing the Mining Code's part on exploitation, international experts participating in the workshop advocated for systematic environmental protection in offshore seabed mining both at project and policy level[14]. Experts also recalled the compelling need for a comprehensive assessment of both the chances and risks of future seabed mining, as well as the obligation to apply a precautionary approach[15]. Major, stricter and more detailed requirements are hence expected with the coming regulations for exploitation.

Current requirements in the Mining Code for exploration include:

- Prevention, reduction and control of pollution and other hazards to the marine environment, applying a precautionary approach. Ecosystem-based management, monitoring and mitigating strategies, and more generally best environmental practices, even though part of the discussion at policy level, remain to be set in further details and standards.
- Gathering of environmental baseline data against which to assess the likely effects on the marine environment of a future seabed mining project.
- Establishment of comprehensive programs for monitoring and evaluating environmental impact.
- Determining of 'impact reference zones' (areas that are sufficiently representative to be used for assessment of impact on the marine environment).

---

[14]Workshop Towards an Environmental Management Strategy for the Area Berlin, 19 to 24 March 2017, the ISA Secretary-General's opening statement (d.d. 13-07-2017): https://www.isa.org.jm/sites/default/files/documents/EN/SG-Stats/remarks.pdf

[15]http://www.umweltbundesamt.de/en/press/pressinformation/deep-sea-mining-germany-lobbying-for-high (d.d. 13-07-2017).

- Determining of 'preservation reference zones' (areas in which no mining shall occur to ensure representative and stable biota of the seabed in order to assess any changes in marine biodiversity).
- Preparation of prior EIAs before any test mining such as large scale extraction or equipment trials; small scale test mining is considered as part of exploration activities and is hence included in the scope of an exploration contract.

The role of the Sponsoring State is to guarantee that the contracting entity will respect the ISA rules, regulations and procedures. In other words, the sponsoring State ensures that the relevant rules of international law apply to public and private entities that are not States. To achieve this, the Sponsoring State has the obligation to adopt national measures, in the form of legally binding instruments. The current state of legislation of EU states is summed up in Table 3.2.

3.4.2.2 Domestic law

Within national jurisdiction, Coastal States are sovereign and can regulate seabed mining occurring on their continental shelf. However, in doing so, they also have to respect the international obligations deriving from global and regional treaty law, including the standards set or to be set by the ISA. While there are thus a variety of different legislations and approach already in place, they tend to be derived from the same principles. For example, the Secretariat of the Pacific Community has developed a framework based on sound legal principles and practice to aid Pacific States in adopting their own legislation[16], while ensuring a high level of requirements and harmonization of the law in the region (Makgill & Linhares, 2015).

In addition, EU member states also have to abide by the relevant and applicable regional conventions and EU law. Environmental rules and procedure in Europe are particularly developed and might add to the ISA requirements, even though EU law is often a form of implementation of international obligations. For instance, the EIA directive (85/337/EEC) and the environmental liability directive (2004/35/EC) can be applied, as well as the EU maritime safety directives and regulations aimed at ensuring safety and environmental protection by EU flag states[17]. An overview of the status of EU states' laws both within and beyond national jurisdiction is provided in Table 3.2.

---

[16]http://dsm.gsd.spc.int/public/files/2014/RLRF2014.pdf (d.d. 13-07-2017).

[17]https://ec.europa.eu/transport/modes/maritime/safety/actions_en (d.d. 13-07-2017).

Table 3.2 Oversight of the national legislation for offshore mining in Europe. CS: Continental Shelf

State	Legislation Adopted – Relevant Acts	Area Draft	In Force	EEZ/CS
	EU Sponsoring States			
Belgium	Belgian Act related to prospecting, exploration and exploitation of the resources of the deep seafloor and subsoil thereof beyond national jurisdiction (17th August 2013)		X	
Czech Republic	Act No. 158/2000 on Prospecting, exploration for, and exploitation of mineral resources from the seabed beyond limits of national jurisdiction (18th May 2000)		X	
France	Mining Code of 20th January 2011			X
	Ordinance No. 2016-1687 of 8 December 2016 relating to the maritime areas under the sovereignty or jurisdiction of the Republic of France	X		X
Germany	Seabed Mining Act (6th June 1995, amended in 2010)		X	
UK	Deep Sea Mining Act 2014, amending Deep Sea Mining (Temporary provisions) Act 1981 (14th May 2014)		X	X
	Other EU Member States – Not Sponsoring			
Denmark	Mining Code Act of 24th September 2009			X
Malta	Malta Resources Authority Act nr XXV of 2000; Continental Shelf Act of 8th August 2014			X
Netherlands	Mining Act of 2002			X
	Note verbale dated 26 March 2013 from the Permanent Mission of the Netherlands to the United Nations.	X		
Portugal	Decree-Law on research and exploitation of minerals, 15th March 1990 (on-going amendment)	X		X
Spain	Law on Mines of 21st July 1973 (last amendment 2014)			X

3.4.3 Societal Impacts and Concerns

Exploration and exploitation of offshore resources could also have serious societal impacts, such as consequences for the livelihoods and well-being of coastal communities in particular for nearshore mining projects. So far no exploitation activities have taken place, which poses uncertainty with respect to the actual impacts of offshore mining.

3.4.3.1 Possible societal impacts

When it comes to offshore mining, the most relevant social impacts will likely be associated with several key changes during mining life cycle, which is likely to be relatively long (20–30 years) and may apply to different stakeholder groups at household, local, regional, national, and international level. Exploration is already occurring in different regions where the absence of conservation areas to protect the unique and little known ecosystems of the deep-sea, and sometimes the lack of an adequate regulatory regime, is striking. Public and local communities participation is also frequently lacking from the project's process and the authorities' decision-making (Franks, 2011; SPC, 2013d; EPRS, 2015, Baker et al., 2017), although trends in legal developments around the globe seem to be heading towards more transparency.

Table 3.3 below presents the potential societal impacts due to offshore mining built upon examples from terrestrial mining as a proxy (EPRS, 2015).

Table 3.3   Overview of potential societal impacts of offshore mining (EPRS, 2015, adapted from SPC, 2013d)

Category	Benefits	Disadvantages
Socio-Political	• Health and safety, • Working conditions, • Remuneration ... • Opportunities for other development options, • Strategic position of metal providers in the global arena	• Social inequalities at local scale • Political and strategic conflicts or inequalities: land-based mining vs offshore mining policies.
Economic	• Employment, • Flow of money, • Training, • Local business expansion, • Community development and social programs, • Equitable distribution	• Change in industrial landscape and composition, • Dominance of foreign entities
Socio-environmental	• Compensatory measures in favour of local communities • Compensatory measures in favour of the scientific world • Increased knowledge of habitat and ecosystem through data, surveys and trials' results	• Access to Marine Resources and competition between users of the sea • Fisheries • Cultural practices, • Environmental damage

Table 3.3 solely lists the societal impacts applicable to offshore mining that are considered likely to have a significant effect as things currently stand. It is an attempt at balancing positive and negative effects.

From a socio-political point of view, impacts can be both positive and negative. For example, labour features as described in Section 3.4.1 may increase remuneration in a given locality because of the higher skilled workforce, as well as ensure good working conditions and health and safety standards. However, this may also increase the social inequalities especially when a project happens in a developing state where communities depend on lower skilled jobs. In Papua New Guinea, one of the main concerns of local communities was the impact that the Solwara 1 project would have on fisheries.

On a bigger political and strategic scale, seabed mining represents opportunities for states or regions in terms of direct growth, but also indirectly through the development of other industries and sectors (e.g. development of industries using the produced metals, service providers, R&D ...). For the reasons earlier explained in Section 3.2, this could also help global strategies and alliances between states or regions. On the other hand, the rise of big and small, new offshore players might affect the economy and political stability of players depending on land-based mining, potentially creating tensions or conflicts.

Economically speaking, while there are a number of benefits directly arising from the sector's growth (e.g. employment, cash flow, community development ...), it might also bring some challenges requiring adaptation. The rise of seabed mining in a state that may be new to the sector, or that already has land-based mining activities occurring on its territory to balance it with, will change the industrial organisation of its system. Inevitably, this will require an adaptation phase, potentially with a new organisation of the sector or even the broader economical balance. In developing states, one of these changes will most likely occur from the arrival or increase in foreign entities joining an economical system, potentially disrupting a pre-established balance.

The socio-environmental impacts of seabed mining are perhaps more difficult to balance. It appears rather obvious what ecological concerns might mean for people worldwide and even so for local communities: lesser access to marine resources for those competing users of the seas (shipping routes, cables, scientific campaigns ...), impact on fisheries, or more generally, the environmental damage undeniably arising out of seabed mining (see Section 3.4.4 for more details). However, as compensatory measures, the sector has the potential to offset those impacts by bringing value back to society, through scientific opportunities on-site, the gathering of data and knowledge on these poorly known environments, or direct benefits to local communities through funding of local infrastructures, training opportunities, etc.

3.4.3.2 Societal impact relevant for the EU

Due to the increasing importance of the topic in the immediate future and the necessity for the EU yet to define a policy on this matter, the European Commission launched a Stakeholder Consultation (including civil society, NGOs, Member States and some private and public consultancies) on offshore mining. The main outcomes of this consultation showed that[18]:

- Commercial mining is not an option unless regulations are in place.
- The drafting and adoption of regulations must be transparent and participatory and any benefits widely shared.
- More emphasis on reuse and recycling of materials rather than on offshore mining is required.

On the other hand, the interviews with industry stakeholders point out the fact that before making any conclusions, the opponents of offshore mining, scientists and governments should look at the overall risk and impact of offshore mining vis-à-vis terrestrial mining, and allow things to go forward.

As further explained in Section 3.4.5, the land, nearshore and offshore mining impacts can be compared and they also diverge for clear reasons. High risk and actual damages already occur on land, and the sole recycling and reuse of metals will not satisfy the increasing need for mineral resources. Hence, it is important to weigh the pros and the cons and to balance environmental risks with the potential for benefits, all the while making sure that the right framework is in place to enable sustainability (EPRS, 2015).

3.4.3.3 Mitigation of societal impacts

Lessons learnt from terrestrial and nearshore mining are provided below together with past relationships between mining companies and Pacific Island communities that have been characterized by complexities, tensions and contradictions (Franks, 2010; SKR Consulting, 2010; EPRS, 2015):

- Use ecological (systematic) approach.
- Be aware that legal limits and scientific data may not be aligned with community expectations.

---

[18]http://ec.europa.eu/dgs/maritimeaffairs_fisheries/consultations/seabed-mining/index_en. htm (d.d. 13-07-2017).

- Societal changes can be indirect, often of economic/political in nature.
- Socio-environmental concerns are very important (use of coastlines, deep-water pollution and disturbance).
- Access to, use of and ownership of land are also important (e.g. issues of fishing or cultural practices).
- Government institutions are crucial to balance environmental preservation against economic gain.
- Corporate governance, corporate social responsibility and transparent procedures need to be established before mining takes place.
- Social scientific research is needed to understand communities' positions.

3.4.3.4 Safeguarding financial revenues for the future

An offshore resource (like any other resource) only has a limited stock and has an end to exploitation at a given time point. After exploitation ends, so does the source of substantial income. As far as international seabed mining in the Area is concerned, the LOSC provides that all mining activities (whether at the exploration or the exploitation phase) shall be carried out for the benefit of mankind as a whole. Hence, some of the provisions in Part XI of the LOSC ensure benefit-sharing in several forms, including non-monetary, particularly in favour of developing States. The sharing of financial and other economic benefits is one of them, although it has not really been implemented yet since exploitation has not started. The LOSC does not give much detail as to how this benefit-sharing should be operationalized, but it does prescribe that a contractor's payment to the ISA shall not be higher than the rates in land-based mining in order to avoid inequalities in the sector. Major discussions are currently talking about rates of 4–6% of the potential revenues (ISA, 2016).

Before granting exploitation concessions, mineral funds should be considered and set up, especially considering that the Area represents the Common Heritage of Mankind and thereby of all nations. Countries like Alaska, East Timor, Norway and São Tomé et Príncipe offer examples and inspiration for their structure and organisation (SPC, 2013d).

3.4.3.5 Safeguarding scientific revenues for the future

Other than financial benefits, the LOSC also provides for the dissemination of marine scientific research results, cooperation with developing states in research programs and training, technology transfer, access to reserved areas of exploration reserved for developing States at lesser costs. During the Berlin workshop in 2017[15,16], it was urged to combine all ecological and physical non-sensitive data and to make it publicly available. This is a matter that should be, and that is to a certain extent, regulated, but more importantly that should be effectively implemented in the future.

3.4.4 Ecological Concerns

3.4.4.1 Potential direct ecological impacts

Various studies emphasize that the ecological impacts of offshore mining are a point of major concern. Amongst others, SPC (2013a, b, c), Ortega (2014), and Sharma (2017) concluded that the impact of offshore mining is expected to be in the various forms of:

- loss of substrate,
- loss of benthic communities,
- loss of biodiversity,
- sediment plumes on the seafloor,
- increased turbidity in the water column, and
- addition of bottom sediments to the surface.

Such impacts would result in changes to the food chain and thus to the marine ecosystem, but impacts on the surface as well, owing to collection, separation, lifting, transportation, processing and discharge of effluents should not be overlooked. Oebius et al. (2001), Ramirez-Llodra et al. (2015) and others described the impact of sediment clouds as a result of other human activities, providing clues and background knowledge from which the impact of seabed mining plumes could be extrapolated. Boschen et al. (2013) describe more specifically the impact on a range of habitats and time scales for SMS deposits.

Mining nodule areas seem especially sensitive, since these deep areas are cold and hardly receive energy input: a standstill world with high and complex biodiversity. The nodules themselves harbour an epiphytic biota distinct from the surrounding sediments. In one CCZ locality, roughly 10 per cent of exposed nodule surfaces were recorded as being covered by sessile, eukaryotic organisms (mostly foraminiferan protozoans) carrying an unique mini-ecosystem themselves (SPC, 2013b; Vanreusel et al., 2016).

The seamount areas of cobalt-rich crusts host biodiversity rich habitats such as deep water coral reefs. Water currents are enhanced around seamounts, delivering nutrients that promote primary productivity in surface waters, which in turn may promote the growth of fish and animals such as corals, anemones, stars and sponges, but also creates an oxygen-minimum zone that inhibits the growth of some organisms (SPC, 2013c). FAO designates seamounts as Vulnerable Marine Ecosystems, a protective status for fishing activities[19].

Hydrothermal vents and SMS deposits are associated ecosystems composed of an extraordinary array of animal life. Chemosynthetic bacteria, which use hydrogen sulphide as their energy source, form the basis of the vent food web, which is comprised of a variety of giant tubeworms, crustaceans, molluscs and other species, with composition depending on the location of the vent sites. Many vent species are considered endemic to vent sites and hydrothermal vent habitats are thus considered to hold intrinsic scientific value (Van Dover, 2008; SPC, 2013a).

Technical and scientific studies have found that there is a general lack of data to make thorough environmental impact assessments (SPC 2013a, b, c, d; Lange et al., 2014; Ecorys, 2014; Rogers et al., 2015). Phosphorite mining examples show how uncertainties and gaps in knowledge and data actually lead to major delays in project development (Baker et al., 2017), in particular due to major discussion on EIAs, potential economic impacts, government shares and social acceptance. Societal protest is due to the fact that phosphorite mining can be nearshore, within the range of fisheries and rich biodiversity (see e.g. Benkenstein, 2014; EPRS, 2015; Baker et al., 2017; Sharma, 2017).

3.4.4.2 Potential indirect ecological impacts

On a more general level, one could state that offshore mining hampers the evolution towards a circular economy (recycling, eco-design, sharing, repairing, etc.), since new resources are reclaimed instead of recycling discarded products. On the other hand, Ecorys (2014) indicated that recycled contents remain rather low, not fulfilling the needs. It also shows that offshore mining can provide a part of the additional new ores that will be needed on the market.

Gas hydrates are thought to influence ocean carbon cycling, global climate change, and coastal sediment stability (issue under serious debate, e.g. Bosswell & Collett, 2011; Lange et al., 2014). In addition the mobilization of gas hydrates as a new, potentially cheap energy source will contribute to

---

[19]http://www.fao.org/in-action/vulnerable-marine-ecosystems/en/ (accessed 13 July 2017).

additional CO_2 in the atmosphere, a cheap new source can also hamper the development of renewable techniques.

3.4.4.3 Mitigation of ecological impacts

Concerns about the ecological impact of offshore mining are recognized by the ISA, who have subsequently taken various actions to describe and support good practices. This includes training – including biodiversity monitoring and development of environmental management systems.

Integrated governance based on the ecosystem approach will be necessary in developing deep-sea mineral policies. Ecosystem-based oceans management strategies, laws, and regulation for seabed mining would include provisions for (SPC, 2013d, ISA):

- Collecting adequate ***baseline information*** on the marine environment where mining could potentially occur.
- Establishing ***protected areas*** where there are vulnerable marine ecosystems, ecologically or biologically significant areas, depleted, threatened, or endangered species, and representative examples of deep sea ecosystems.
- Adopting a ***precautionary approach*** that, in the absence of compelling evidence to the contrary, assumes offshore mining will have adverse ecological impact and that proportionate precautions should be taken to minimize the risks.
- Applying ***adaptive management*** in which different hypotheses on exploitation and impacts are formulated and tested during exploitation in order to switch to different management strategies.

Processes at the deep seafloor require lots of energy, e.g. for transport of raw material to the surface and for processing and transport on board of vessels and platforms. Therefore, the use of on-site renewable energy sources may be considered to reduce the supply and costs of fuels, and emissions of CO_2. Especially when combined with floating or fixed platforms, wave energy and wind farms could possibly be used. To this end, innovation and R&D in the seabed mining sector is a crucial and on-going step.

3.4.5 Comparing the Impacts of Land-based Mining versus Offshore Mining

Aside from sediment plumes being dispersed in the water column at different depths with different consequences, seabed mining will also undoubtedly destroy the habitats and biodiversity locally and in the case of nodules most likely permanently, on the sites where the mining occurs. However, these two impacts (plumes and habitat destruction) need to be relativized when compared to land-based mining's social and environmental footprint.

On land, mining tailings could be the equivalent of sediment plumes. Mining tailings are often dumped directly in the surrounding environment, may it be grounds or rivers, and are more often than not charged with chemical and heavy metals remaining from minerals processing into commercially exploitable metals. While the dumping of sediment tailings has significant effects on the surrounding environment comparable to the ones of underwater plumes, contaminated tailings flowing into the water cycle – groundwater, watercourses and eventually the sea – is quite worrisome, to say the least (Hein et al., 2013; Ramirez-Llodra et al., 2015; Rogers et al., 2015).

With seabed mining, contaminated sediments plumes in the water column are not only unlikely because on-board processing methods differ, but they are also legally forbidden. Not only the LOSC and ISA standards do not and will not allow it, but maritime practice and customary rules built upon the relevant IMO conventions have long been applied, monitored and effective (IMO, 1972; IMO, 1996). Even though to date IMO conventions are not directly applicable to seabed mining, the ISA, following the International Tribunal for the Law of the Sea's advice, is taking steps to avoid the emergence of "sponsoring States of convenience" in the seabed mining sector, meaning that States will be treated equally irrespective of their status or capacity when it will come to compliance with Part XI of the LOSC and the ISA Mining Code (ITLOS, 2011). This is also of relevance considering that land-based mining often occurs in places where, even when environmental safeguards are in place, their effective application is often lacking. Indeed, major extracting activities happen on the territory of developing States that are at best, unable to monitor and enforce and at worse unstable and corrupted (e.g. China, Congo).

Hence, even though the geographical scope of sediment plumes is likely to be larger than onshore mining due to the size of exploitation areas and oceanic currents and dynamics, measures to maintain turbidity at an acceptable level and to prevent the use of contaminants will be effectively applied and monitored by several levels of authorities (sponsoring States, ISA, IMO). A major concern is still the definition of what is acceptable and what is harmful impact in this offshore environment[8].

Comparison of habitat destruction onshore and offshore bears different concerns. Seabed mining is likely to occur on areas much larger than typical land-based mine sites. The exploration area of GSR is three times the size of

Belgium[5]. Even if they are likely to actually exploit only a small proportion of it, that could represent up to a third of said country. However, the direct impact of extraction will have a superficial impact on the seafloor (Hein et al., 2013). Indeed, whichever mineral is targeted (nodules, sulphides or crusts) will not require deep-cutting excavation methods. Because of the formation of such minerals either from superjacent water deposit or subsoil volcanic and geologic activity which are specific to their oceanic environment, mineral extraction does not require much more than scraping the seafloor's surface of a few meters deep only (Hein et al., 2013). When compared to land-based mining, where entire mountains can be taken down or underground mining can go too deep as to weaken stability and provoke slides (e.g. Chile), seabed mining's negative impact on the seafloor habitat may appear minor from a geological point of view. From an ecological point of view, habitat, biodiversity, genetic information and ecosystems concerns are not fully addressed and compared yet (SPC, 2013a, b, c, 2016; Ecorys, 2014; Rogers et al., 2015[20]).

Last but not least, working conditions on board will without a doubt be a lot better than conditions of onshore mine workers. Indeed, the technicality of seabed mining operations and the restrictions of having to sail on the high seas require limited and higher trained workers and seafarers, as opposed to the potentially terrible conditions of miners' populations often abused by corporates and governments (e.g. Congo) in terms of salary, health and safety rules[21].

In summary then, it is difficult to compare the environmental impacts of land based mining with seabed mining because one is a mature, large scale, destructive industry and the other has only limited information. As a consequence in all individual cases decision makers would need to evaluate independently – taking into consideration the market and environmental conditions of the individual minerals at the moment of deciding and in the future – whether the integrated economic, social and environmental footprint of seabed mining is acceptable and preferable or that land-based mining provides a better solution to meet the standards for the integrated economic, social and environmental footprints.

---

[20]http://dosi-project.org/ (d.d. 13-07-2017).

[21]http://www.economist.com/news/middle-east-and-africa/21705860-can-ambitious-mine-make-difference-eastern-congo-richest-riskiest (d.d. 13-07-2017).

3.5 Innovation

Innovation needs are firstly introduced using the guiding principles of LCA and the value chain and next described in more detail. The value chain for offshore mining, irrespective of the specific resource, can be considered to include six main stages (Figure 3.10; Ecorys, 2014):

1. Exploration;
2. Resource assessment, evaluation and mine planning;
3. Extraction, lifting and surface operations;
4. Offshore and onshore logistics;
5. Processing stage;
6. Distribution and sales (this stage is not included in this study's analysis).

The current state of technology can be assessed on the basis of Technology Readiness Levels (TRL). The TRL levels for offshore mining value chain have recently been assessed and reported (Ecorys, 2014), and this section builds on the results of this study that was commissioned by the EU, and by the SPC study (2016).

3.5.1 Lifecycle Stages

The concept of business lifecycle considers how industries and firms were not in a steady state and appeared to evolve over time. The general value chain of nearshore and offshore mining is given Figure 3.10. For each sector and segment information for the LCA is given throughout the document. Per subsector more information on the most conspicuous features is given in the next sections. The most dominating aspect at the moment is the interpretation of exploration results, extraction and ore processing (steps 2, 3 and 5 in Figure 3.10) where the TRL of most aspects is still fairly low.

3.5.1.1 LCA of nodules, SMS deposits and cobalt crusts

The LCAs of nodules, SMS deposits and cobalt crusts are discussed in combination since they experience the same driving forces. The main drivers of the interest in offshore mining of metals seems to be the high market prices of the resources at stake at a certain moment in combination with the high exploitation costs vs geopolitical concerns on flows of essential ores.

Typically, TRL levels are lower (range 1–4) for technologies required on the seabed (collectors like cutters and scavengers) and for vertical transport (lifters). The on-board processing of ores for metal extraction -in order to reduce material loads to be transported- also needs to be improved. Technologies required at sea level (ship/platform and associated equipment, logistics) and onshore are more mature as they have similarity to applications in other sectors already existing. In addition the refined metals have their long established markets (Ecorys, 2014).

Innovation is expected to reduce the exploitation costs. Since these prices are highly dynamic and innovation costs are high and time consuming, major developments in activities are not expected at the moment (except for a few exceptions with high concentrations of resources).

3.5.1.2 LCA of phosphorites

A first remark is that extensive reviews are scarce on marine phosphorite mining. Only limited information is available. Most informative are websites. Given the high potential of this resource a more elaborate study is welcome.

Contrary to the metals, phosphorites can have valid business cases in the three projects in Namibia (two companies), Don Diego, Mexico and Chatham rise New Zealand. Several aspects make these business cases alive:

1. The large local demand for phosphates (Don Diego, 2015);
2. High global market prices (Figure 3.1);
3. Reasonable exploitation costs (Table 3.4);
4. Potential export and a share in the global market (Benkenstein, 2014).

Whereas they are imported now, rich relative shallow concessions are available and investors are willing to make the necessary high start-up investments. Amongst other reasons, problems with land-based ore qualities, increased demands, and geopolitical concerns (de Ridder et al., 2012), a more stabilized higher price and presumably technological developments will have altered the business case.

For phosphorites the business case seems more viable: large concessions can be found in the easily reachable nearshore and the shallow offshore. This enables the use of standard equipment what only has to be adapted to a minor extend (Schulte, 2013). As a result preparations have been made to exploit the resources with substantial interest expected (like being able to deliver 10% of the global market for phosphates). Environmental considerations have blocked the actual exploitation until further evaluations partly due to the fact that this type of bottom destruction in this zone has not been attempted before and e.g. impacts on bottom-life and associated fish communities are feared (Benkenstein, 2014; Rogers et al., 2015; Baker et al., 2017, Sharma, 2017)[22].

---

[22]See e.g. the continuing discussion on the Namibian NMP Sandpiper project (accessed 13-07-2017): https://southernafrican.news/2016/11/07/namibia-u-turn-on-phosphate-mining/

Table 3.4 CAPEX, OPEX costs and IRR of metal ores (Ecorys 2014) and an example phosphate project[7]. The metals used for calculations and the relative contribution to price: SMS: copper, gold, silver (70:28:2). Nodules: copper, cobalt, nickel (25:11:63). Including processing, however assumed to exclude processing of manganese. Cobalt crusts: copper, cobalt, nickel (4:63:33). Including processing, however assumed to exclude processing of manganese

	SMS Deposits	Polymetallic Nodules	Cobalt Crust	Phosphate
CAPEX ($)	1,000,000,000	1,200,000,000	600,000,000	400,000,000[a]
Years of operation:	15	20	20	20
Linear depreciation $/yr:	66,666,667	60,000,000	30,000,000	
Yearly production (tonnes crude ore):	1,300,000	2,000,000	450,000[1]	3,000,000
CAPEX per tonne crude ore($):	51	30	37	7
OPEX Cost excluding processing $/tonne crude ore:				60
OPEX Cost including processing $/tonne crude ore:	70–100	85–300	95–310	
Revenue (excluding manganese) $/tonne crude ore:	718	306	216	
Total OPEX per project	3,315,000,000	7,000,000,000	3,200,000,000	3,600,000,000
Total Revenues per project	14,001,000,000	12,240,000,000	3,456,000,000	7,500,000,000
Net profit per project	9,686,000,000	4,040,000,000	–344,000,000	3,500,000,000[b]
IRR % (excluding manganese)	68	2	no positive cash flow over period	23.6
IRR % (including manganese)		109	46	

[a]Estimated for total project.
[b]at $125/tonne.

3.5.1.3 LCA of gas hydrates

According to current estimates, global hydrate deposits contain about 10 times more methane gas than conventional natural gas deposits. There is a strong urge to make the exploitation of gas hydrates viable. In particular, highly developed countries without their own sources of energy are investing in this sector. The technology needs to be developed since it is a whole new substance type for exploitation. There are some doubts whether it can be exploited in a profitable approach. It remains to be seen whether hydrate extraction at great depths is economically viable at all.

Continuing on the specific stages in the value chain (Figure 3.10):

3.5.2 Resource Assessment

During the last decade, stage 1 has been developed up to a reasonable level to proceed with the actual exploitation phase. In stage 2 Planning, deep-sea geotechnical site investigation and evaluation methods and procedures for pit design, including slopes and ground conditions as well as for predicting extraction efficiencies are the subject of current R&D projects[23]. However, the extraction methodology still needs to be validated in lab and real environments. Since offshore and onshore logistics are already well developed, the critical stage in the value chain from a technology perspective is stage 3. It will be considered here.

3.5.3 Extraction

No commercial offshore mining operations have taken place yet, and especially the extraction techniques required on the seabed are not operational yet (Ecorys, 2014, SPC, 2016). The technology to be used depends mainly on the type of deposit. The extraction process for deep-sea minerals starts with the excavation. For nodules the proposed technique for excavation is by making use of collectors, while for SMS deposits crusts cutters are being developed. Some processing may also take place on the seabed. The TRL for proposed extraction technologies is scored low, ranging from TRL2 (formulated concept) to TRL 5 (technology validated in relevant environment). Hence, more development should take place before exploitation from the deep-sea bed can take place. In the summer 2017, GSR[5] will test a nodule harvesting tool for the first time in the Area, showing great improvements in the design and

---

[23] See for instance the Blue Mining project www.bluemining.eu; or the Blue Nodules project www.blue-nodules.euor MIDAS: www.eu-midas.net (d.d. 13-07-2017).

preparation of a future exploitation project. Nautilus will also be undertaking submerged trials in PNG[24].

The availability of the operational gear is a crucial aspect (SPC, 2016). At the time of writing, the most advanced (and applied) technique to raise crude ore from the seabed appears to be the technique as developed by the diamond industry to recover eroded diamonds, deposited on the (nearshore) ocean floor by land runoff and fluvial systems. The maximum reported commercial and full-scale operative removal depth reported so far is limited to – 140 m (ROVs and scraping & vertical lifting[2]). This basic technology – as developed for this given mining environment – may amongst others also be applicable to the environments under consideration in the present chapter, although we are here dealing with depths ranging from a few hundred meters for seamount crusts down to –4000 m for nodules. In addition, for the phosphorite concessions of Don Diego and Chatham Rise, an adapted TSHD with trailing technique will most likely be developed to remove ores till depths of –450 m[3]. Cutters need to be tested and optimised for SMS deposits and crusts.

3.5.4 Vertical Lifting

Vertical lifting is another critical part of the mining process. Air lift systems and especially hydraulic systems seem most applicable for use in deep sea mining operations. However, TRL levels for proposed lifting systems is at 5 at the highest, and therefore further development is required. Both require high power input and are so far sensitive to unstable flows, which again given the depths at stake is one of the most critical aspects from a technological and technical point of view. Possibly, techniques being used in the offshore oil and gas sector (transport of drill cuttings and mud) could be adjusted for use in ore transport. For nodules the ROVs or AUVs seem the most promising technique. These have to be tested for operation at depth of 6000 m (real operating environment).

3.5.5 On Board Processing

Once raw material is transported to the surface, a working platform is required for further handling. Support vessels or platforms are proposed

---

[24]See http://www.tijd.be/nieuws/archief/Knollen-rapen-op-de-zeebodem/9854214?ckc=1& ts=1491349056 and http://www.nautilusminerals.com/irm/PDF/1893_0/NautilusMineralsSea floorProductionToolsarriveinPapuaNewGuinea (d.d. 13-07-2017).

as dispatching system, storage facility, dewatering and on-board processing facility. Simple dewatering systems can easily be applied on board of vessels and platforms, but further processing on board like concentrating ore and application of metallurgical processing requires further development. Fixed platforms offer better opportunities for processing than ships because comminution, the grinding to smaller particles, is performed by large and heavy equipment.

For efficient use of ships and equipment, use of a platform in a central place with respect to the mining locations should be envisaged. Platforms are very stable, and instability issues like on ships are not important. The technology for such platforms in deep sea is well established in the oil-industry. In addition one can think of floating platforms, as well as of the installation of renewable energy structures in order to reduce the energy costs and carbon footprint.

A central platform located nearby the mining site or halfway to port, and where most of the processing would be carried out, could be more efficient than carrying-out processing on the ship. The ship could transport the retrieved minerals to the platform rather than sailing all the way to port every time, which is particularly relevant in remote cases like the high seas and the Area. Processing of the ore can proceed on the platform and concentrates can then be shipped to on-shore locations (Ecorys, 2014, SPC, 2016).

3.5.6 Final Processing

Due to the large quantities of ore, and – in some cases – complex chemical process involved, the final processing will most likely take place on-shore in dedicated facilities. In general two techniques have been tested: hydrometallurgy, where the metals are separated with acids (hydrochloric or sulphuric) or basic reagents (ammonia), and smelting. Some ores, especially manganese and cobalt (to a lesser extent) still pose problems and require extensive energy input or use of aggressive chemicals (by methods still in optimisation phase, SPC, 2016).

However, most developments that are currently taking place focus on adapting available techniques to deep-sea environments rather than developing novel techniques and processes specifically suitable for deep-sea deployment. It seems therefore that higher operating expenses (OPEX) are accepted to avoid higher capital expenditures (CAPEX), e.g. for lifting (Schulte, 2012[3]).

3.6 Business Economics and Investment

From a commercial perspective, seabed mining is a small sector, with only few active companies. This section reviews available information on the economic performance of seabed mining, addresses the current status of investment and identifies key concerns among investors.

3.6.1 Economic Climate for Offshore Mining

Land-based mining was developed over a long period of time. Starting with small-scale mining of easily accessible deposits, this sector is gradually increasing in size of operations and targeted less accessible depots. As a consequence, knowledge and investments increased gradually. This development pathway is not foreseen for seabed mining where – particularly for resources at depths exceeding 200 metres – investors need to be fully committed with high initial CAPX costs (near $1,000 million starting, Clark et al., 2013, Rozemeijer et al., 2015, Table 3.4).

Despite of the availability of a lot of documents on the subject, it is hard to dig into the details of the costs involved in offshore mining in order to pinpoint a target for innovation on the basis of CAPEX or OPEX. This is due to the lack of uniformity in the data provided by different authors concerning CAPEX and OPEX. When considering the cost and revenues it is important to remain cognizant of the fact that all costs are based on technology that has been piloted but not proven at the commercial level of operation. Cost estimates are highly uncertain and may change significantly depending on the mining technology that is in place at the time of full-scale commercial operation.

From the assessments described below, the following general picture emerges. Phosphorites exploitation can be profitable at this time. Offshore mining of SMS deposits seem economically profitable, when enough resources can be found clustered to support 15 years of continuous operations. Nodules revenue estimates are subject to serious debate and exploitation of crusts is far from profitable. Gas hydrates are even further from actual exploitation (see Table 3.4).

3.6.1.1 Market price for key resources

It is interesting to study long-term trends of metal prices as the fluctuation in price correlates to the interest in offshore mining. The first wave of offshore mining development took place in the 1970s when resource prices where high

(Figure 3.1, Section 3.2). Likewise, interest in the first decade of the 21st century can be related to high resource prices.

In the mid-2000s, prices for these metals rapidly increased, but then started to decline around the year 2010 (Figure 3.1). Current prices are somewhere in between pre-2000 prices and the highest recorded prices[25].

3.6.1.2 Costs and revenues of SMS deposit mining

SMS deposit mining requires a very high initial investment to start the operations. Initial investments (CAPEX) are estimated at around $300M–400M for a typical seafloor SMS deposits operation (Birney et al., 2006; Yamazaki, 2008). However based on actual costs developments for the Nautilus Solwara 1 operation, actual CAPEX is likely to be much higher. In practice total CAPEX, including exploration costs, is estimated to be closer to $1,000M (Table 3.4; Ecorys, 2014; EPRS, 2015). The OPEX of seabed mining, including transport to shore, are estimated to be between $70–140/tonne crude ore based on the above sources. Necessary processing costs increase total OPEX to $150–260/tonne crude ore.

Boschen et al. (2013), backed by the studies of Ecorys (2014) and SPC (2016), estimated that SMS deposits will be profitable due to the high content in currently highly priced copper, gold and silver (with copper contributing the most ~2/3). In addition REEs and other metals will contribute also to the revenues. The Ecorys study calculated a potential internal rate of return (IRR) of 68% of total investment. Total revenues are $14,001,000,000 and net profit $9,686,000,000 (Table 3.4).

However, there is considerable uncertainty regarding SMS deposits as it is assumed that an operation of 15 years is needed to generate returns on investment, whereas most resources and proven reserves seem to point to smaller sizes, and a strain of operations on different locations needs to be established. The Solwara1 project seems to have only a limited amount of deposits (2 years) (SRK Consulting, 2010; Ecorys, 2014; EPRS, 2015; SPC, 2016). In addition, when comparing all different sources, different values for CAPEX and OPEX are encountered every time (Rozemeijer et al., 2015).

3.6.1.3 Costs and revenues of nodule mining

Nodules mining is expected to be more capital intensive than SMS deposit mining due to the larger depths and more widespread distribution over the seafloor. An initial estimated CAPEX of $1,200M seems realistic to start

---

[25]www.infomine.com (d.d. 13-07-2017).

operations (Yamazaki, 2008, Clark et al., 2013). A more detailed estimate–as described in EPRS' study (2015) indicates a CAPEX cost of almost $1,800M. Still according to ERPS, almost half of these capital investments come from investments in a processing facility. Estimates of nodule mining OPEX range between $85-500/tonne, of which costs related to processing form an important component.

Considering copper, cobalt and nickel, Ecorys' study estimated the IRR at 2% (Table 3.4) with nickel being the main contributor ($\sim$1/2). Manganese was excluded from their calculations. Including manganese IRR increases to 102%. For manganese no efficient extraction method is yet available. The manganese residuals could of course be stored till further developments enable costs effective isolation (SPC, 2016). The conclusion of Ecorys (2014) is not consistent with other sources that consider nodules as the most attractive deposits economically (EPRS, 2015). According to SPC (2016), nodules were profitable only in 60% of various scenarios with different CAPEX, OPEX and revenues. Clark et al. (2013) give an IRR range of 6–38%. Note that Martino & Parson (2012) propose that a lower IRR of 15–20% could be advocated since seabed mining is less risky than onshore mining (IRR > 30%).

Rozemeijer et al. (2015) calculated based on different scenarios with copper, cobalt and nickel prices of 2015 and were not able to show profitable exploitation. The assumptions taken on e.g. equipment efficiency and costs are very important in the calculations and vary highly between authors (Rozemeijer et al., 2015; SPC, 2016).

3.6.1.4 Costs and revenues of cobalt crust mining

Only the costs and revenues of a single cobalt crust source have been assessed. Yamazaki (2008) has estimated the CAPEX and OPEX of cobalt crusts based on nodule mining. CAPEX is expected to be some 50% of nodule mining and OPEX stand at 45%. However, assumed production volumes (dry) in these estimate for cobalt crusts stands at some 40% of manganese nodules which makes the CAPEX and OPEX per tonne some 25% resp. 12.5% higher than for manganese nodules. Based on calculations by Ecorys (2014), Rozemeijer et al. (2015) and SPC (2016), it is concluded that under current market prices, there is no viable business case for cobalt crusts mining.

3.6.1.5 Costs and revenues of phosphorites

Namibian Marine Phosphate (NMP) estimates the further CAPEX for a whole project on phosphorite mining will amount to approximately $326M.

In addition, $50 million OPEX will be spent on the project. The mining licence of NMP has been granted for an initial period of 20 years. Approximately 3 million tonnes of dry product for export are expected to be processed starting from year three, at a price of $125 per tonne (Table 3.4). This is approximately $7–7.5 billion for 20 years. It is claimed to be very profitable. The IRR for Namibian Marine Phosphate project Sandpiper is estimated at 24%[7].

Leviev's private company LL Namibia Phosphate (LLNP) plans investing $800 million in building a mining facility to produce about two million tons annually from an estimated two billion tons at a depth of 300 meters. At a selling price of an estimated $125/tonne, the revenues are about $250M a year. Chatham Rock Phosphate expects yearly revenues of $280M and a yearly profit of $60M (Schilling et al., 2013).

3.6.1.6 Concerns and uncertainty about economic viability

Doubts can be raised on the economic viability of offshore mining of metal ores. Ecorys (2014) examined the estimated CAPEX, OPEX and market price for metals of seabed mining and concluded that SMS deposits are likely to have the highest commercial viability (to be treated with caution as no actual operations have taken place yet). This is due to the fact that in SMS deposits, copper can be extracted in large amounts from these resources at a moderate market costs. Furthermore, it is possible to extract gold from these reserves (Boschen et al., 2013).

In the calculations presented above, exploitation of nodules and crusts is not commercially feasible. This finding is not consistent with the answers given by some interviewees (EPRS, 2015), who mentioned nodules as the most attractive deposits commercially. This can be due to the fact that mining companies assume an operation of 15 years (20 years for nodules and cobalt crusts) to generate returns on investment, while key uncertainties exist in case of SMS deposits about the resources and reserves which seem to point to smaller sizes (SRK Consulting, 2010; Ecorys 2014). This has been confirmed by the industry stakeholders, mentioning that it is challenging to find and extract SMS deposits as they are more difficult to spot and are relatively small, while the operations are usually calculated with a proven resource for 20 years.

3.6.2 Government Support

Government support for development and commercialisation of offshore mining can take different forms. In a basic form, governments can be catalysts via their membership of the ISA, enabling their national companies to obtain exploration and exploitation contracts with the ISA.

Funding can also stimulate innovation. Offshore mining could use a boost in order to exploit at less energy costs (cheaper) and with less environmental impact, making it economically viable. To this end, national and EU publicly funded research projects related to offshore mining and offshore exploration technologies are carried out. Research is often supported by engineering firms and technology providers, which themselves work closely together with research institutes and universities. Three important EU projects aiming at deep-sea resource extraction are Blue Mining and MIDAS[23]. Blue Mining explores the needs for developing the technologies required for nodule and SMS mining, while MIDAS focuses on environmental impacts from deep-sea activities. Other research efforts are linked with seabed mining, but have a wider scope.

An important programme is the European Innovation Partnerships (EIP). The EIP aims to reduce the possibility that a shortage of raw materials may undermine the EU industry's capacity to produce strategic products for EU society. The EIP on Raw Materials is not a new funding instrument. It aims to bring stakeholders together to exchange ideas, create and join partnerships in projects that produce concrete deliverables. In 2014, 80 commitments were recognized as 'Raw Material Commitments', out of which, six are related to seabed mining (Ecorys, 2014).

3.6.3 Status of Investment in Seabed Mining

The recent history of deep-sea mining is not a story of great commercial success. A number of companies have succeeded in getting listed on various stock markets, including well-known companies such as Nautilus Minerals (Toronto Stock Exchange), Neptune Minerals, Chatham Rock Phosphate (New Zealand) and Odyssey Marine Exploration. However, where common stock-market indexes have risen considerably in the last years, the stock-market value of these companies has dropped sharply between 2013–2015 and is consistently low since then. Traded volumes are also low (see Figure 3.11).

Research on identification of investors in seabed mining, and their interests – carried out under the EU Maribe-project and reported in van den Burg et al. (2017), is illustrative of the low interest of investors in seabed mining.

The inventory of investors active in the various Blue Growth and Blue Economy sectors (van den Burg et al., 2017) identifies 31 investors in seabed

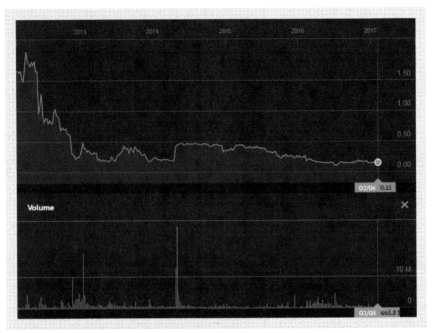

Figure 3.11 Stock market price of Nautilus, listed on Toronto stock exchange https://www.bloomberg.com/quote/NUS:CN (06022017).

mining (out of 244 total investors). The majority of these investors are so-called internal investors; these are companies that invest in R&D in seabed mining, as this can be a future market for their products or services. Examples include shipbuilding companies (Damen, Royal IHC), generic maritime service companies (Kongsberg Maritime, Heerema) and dredging companies (such as Boskalis). Notably absent are private equity investors, business angels and banks. These investors are generally from the UK, USA or the Netherlands, with a few exceptions.

In a survey, investors were asked how important the different Blue Growth sectors are for them (see van den Burg et al., 2017). Seabed mining scores considerable lower than the other sectors, with an average score of only 1.82 (1 = not important, 4 = highly important), see Figure 3.12.

3.6.4 Factors Hampering Further Investment

Research into investor behaviour (see van den Burg et al., 2017) pointed out some of the main concerns of investors. The top 5 risks that stand in the

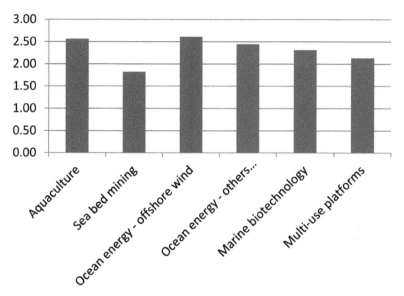

Figure 3.12 Importance of various Blue Growth sectors (from van den Burg et al., 2017).

way of investment in seabed mining are discussed from a seabed mining perspective below.

3.6.4.1 Operational risks
Technologies for seabed mining are under development but in the absence of large-scale mining activities, this remains experimental development. While surveys of deep-sea resources have a long history, the actual mining of these resources has hardly been done before. Uncertainty about the technology to deploy, the risk and associated costs impede investment.

3.6.4.2 Financial risks
Low interest of investors is inextricably linked to doubts about the financial performance of the sector. In Figure 3.1 it was shown that prices of the resources fluctuate significantly over time. In the period 2000–2010, prices for natural resources peaked, increasing interest in the exploitation of deep-sea resources. As mentioned before, copper, gold and silver are the main metals of interest for SMS deposits, cobalt, and nickel (and copper to a very limited extend) for nodules. Crusts seem too costly at this moment. Apart from the overall uncertainty within the assumptions, a specific uncertainty exists regarding potential revenue streams for manganese (Section 3.5.6).

This directly points to the importance of further efficiency increases in mining itself and in processing as well.

3.6.4.3 Regulatory risk

The recent history of seabed mining shows that the risk of (sudden) withdrawal of governmental support is real. This also includes permitting. Chatham Rock Phosphate lost 92% of its stock value in one trading day when it was refused consent to mine from the Chatham rise in 2015. With continued pressure from NGO's and other interest groups to halt further development – also witnessed e.g. in Papua New Guinea – regulatory risks are a key obstacle to investment in seabed mining (Baker et al., 2017, Sharma 2017[3]).

3.6.4.4 Environmental issues

Given the attention the offshore mining industry receives from stakeholders, none of the companies would be willing to add risks to their investment by developing environmentally harming techniques. Before licenses are issued, environmental impact assessments need to be approved, including the techniques and mitigating actions concerning the environment. Therefore, it can be expected that the technologies being developed at the moment are technologies that will mitigate environmental impacts as much as possible. Acting in an environmentally friendly way is a prerequisite for economically attractive operations, as the risk of refusal, suspension or withdrawal of permits is too high. However, standards and protocols for environmentally friendly seabed mining are still under development (sees Section 3.4).

3.6.4.5 Product market risk

Finally, scarcity and increasing prices will have a direct impact on the commercial viability of offshore mining operations, although this will also trigger further terrestrial (including recycling) developments. Offshore mining operations themselves are not expected to directly influence global prices of most metals, except for cobalt and phosphorites. This will limit the number of operations that can be exploited in parallel to crust and nodules, to avoid boom and bust developments.

3.7 Concluding Remarks

Offshore mining is seen as one of the Blue Growth sectors, with a potential contribution to growing the (European) economies. This promise stems from an idea of vast natural resources, available for human exploitation, that are in great demand. The reality is however less bright and shiny. There are not only technological challenges to offshore mining; it is also trapped in a vicious circle of uncertain operations, the need for high capital investments and fluctuating prices for the resources. The target resources for offshore mining are very scarce in the European basins. On a global level, the European sector is of importance though, since the EU has some major operators.

A closer look at the sectors reveals the differences in status and potential. From an economic perspective, the polyphosphate sector is closest to commercial take off, with high enough and stable prices for the products. Offshore mining of metals is less promising, given low resource prices and enormous costs for exploration. Also, the urgency for exploration of new resources has decreased in recent years. Gas methane mining is in the early stages of development and development of this sector is inextricably linked to the development of global energy market. The gas hydrates initiatives are typically lead by governments. These subsectors seems driven by governmental interests for control of strategic ore reserves. The polyphosphate sector seems ready to take the next step in exploitation, licensing and actual exploitation. However, its operations are now hampered by discussions and uncertanties on environmental impact and on impacts on other economic sector activities like fisheries and vulnerable areas like seamounts.

A new balance between sectors with at times conflicting interests has to be found. Governments and international policy makers (such as ISA) will need to develop protocols, guidelines and legislation to settle re-occurring debates. But this is not only a governmental responsibility. In an era of social corporate responsibility and social licences to produce, the nearshore and offshore mining sector needs to justify why marine resources need to be explored and bears responsibility for mitigation of social and environmental impacts.

3.7.1 Moving Forward

Ore prices are the major incentive for market-driven development. When especially nickel and cobalt prices will rise structurally, offshore mining on nodules can be achievable. Further development of the technologies used for mining can strengthen the sector (Figure 3.10, exploration, collecting, lifting, on board handling, on land extraction of ores).

Market driven technological development is hampered by the large uncertainties and ample availability of land-based ores and recycling. Today, most of the exploitation of offshore metal and gas resources is in the hands of governmental related companies. Some private enterprises can be found as well for high grade concessions and for commercially exploitable phosphorites. In view of high investment, technological challenges and economic considerations, private-public cooperation could be an effective means to make offshore mining a success. The EU and individual governments can step in here and stimulate the technological and governmental innovations to achieve lower CAPEX and OPEX and at the same time lower environmental impacts in the sensitive nearshore and offshore seabed and associated systems.

Further support for the development of offshore mining can also be driven by the desire to be front-runner in technological development. Extracting some deposits now, getting acquainted with offshore mining in practice, helps to develop techniques and earn a reputation in this uncertain field. It can be of strategic importance to create a first-move advantage, useful when conditions change and offshore mining becomes profitable.

3.7.1.1 Some considerations

Among the reasons for exploring offshore metal extraction, a potential shortage of natural resources is often mentioned, due to geopolitical reasons or limited availability on land. Resource prices are prone to speculation and not a good indicator for worldwide availability. Various researchers have pointed at the real danger that resources will be scarce in the future, for example for phosphate (Gilbert, 2009), and highlighted different national strategies to deal with future resource scarcity (Bartekova & Kemp, 2016). On the other hand our analysis suggests the contrary, that there are ample land-based stocks of economically minable deposits today for at least 30 years and large stocks which one can expect to become economically minable deposits in due time (Figure 3.3, Table 3.1, see also USGS, 2017). NB; an important intriguing aspect is that reserves (economically mineable amounts) appear constant in amount over time (Arndt et al., 2017). Probably due to reasons of financing prospecting and research, the market gives little consideration to a reserve life sufficient to supply more than 20 to 40 years of present consumption (Arndt et al., 2017). On the other hand there is also the unresolved debate of the differences in impacts of land-based mining vs seabed mining, where land-based impacts are estimated to be substantial. Having at least 30 years of reserves and an immense amount of resources (Figure 3.3, Arndt et al., 2017, USGS, 2017) implies that there is no direct urgency for offshore mining. On the other hand, given the fact that some exploration licenses are ending, bodies like ISA should make steady progress to install the necessary regulations and additional preparations in order to enable seabed mining. Regarding nodules, there appears to be a momentum given a recent workshop[16] despite doubts on profitability.

To add to the discussion, Hannington et al. (2017) pointed out to large ore reserves of all sorts nearshore. Those nearshore reserves could also provide the necessary ores at lesser costs. Considering the fierce environmental discussions about nearshore phosphorites, similar environmental discussions can be anticipated.

Given the fluctuating market prices, technical risk and uncertain environmental impact, private entrepreneurs and companies can be expected to be hesitant to invest significantly. A coherent and stable European policy to support offshore mining can benefit society and the sector. Anticipating now a future need for offshore mining could help in geopolitical stabilisation, be a technological pull for knowledge developments and incentivize European exploiters to further develop technologies for offshore mining.

References

Arezki R., Matsumoto A., Zhao H., Toscani F. (2015). Commodity Special Feature from World Economic Outlook, Developments and Forecasts, with a Focus on Metals in the World Economy. International Monetary Fund World Economic Outlook October 2015.

Arndt, N.T., Fontbote, L., Hedenquist, J.W., Kesler, S.E., Thompson, J.F.H., Wood, D.G., 2017. Future global mineral resources. Geochemical Perspectives 6, 1–166

Baker E., Gaill F., Karageorgis A. P., Lamarche G., Narayanaswamy B., Parr J., Raharimananirina C., Santos R., Sharma R., Tuhumwire J. (2017). Chapter 23. Offshore Mining Industries. In: The First Global Integrated Marine Assessment. World Ocean Assessment I. United Nations (Division for Ocean Affairs and the Law of the Sea, Office of Legal Affairs) Cambridge: Cambridge University Press. Online ISBN: 9781108186148.

Barteková, E., Kemp, R., 2016. National strategies for securing a stable supply of rare earths in different world regions. Resources Policy 49: 153–164.

Benkenstein, A., (2014) Seabed Mining: Lessons from the Namibian Experience, SAIIA Policy Briefing No. 87, April 2014.

Bertram C., Krätschell A., O'Brien K., Brückmann W., Proelss A., Rehdanz K. (2011). Metalliferous sediments in the Atlantis II Deep—Assessing the geological and economic resource potential and legal constraints. Resources Policy 36: 315–329.

Birney, K., Griffin, A., Gwiazda, J., Kefauver, J., Nagai, T. and Varchol, D., 2006. Potential deep-sea mining of seafloor massive sulfides: A case study in papua new guinea. Donald Bren School of Environmental Science and Management Thesis.

Boschen, R. E., Rowden, A. A., Clark, M. R., Gardner, J. P. A., (2013) Mining of deep-sea seafloor massive sulfides: a review of the deposits, their benthic communities, impacts from mining, regulatory frameworks and management strategies. Ocean and Coastal Management 84: 54–67.

Boswell, R., Collett, T. S., (2011). Current perspectives on gas hydrate resources. Energy and Environmental Science 4: 1206–1215.

BP (2016). Statistical review of world-energy 2016 natural gas.

Clark A. L., Cook Clark J., Pintz S. (2013). Towards the Development of a Regulatory Framework for Polymetallic Nodule Exploitation in the Area. ISA Technical Study: No. 11. ISBN 978-976-8241-16-0 (ebk).

Conceição, P., Mirão, J., Madureira, P., Costa, R. (2014). Fe-Mn crusts from the Central Atlantic. Technical Presentations Harvesting Seabed Mineral Resources in Harmony with Nature. UMI 2014, Portugal: p. 4.

Cronan, D. S., (Ed.) (2000). Handbook of Marine Mineral Deposits, CRC Press, London, Boca Raton.

De Ridder M., de Jong S., Polchar J., Lingemann S. (2012). Risks and Opportunities in the Global Phosphate Rock Market. The Hague Centre for Strategic Studies (HCSS) Rapport No. 17 |12 |12. ISBN/EAN: 978-94-91040-69-6.

Devey, C. W., ed and Shipboard scientific party. (2015) RV SONNE Fahrtbericht/Cruise Report SO237 Vema-TRANSIT: bathymetry of the Vema-Fracture-Zone and Puerto Rico TRench and Abyssal AtlaNtic BiodiverSITy Study, Las Palmas (Spain) – Santo Domingo (Dom. Rep.) 14.12.14–26.01.15 GEOMAR Report, N.Ser. 023. GEOMAR Helmholtz-Zentrum für Ozeanforschung Kiel, p. 130.

Don Diego (2015). Feeding the future, Environmental impact assessment. Non-technical executive summary. http://www.dondiego.mx/downloads/ (d.d. 08-07-2015).

ECB (2014). ECB Monthly Bulletin. October 2014.

Ecorys (2014). Study to investigate the state of knowledge of deep-sea mining. Final Report under FWC MARE/2012/06 – SC E1/2013/04.

EPRS (2015). Deep-Seabed exploitation – tackling economic, environmental and societal challenges. European parliamentary research service scientific foresight unit (Stoa) PE 547.401, 92 pags.

Franks, D. M., (2011). Management of the social impacts of mining. Society for Mining, Metallurgy and Exploration, SME Mining Engineering Handbook, 3rd ed., pp 1817–1825, Littleton, Colorado.

Gaspar, L. (1982). Fosforites da Margem Continental Portuguesa. Alguns Aspectos Geoquímicos, Bol. Soc. Geol. Portugal 23: 79–90.

Gilbert, N., 2009. The disappearing nutrient: phosphate-based fertilizers have helped spur agricultural gains in the past century, but the world may soon run out of them. Natasha Gilbert investigates the potential phosphate crisis. Nature 461: 716–719.

Gonzalez, F. J., Somoza L., Hein J. R., Medialdea T., León R., Urgorri V., Reyes J., Antonio Martín-Rubí J.(2016). Phosphorites, Co-rich Mn nodules and Fe-Mn crusts from Galicia Bank, NE Atlantic: Reflections of Cenozoic tectonics and paleoceanography. Geochem. Geophys. Geosyst. 1525–2027: 1–29.

Gurney, J. J., Levinson, A. A., Stuart Smith, H. (1991). Marine Mining of Diamonds off the West Coast of Southern Africa Marine Mining of Diamonds. Gems & Gemology 27(4): 206–219.

Hannington, M. D., Jamieson, J., Monecke, T., Petersen, S., (2010). Modern sea-floor massive sulfides and base metal resources: toward an estimate of global sea-floor massive sulfide potential. Society of Economic Geologists Special Publication 15, 317–338.

Hannington, M. D., Jamieson, J., Monecke, T., Petersen, S., Beaulieu, S., (2011). The abundance of seafloor massive sulfide deposits. Geology 39, 1155–1158.

Hannington, M., Petersen S., Krätschell A. (2017) Subsea mining moves closer to shore. Nature Geoscience 10, 158–159.

Hein J., Mizell K., Koschinsky A., Conrad T. (2013). Deep-ocean mineral deposits as a source of critical metals for high- and green-technology applications: comparison with land-based resources. Ore Geology Review 51: 1–14.

Hein, J. R., Spinardi F., Okamoto N., Mizell K., Thorburn D., Tawake A. (2015). Critical metals in manganese nodules from the Cook Islands EEZ, abundances and distributions, Ore Geol. Rev. 68: 97–116.

IMO (1972). Convention on the Prevention of Marine Pollution by Dumping of Wastes and Other Matter, London 1972.

IMO (1996). Protocol on the Prevention of Marine Pollution by Dumping of Wastes and Other Matter, London 1996, amended in 2006.

ISA (2015). Discussion Paper on the Development and Implementation of a payment mechanism in the Area for consideration by members of the Authority and all stakeholders, published in March 2015 by the ISA, available at https://www.isa.org.jm/files/documents/EN/Survey/DPaper-FinMech.pdf

ISA (2016). Conference Report on the Deep Seabed Mining Payment Regime Workshop (San Diego, 17–18 May 2016). Available at: https://www.isa.org.jm/files/documents/EN/Pubs/2016/DSM-ConfR ep.pdf

ITLOS (2011). Responsibilities and obligations of States sponsoring persons and entities with respect to activities in the Area, Seabed Dispute Chamber's advisory opinion of 1 February 2011.

Lange, E., Petersen, S., Rüpke, L., Söding, E., Wallmann, K. (eds.) (2014). "World Ocean Review 3. Marine Resources: Opportunities and Risks, 165 Pages, ISBN 978-3-86648-221-0.

Laurila, T. E., Hannington M. D., Leybourne M., Petersen S., Devey C. W., Garbe-Schonberg D. (2015). New insights into the mineralogy of the Atlantis II Deep metalliferous sediments, Red Sea, Geochem. Geophys. Geosyst.: 16, 4449–4478.

Liu, J. P., Milliman, J. D., Gao, S., Cheng, P., (2004). Holocene development of the Yellow River's subaqueous delta, North Yellow Sea. Marine geology 209: 45–67.

LOSC United Nations Convention on the Law of the Sea, Montego Bay, 1982 http://www.un.org/depts/los/convention_agreements/convention_overvie w_convention.htm

Makgill, R., Linhares, A. P., (2015) Deep Seabed Mining: Key Obligations in the Emerging Regulation of Exploration and Development in the Pacific", in Warner, R., Kaye, S., (eds) Routledge Handbook of Maritime Regulation and Enforcement, Routledge Handbooks.

Marques A. F. A., Scott S. D. (2011). Seafloor massive sulfide (SMS) exploration in the Azores. Society of Mining Engineers Annual Meeting, Denver, Preprint 11–306, p. 3.

Martino S., Parson, L. M. (2012). A comparison between manganese nodules and cobalt crust economics in a scenario of mutual exclusivity. Marine Policy 36: 790–800.

Mudd, G. M., Weng, Z., Jowitt, S. M., (2013). A detailed assessment of global Cu resource trends and endowments. Economic Geology 108: 1163–1183.

Muiños, SB, Hein, JR, Frank, M, Monteiro, JH, Gaspar, L, Conrad, T, Pereira, HG, Abrantes, F (2013). Deep-sea Fe-Mn Crusts from the Northeast Atlantic Ocean: Composition and Resource Considerations. Marine Georesources & Geotechnology 31: 40–70.

Oebius, H. U., Becker, H. J., Rolinski, S., & Jankowski, J. A. (2001). Parametrization and evaluation of marine environmental impacts produced by deep-sea manganese nodule mining. Deep Sea Research Part II: Topical Studies in Oceanography 48: 3453–3467.

Ortega, A. (Ed.)(2014). Towards Zero Impact of Deep Sea Offshore Projects – An assessment framework for future environmental studies of deep-sea and offshore mining projects. Report IHC Merwede.

Petersen S., Krätschell A., Augustin N., Jamieson J., Hein J. R., Hannington M. D. (2016). News from the seabed – Geological characteristics and resource potential of deep-sea mineral resources. Marine Policy 70: 175–187.

Ramirez-Llodra, E., Trannum, H., Evenset, A., Levin, L., Andersson, M., Finne, T., Hilario, A., Flem, B., Christensen, G., Schaanning, M., Vanreusel, A. (2015). Submarine and deep-sea mine tailing placements: a review of current practices, environmental issues, natural analogs and knowledge gaps in Norway and internationally. Mar. Pollut. Bull. 97: 13–35.

Rogers, A. D., Brierley, A., Croot, P., Cunha, M. R., Danovaro, R., Devey, C., Hoel, A. H., Ruhl, H. A., Sarradin, P-M., Trevisanut, S., van den Hove, S., Vieira, H., Visbeck, M. (2015) Delving Deeper: Critical challenges for 21st century deep-sea research. Larkin, K.E., Donaldson, K. and McDonough, N. (Eds.) Position Paper 22 of the European MarineBoard, Ostend, Belgium. p. 224. ISBN 978-94-920431-1-5.

Rozemeijer M. J. C., Jak R. G., Lallier L. E., van Craenenbroeck K., van den Burg S. (2015). WP4.2 Sector: Nearshore and Offshore Mining A sector context review is to identify the key socio-economic features of nearshore and offshore mining. In: Johnson K. R., Legorburu I. et al. Eds (2015) WP 4: Socio-economic trends and EU policy in offshore economy Socio-economic drivers and trends- Historic Life. MARIBE D-4.1 report.

Schilling, C., Clough, P., Zuccollo, J., (2013). Economic assessment of Chatham Rock Phosphate. Input to the Environmental Impact Assessment. Final report to Chatham Rock Phosphate Ltd.

Schulte, S. A., (2013). Vertical transport methods for Deep Sea Mining. Delft University of Technology, Section of Dredging Engineering. Thesis version 2.0 June 19.

Scott, S., Atmanand, M. A., Batchelor, D., Finkl, C., Garnett, R., Goodden, R., Hein, J., Heydon, D., Hobbs, C., Morgan, C., Rona, P., Surawardi, N., (2008). Mineral Deposits in the sea: Second report of the ECOR (Engineering Committee on Oceanic Resources) Panel on Marine Mining, London, October 2008, p. 36. (http://oceanicresources-ecor.amc.edu.au).

Sharma R. Editor (2017). Deep-Sea Mining: Resource Potential, Technical and Environmental Considerations 2017. Springer International Publishing AG 2017. ISBN 978-3-319-52557-0.

Siesser, W. G., Dingle, R. V. (1981). Tertiary sea-level movements around southern Africa. Journal of Geology 89: 83–96.

SPC (2013a). Deep Sea Minerals: Sea-Floor Massive Sulphides, a physical, biological, environmental, and technical review. Baker, E., and Beaudoin, Y. (Eds.) Vol. 1A, Secretariat of the Pacific Community.

SPC (2013b) Deep Sea Minerals: Manganese Nodules, a physical, biological, environmental, and technical review. Baker, E., and Beaudoin, Y. (Eds.) Vol. 1B, SPC Secretariat of the Pacific Community.

SPC (2013c) Deep Sea Minerals: Cobaltrich Ferromanganese Crusts, a physical, biological, environmental, and technical review. Vol. 1C, SPC Secretariat of the Pacific Community.

SPC (2013d). Deep Sea Minerals: Deep Sea Minerals and the Green Economy. Baker E., Beaudoin Y. (Eds.) Secretariat of the Pacific Community Vol. 2.

SPC (2016). An Assessment of the Costs and Benefits of Mining Deep-sea Minerals in the Pacific Island Region. Deep-sea Mining Cost-Benefit Analysis. Prepared by Cardno for the Secretariat of the Pacific Community. Suva, Fiji. SPC Technical Report SPC00035, ISBN: 978-982-00-0955-4.

SRK Consulting (2010). Offshore Production System Definition and Cost Study. SRK Project NAT005. Document No: SL01-NSG-XSR-RPT-7105-001.

Thiel H, Angel MV, Foell EJ, Rice AL, Schriever G (1998). Environmental risks from large-scale ecological research in the deep sea: a desk study. European Commission: Marine science and technology. Office for Official Publications of the European Communities XIV, p. 210, ISBN 92-828-3517-0.

Thiel H., Karbe L., Weikert H. (2013). Environmental Risks of Mining Metalliferous Muds in the Atlantis II Deep Red Sea. Conference Paper February 2013, Springer Berlin Heidelberg, Berlin, Heidelberg.

Tully, S. (2006). "Welcome to the Dead Zone". Fortune (May 5, 2006). http://money.cnn.com/2006/05/03/news/economy/realestateguide_fortu ne/ d.d. 09-11-15.

USGS (2015), Mineral commodity summaries 2015: U.S. Geological Survey, p. 196, ISBN 978-1-4113-3877-7.

USGS (2017). Mineral commodity summaries 2017: U.S. Geological Survey, p. 202, ISBN 978-1-4113-4104-3.

van den Burg SWK, Stuiver M, Bolman BC, Wijnen R, Selnes T and Dalton G (2017) Mobilizing Investors for Blue Growth. Front. Mar. Sci. 3:291.

Van Dover, CL. (2008). Hydrothermal Vent Ecology. In Encyclopedia of Ocean Sciences: Second Edition, 151–158.

Vanreusel, A., Hilario, A., Ribeiro, P. A., Menot, L., Martinez Arbizu, P. (2016). Threatened by mining, polymetallic nodules are required to preserve abyssal epifauna. NPG Scientific Reports 6, p. 6.

Worldbank (2012). Global economic prospects: Uncertainties and vulnerabilities. The International Bank for Reconstruction and Development. The World Bank, Washington, DC.

Worldbank (2016). Precautionary Management of Deep Sea Mining Potential in Pacific Island Countries. Pacific-Possible. World Bank Draft.

Yamazaki, T., 2008. Model mining units of the 20th century and the economies. In Technical paper for ISA Workshop on Polymetallic Nodule Mining Technology-Current Status and Challenges Ahead: 18–22.

4

Ocean Energy – Wave and Tide

Gordon Dalton

University College Cork, Ireland

4.1 Introduction

4.1.1 Policy and EU Strategy Initiatives Overview for the Ocean Energy Sector

Our seas and oceans have the potential to become important sources of clean energy. Marine renewable energy, which includes both offshore wind and ocean energy (wave and tidal energy), presents the EU with an opportunity to generate economic growth and jobs, enhance the security of its energy supply and boost competitiveness through technological innovation. Following the 2008 Communication on offshore wind energy (European Commision 2008), the European Commission (EC) considered the potential of the ocean energy sector to contribute to the objectives of the Europe 2020 Strategy (European Commision 2010) as well EU's long-term greenhouse gas emission reduction goals. It also looked over the horizon at this promising new technology (Blue Growth) and outlines an action plan to help unlock its potential.

In 2008, the European Commission stated that "Harnessing the economic potential of our seas and oceans in a sustainable manner is a key element in the EU's maritime policy" (European Commision 2007). The ocean energy sector was highlighted in the Commission's Blue Growth Strategy (European Commision 2012) as one of five developing areas in the 'Blue Economy' that could help drive job creation in coastal areas. Other Commission initiatives were the Communication on Energy Technologies and Innovation (European Commision 2013) and the Atlantic Action Plan (European Commision 2013). The Atlantic Action Plan recognised the importance of ocean energy and aimed to encourage collaborative research and development and cross-border cooperation to boost its development and published two key reports on Ocean

Feb 2014	Dec 2014	Feb 2015	Sept 2015	Sept 2016	Nov 2016
Ocean Energy Communication (COM/2014/08) Setup Ocean Energy Forum Roadmap expected end of 2016	Towards an Integrated Roadmap: Research and Needs of the EU Energy System 13 Actions in 3 different programmes for the uptake of Ocean Energy in EU	Energy Union ((COM/2015/80) Retain Europe's leading role in global investment in renewable energy	SET-Plan Communication (COM/2015/6317) Reduce the cost of key technologies Increase regional cooperation, in the Atlantic area for ocean energy	SET-Plan Declaration of intent, defining LCOE targets for tidal and wave energy	Publication of the Ocean Energy Strategic Roadmap developed by the Ocean Energy Forum and supported by DG Mare

Figure 4.1   The history of Ocean Energy Policies at EU level. Image from JRC report 2016 (Magagna, Monfardini et al. 2016).

Energy development: "Blue Growth, opportunities for marine and maritime sustainable growth" (Altantic Action Plan 2013), and "Action Plan for a maritime strategy in the Atlantic area" (Atlantic Action Plan 2013). In 2014, the European Commission summarised all the initiatives in its COM/2014/08 final report "Blue Energy Action needed to deliver on the potential of ocean energy in European seas and oceans by 2020 and beyond" (European Commision 2014).

In 2014, the Strategic Initiative for Ocean Energy (SI Ocean)[1], released a report (SI Ocean 2014) detailing four main barriers to widespread wave and tidal energy deployment in Europe, namely:

1. Financial risks: market stresses, public support mechanism fluctuations, reduced investor confidence.
2. Technology risks: lack of commercially ready prototype devices, TRL8 or higher, due to failure of technology developers to overcome technology barriers. Insufficient cost reduction has been demonstrated as technology moves to higher TRL.
3. Regulatory and consenting barriers still exist in most jurisdictions with slow progress on their resolution. On the other hand, environmental impact requirements are increasing, delaying consents and increasing costs.
4. Grid connection, both adequate and sufficient, still remains a huge non-technical barrier, mainly due to the remote nature of most ocean energy resource areas, and lack of existing infrastructure. Lack of grid infrastructure could posing real risk to large scale deployment once technical barriers are overcome.

---

[1] https://ec.europa.eu/energy/intelligent/projects/en/projects/si-ocean

The report offered recommendations for addressing those barriers, as part of its market deployment strategy. SI Ocean presented a vision of Europe reaching 100 gigawatts (GW) of installed wave and tidal energy capacity by 2050, the report's subsequent chapters focus on finance, technology development, regulatory regimes and the grid. Each chapter identifies the challenges these risk areas present, offers goals to remove barriers and recommends way to meet those goals. The report suggests that regulators incorporate wave and tidal energy projects into long-term grid development plans.

In 2014, the Ocean Energy Forum[2] was created by the European Commission, under the stewardship of Ocean Energy Europe[3]. The Forum brought together more than 100 ocean energy experts over two years. Ocean Energy Europe created TP Ocean (Ocean Energy Europe 2014) initiative, called the European Technology and Innovation Platform for Ocean Energy. TP Ocean identified six essential priority areas to be addressed to improve ocean energy technology and decrease its risk profile:

1. *Testing* sub-system components and devices in real sea conditions.
2. Increasing the *reliability and performance* of ocean energy devices allowing for future design improvements.
3. Stimulating a dedicated *installation and operation and maintenance* value chain, to reduce costs.
4. Delivering *power to the grid*, with hubs to collect cables from ocean energy farms and bring power to shore.
5. *Devising standards and certification*, to facilitate access to commercial financing.
6. *Reducing costs and increasing performance* through innovation and testing.

In November 2016, the Ocean Energy Forum created the 'Ocean Energy Strategic Roadmap' (Figure 4.1) (Ocean Energy Forum 2016).

The Roadmap puts forward four key Action Plans focused on maximising private and public investments in ocean energy development by de-risking technology as much as possible, ensuring a smoother transition from one development phase to another on the path to industrial roll-out and a fully commercial sector.

The second initiative of the Ocean Energy Forum was Strategic Research Agenda for Ocean Energy developed by Technology and Innovation Platform

---

[2]https://www.oceanenergy-europe.eu/en/policies/ocean-energy-forum
[3]https://www.oceanenergy-europe.eu/en/

for Ocean Energy (TP Ocean 2016). The ocean energy sector has identified 12 priority research areas and 54 research and innovation actions. The research areas have been attributed indicative budgets that industry, national authorities and the European Commission need to commit to finance the RD&I programmes. Rolling-out the actions of this Agenda would generate around €1 bn in investment over 4 to 5 years. The outcomes for the ocean energy sector would be the improvement of current technologies and the identification of novel financial instruments to sustain the critical phase of moving to demonstration projects.

4.1.2 Tidal Energy Development Demographics

Tidal energy is predictable up to 100 years in advance (Alcorn, Dalton et al. 2014), making tidal energy attractive to grid operators by adding more predictable and consistent sources of renewable energy which has the effect of smoothing out the overall power supply from renewables. In tidal energy, there has been a general convergence of the technologies, with several developers testing full-scale prototypes and plans for commercial deployments.

Worldwide, many companies are currently developing tidal energy devices with most (about 52%) being based in the EU (Magagna, Monfardini et al. 2016). In Europe, the country with the highest level of development is the United Kingdom, followed by the Netherlands, and France. The United States and Canada are the major non-EU players (Figure 4.2).

The development of tidal technology is taking place in countries with the major tidal energy resources: UK, France, and Ireland (OES 2016). Other active countries, with more limited resources include Germany and Sweden.

4.1.3 Wave Energy Development Demographics

Wave energy is highly predictable days in advance and compliments wind energy by generally achieving its peak energy after wind energy has reached its maximum (Alcorn, Dalton et al. 2014). Therefore wave energy is a further alternative for grid operators seeking to smoothing out the overall power supply from renewables. By 2016 about 70 different design concepts were under development (OES 2016), Unlike wind energy (or even tidal current), designs for wave energy devices have not converged around a standard technology solution (more likely that wave energy will converge on a number of standard technologies), and relatively few have made it to full scale prototype testing, and there are no current plans for commercial arrays. The majority of

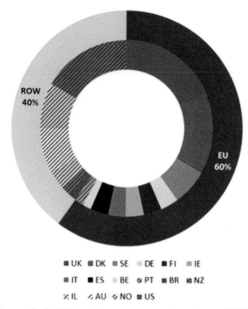

Figure 4.2 Global spread of tidal development companies. Image JRC Ocean Energy Status Report 2016. (Magagna, Monfardini et al. 2016).

companies developing wave energy devices are based in the EU (Magagna, Monfardini et al. 2016) (Figure 4.3). The United Kingdom has the highest numbers of developers, followed by Denmark. Outside the EU, countries with a larger number of wave energy developers are USA, Australia, and Norway. Globally, about 57 wave energy developers have tested their devices in open waters or will do so in the near future.

See Section 4.5 'Innovation' for details on wave and tidal companies and their lifecycle stage.

4.2 Market

There are potentially enormous exploitable energy resources available in the world's oceans. This would suggest significant potential markets for the sale of ocean energy as well as opportunities for supporting industries and services involved in the development, manufacturing, construction, installation and operation (Alcorn, Dalton et al. 2014). However, uncertainty in future costs makes it difficult to estimate the scale of the opportunity and the size of the long term potential market.

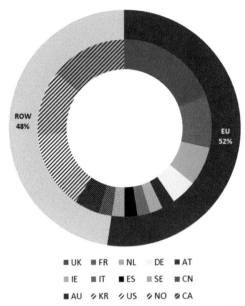

Figure 4.3   Global spread of wave development companies. Image JRC Ocean Energy Status Report 2016 (Magagna, Monfardini et al. 2016).

4.2.1 Global Ocean Energy Resources and Potential Economic Return

The total theoretical energy contained in the seas is estimated to be 32,000 TWh/yr for wave (Mork, Barstow et al. 2010) and 7,800 TWh/y for tides (IEA-OES 2011). It is this potential scale that justifies the drive for its development (Alcorn, Dalton et al. 2014, Magagna, Monfardini et al. 2016). Wave energy devices derive energy from the three dimensional movement of ocean waves. Tidal energy devices harnesses the bodily movement of water resulting from the environmental pull between the moon and the earth. The efficiencies of future ocean energy technologies will dictate how much of this resource can be usefully harnessed. The technically exploitable energy of wave energy devices is estimated to be 5,500 TWh/yr (Lewis 2011), which is approximately 30% of world electricity demand. Whilst currently under development, the Ocean Energy Forum goal is to install 100GW of wave and tidal by 2050. This equates to 350 TWh of exploitable electricity and opens up a global market for investment, jobs and growth. This would meet 10% of the power demands of the EU, a significant component in the transition to a low carbon clean economy.

In 2009 the Renewables Directive 2009/28/EC (European Commision 2007) set binding targets for all EU Member States, such that the EU will reach a 20% share of energy from renewable sources by 2020 and a 10% share of renewable energy specifically in the transport sector. The primary production of renewable energy within the EU-28 in 2014 was 196 million tonnes of oil equivalent (toe) – a 25.4% share of total primary energy production from all sources (Eurostat 2016).

For Europe to meet its objective of reducing greenhouse gas emissions to 80–95% below 1990 levels by 2050 (European Commision 2011), Ocean Energy is needed in a diversified low carbon and renewable energy portfolio. Investment wise the global market between now and 2050 is estimated to be worth €653 bn (Ocean Energy Forum 2016) (cumulative, undiscounted) which would bring great benefit to European and world economies. Tidal energy is going strongly in its development and some niche opportunities are expected, whilst wave energy has suffered some setbacks in investment in 2015 in the EU. The World Energy Council estimates the global capital expenditure for wave energy projects to be more than £500 billion, based on a technically exploitable wave resources of 2,000 TWh/year (World Energy Council 2007). So far, over the past 10 years the ocean energy industry has invested an estimated €1 bn in capital to move concepts from the drawing board to deployment in EU waters (OEE 2016 (Ocean Energy Forum 2016)).

4.2.2 Installed Capacity and Consented Capacity for Wave and Tidal

This section presents the target deployment predictions of the major policy agencies reviewing ocean energy. There was great optimism in the early 2000's and accordingly ambitious targets. Successive reviews for both near term, 2015, and far term, 2050, were revised downwards, as real deployments failed to materialise. It is likely that the current 2050 projections will be revised down in subsequent reviews.

<u>2020 deployment predictions</u>

JRC and European Commission in 2010 (European Commission 2010) set European targets for wave and tidal of 1.9 GW by 2020. In 2015, OEE downsized the prediction for ocean energy deployment, reaching a cumulative capacity of 850 MW by 2020 (OEE 2015 (Ocean Energy Forum 2016)).

2050 deployment predictions

In 2007, the IEA-Ocean Energy Systems Implementing Agreement (IEA-OES), predicted combined wave and tidal deployment of **337 gigawatts (GW)** of capacity worldwide by 2050 (IEA-OES., Khan et al. 2008). (By comparison, the capacity of the much more developed wind energy sector reached the same figure – 336 GW – by the end of June 2014).

Current estimates from 2014 for 2050 deployments, as quoted by SI Ocean (SI Ocean 2014), currently stand at **100 GW** of combined *wave* and *tidal* capacity installed (elaborated by Magagna (Magagna and Uihlein 2015)).

Table 4.1 represents more detailed breakdown provided by OES 2015 Annual Report for Ocean Energy up to 2020 (OES 2015):

- current installed capacity
- consented capacity.

Current capacity (2015) installed for tidal energy exceeds wave energy by a factor of 5, at 2.4 MW for wave energy and 14 MW for tidal.

The current predictions for wave energy deployment was optimistic (consented capacity in Table 4.1), requiring a sizeable increase in deployment

Table 4.1   Table from Ocean Energy Systems Data taken from OES 2015 report (OES 2015)

Basin	Country	Installed Capacity MW 2015		Consented Capacity	
		Wave	Tidal Stream	Wave	Tidal Stream
Atlantic	UK	0.96	2.1	40	96
	Portugal	0.4	–	5	–
	Spain	0.3	–	–	–
	France	–	2.5	–	21.5
	Ireland	–	–	–	–
Baltic	Sweden	0.2	8	10.6	
	Belgium	–	–	20	–
	Netherlands	–	1.3	–	2.2
	Norway	–	–	0.2	–
	Denmark	–	–	0.05	–
Caribbean	Inactive	–	–	–	–
Mediterranean	Inactive	–	–	–	–
Rest of World	Canada	0.09	–	–	20
	China	0.45	0.17	2.7	4.8
	United States	–	–	1.5	1.3
	Korea	0.5	1	0.5	1
Total	–	2.4	14.07	80.05	145.8
			16.47		225.85

of 4000% in MW deployed, from current 2 MW up to 80 MW. UK, Sweden and Belgium plan to take the lead, with approx. 20–40 MW deployments in each jurisdiction. Tidal energy also has optimistic deployment gains, although more modest, with a 10 fold increase in MW deployed from 14 MW to 145 MW. Deployments in the remainder of the world are currently modest, with no major plans for increases. The exception is Canada, where tidal energy is predicted to reach 20 MW installed by 2020.

In summary, current capacity deployments to date (2016) of 16.7 MW will make it highly unlikely that the OEE target of 850 MW by 2020 will be reached (OEE 2015 (Ocean Energy Forum 2016)).

However, the global potential market identified by SI Ocean (SI Ocean 2014) of 100GW by 2050 is substantial, with very large capital expenditure. These investments would add significantly to Europe's strategic goals of jobs and growth for the European Area.

4.2.3 Capital Expenditure (Capex/MW or €/MW)

Chozas et al., conducted a comprehensive literature review of published data on historical costs, planned projects and reference reports that estimate capital expenditure (Capex costs/MW) for both wave and tidal (Figure 4.4) (Chozas, Wavec et al. 2015). They state that there is a significant variability of CAPEX values for the first pilot projects (up to 1 MW) installed worldwide, ranging from €10–50 M/MW for wave energy, and a much lower €5–20 M/MW for tidal energy. The trends for both technologies were relatively similar as they progressed to commercial stage, converging to €3–6 M/MW for both wave and tidal energy. Other reviews of Capex/Mw for ocean energy are conducted by Dalton et al. (Dalton, Alcorn et al. 2009, Dalton 2010, Dalton,

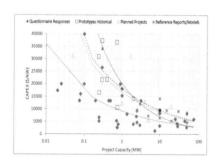 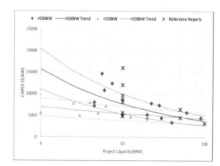

Figure 4.4 CAPEX cost per kW installed for 1: wave 2: tidal, relative to project deployed capacity. Image taken from Chozas et al., (Chozas, Wavec et al., 2015).

Alcorn et al. 2010, Dalton, Alcorn et al. 2010, Dalton 2011, Dalton and Lewis 2011, Dalton, Alcorn et al. 2012, Dalton, Allan et al. 2016, Dalton, Allan et al. 2016).

4.2.4 Prices – Cost of the Product – Levelised Cost of Electricity LCOE

The Levelised Cost of Electricity (LCOE) is one of the most commonly used financial indicators to compare the cost of energy projects. Magagna et al. (Magagna and Uihlein 2015) published a comprehensive report in 2015 on the business cases for wave and tidal. Figure 4.5 compares wave and tidal LCOE to other renewable technologies as well as fossil fuels. LCOE for wave has a range of €500–650/MWH and Tidal a range of €350 to 450/MWh. Their forecast for cost reductions and learning for both however are optimistic, with Wave LCOE dropping to €80/MWH and Tidal €60/MWH, competitive to all other renewables and fossil fuels.

The JCR report, authored again by Magagna (Magagna, Monfardini et al. 2016), approached LCOE reduction from a cumulative installed prospective and in Figure 4.6, also insert timeframe benchmarks. By 2030, they predict

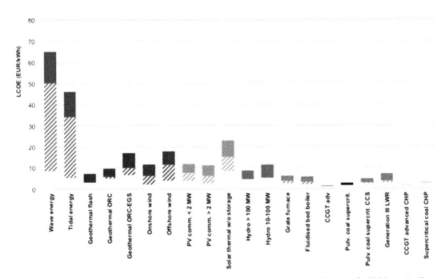

Figure 4.5 LCOE for alternative and conventional energy technologies. Solid bars indicate current cost ranges, while shaded bars indicate expected future cost reductions. Image taken from Magagna (Magagna and Uihlein 2015).

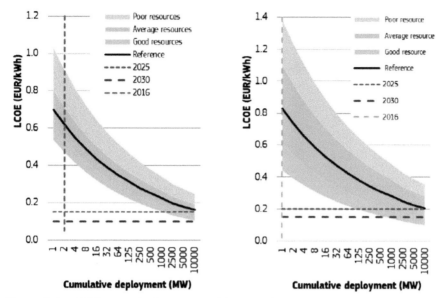

Figure 4.6 LCOE cost reduction ranges with cumulative deployments; 1. Tidal 2. Wave. Image taken from (Magagna, Monfardini et al. 2016).

cumulative installed capacity for both technologies will reach 10 GW each, and that LCOE for both technologies will drop to €100–120/MWH.

Another JCR report (Global CCS Institute 2013), conducted by Global CCS, has a longer time span projection to 2050, also predicting that wave and tidal LCOE cost will reduce to approximately €80/MWH (Figure 4.7).

A more detailed review and modeling of LCOE of Wave and Tide was published by Chozas (Chozas, Wavec et al. 2015). Table 4.2 is taken from that report, and presents LCOE results for the various stages of commercialization for both technologies, however not specifying size of deployment, cumulative installed capacity or timeframe specified. At full commercial scale, Chozas predicts a tidal LCOE of €130/MWH and most unusually, wave lower than tidal at €120/MWH.

Chozas (Chozas, Wavec et al. 2015) also presents LCOE modeling based on learning curves, as does Dalton (Dalton, Alcorn et al. 2012). Other reviews of LCOE for ocean energy include Dalton et al. (Dalton, Alcorn et al. 2009, Dalton 2010, Dalton, Alcorn et al. 2010, Dalton, Alcorn et al. 2010, Dalton 2011, Dalton and Lewis 2011, Dalton, Alcorn et al. 2012, Dalton, Allan et al. 2016, Dalton, Allan et al. 2016).

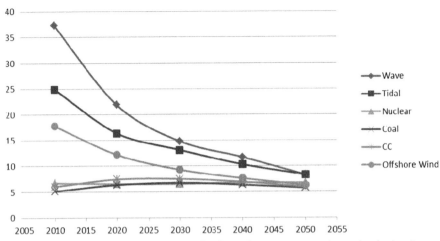

Figure 4.7 LCoE (€/kWh) projections for the main power generation technologies. Image taken from JCR Report (Global CCS Institute 2013).

Table 4.2 LCOE of wave and Tidal, for 3 stages of development: First array, second array and Commercial. Table taken from Chozas et al. (Chozas, Wavec et al. 2015)

Deployment Stage	Variable	Wave		Tidal	
		Min	Max[1]	Min	Max
First array/First Project[2]	Project Capacity (MW)	1	3[3]	0.3	10
	CAPEX ($/kW)	4000	18100	5100	14600
	OPEX ($/kW per year)	140	1500	160	1160
Second array/ Second Project	Project Capacity (MW)	1	10	0.5	28
	CAPEX ($/kW)	3600	15300	4300	8700
	OPEX ($/kW per year)	100	500	150	530
	Availability (%)	85%	98%	85%	98%
	Capacity Factor (%)	30%	35%	35%	42%
	LCOE ($/MWh)	210	670	210	470
First Commercial-Scale Project	Project Capacity (MW)	2	75	3	90
	CAPEX ($/kW	2700	9100	3300	5600
	OPEX ($/kW per year)	70	380	90	400
	Availability (%)	95%	98%	92%	98%
	Capacity Factor (%)	35%	40%	35%	40%
	LCOE $/MWh)	120	470	130	280

4.2.5 Funding Support Schemes

4.2.5.1 History of EU funding programme support schemes for ocean energy

In 2007 EU approved of the Strategic Energy Technology Plan (SET-Plan) (Europeaan Commision 2015), with aims to develop technologies in areas including renewable energy, energy conservation, low-energy buildings, fourth generation nuclear reactor, coal pollution mitigation, and carbon capture and sequestration (CCS).

In order to implement the research required for the SET-Plan, the European Energy Research Alliance (EERA)[4] was founded by more than 175 research centres and universities in the European Union (EU). The aim of EERA is to expand and optimise EU energy research capabilities through the sharing of world-class national facilities and the joint realisation of national and European programmes, and builds on national research initiatives.

The following are the list of EU funded programs for ocean energy:

1. Within the EERA, a joint programme for investment in ocean energy has been set up. NER 300 is an example of one of the EERA initiatives (see NER 300 described below under push mechanisms). Three ocean energy projects were awarded around €60 million in total under the first round of the NER 300 programme, which will enable the demonstration of arrays from 2016 (European Commision 2014).

2. The development of ocean energy has been highlighted in the recent Commission Communication entitled "Action Plan for the Atlantic Ocean area" (Atlantic Action Plan 2013, European Commision 2013) which encouraged national and regional governments to consider how they could use EU structural and investment funds as well as research funds or European Investment Bank funding to support the development of the sector.

3. Research Framework Programmes (FP4,5,6,7) and the Intelligent Energy Europe Programme provided an amount of up to €90 million for ocean energy development since the 1980s (European Commision 2014). (Ocean Energy Europe[5] reports €124 m to ocean energy projects between 2005 and 2014, almost €14 m per year).

4. Horizon 2020[6], the EU's research and innovation programme, will aim to address important societal challenges including clean energy and

---

[4]https://www.eera-set.eu/

[5]http://www.oceanenergy-europe.eu/en/14-policy-issues

[6]https://ec.europa.eu/programmes/horizon2020/

marine research. As such, it is a powerful tool that can drive the ocean energy sector towards industrialisation, creating new jobs and economic growth. Between 2014–15, H2020 programme has funding over EUR 60 million (Magagna, Monfardini et al. 2016) of R&D projects in wave and tidal energy. €30 M[7] in demonstration funding was awarded (LCE3 and 12). For 2016–17, total of €22.6 M will be awarded for ocean energy specific calls, 9.8% of LCE budget. A further €35 M was allocated to Blue Growth and Co-Funded calls, which include ocean energy.

5. Other funding instruments available in Europe are InnovFin[8] (a series of integrated and complementary financing tools and advisory services offered by the European Investment Bank Group together with the European Commission) and the European Regional Development Fund (ERDF)[9]. These funding mechanisms are supporting the deployment of demonstration projects. Collaboration initiatives at regional level are catalysing the formation of marine energy clusters to consolidate the European supply chain.

There are two types of support type mechanisms.

1. Push: = grants and equity
2. Pull: = tariff and other revenue mechanisms

4.2.5.2 Pull support schemes – Feed-in Tariff

Market pull mechanisms for wave and tidal sectors include financial supports mechanisms such as feed-in tariff and renewable obligations.

Feed-in tariffs (FIT) are the most common support mechanism, and are also currently the most popular and sought after mechanism by investors.

A feed-in tariff (FIT, FiT, standard offer contract, advanced renewable tariff, or renewable energy payments) is a policy mechanism designed to accelerate investment in renewable energy technologies (CfD described below separately). It achieves this by offering long-term contracts to renewable energy producers, typically based on the cost of generation of each technology. Ocean energy technologies such as wave and tidal power are offered a higher FIT price, reflecting costs that are higher at the moment. Table 4.3 presents a range of market pull mechanisms.

---

[7]http://maritimebrokerageevent2015.eu/media/sites/11/dlm_uploads/2015/11/Ocean-Energy-presentation.pdf

[8]http://www.eib.org/products/blending/innovfin/

[9]http://ec.europa.eu/regional_policy/en/funding/erdf/

Table 4.3 European Market support 'pull' mechanisms. Information adapted from JRC Ocean Energy Status Report 2016 Edition (Magagna, Monfardini et al. 2016)

Country	Tariff Support Scheme
Denmark	Maximum tariff of 0.08 EUR/kWh for all renewables including ocean energy
France	Feed-in Tariff for renewable electricity. Currently 15 cEUR/kWh for ocean energy.
Germany	Feed-in Tariff for ocean energy between EUR 0.035 and 0.125 depending on installed capacity
Ireland	Market support tariff for ocean energy set at €260/MWh and strictly limited to 30 MW
Italy	For projects until 5 MW 0.3 EUR/kWh For projects >5 MW 0.194 EUR/kWh
Netherland	The SDE+ (feed-in premium) supports ocean energy with a base support of 0.15 EUR/kWh minus the average market price of electricity in the Netherlands (support is given for a 15 year period). Total budget for SDE+ capped (EUR 8 billion in 2016)
UK	Renewable Obligation (RO) Scheme. Renewable Obligation Certificates (ROCs) price set to 44.33 GBP in 2015/16. Replaced by a Contract for Difference (CfD) scheme in 2017. Wave and tidal energy technologies will be allowed to bid for CfDs, however they are currently expected to compete with other technologies (e.g. Offshore Wind) to access CfD.

In addition, feed-in tariffs may include "tariff degression", a mechanism whereby the price (or tariff) ratchets down over time. This is done in order to encourage technological cost reductions. The goal of feed-in tariffs is to offer cost-based compensation to renewable energy producers, providing price certainty and long-term contracts that help finance renewable energy investments.

The disadvantage of Feed-in tariff support schemes is that they are only beneficial in stimulating investment when the technologies are near commercial (at TRL9[10]). They have benefited the tidal developments to some extent, but have not provided a benefit to wave energy prototypes. The advertised tariffs for wave energy could be viewed as purely theoretical, as the funds allocated have never been drawn-down. Moreover, many studies for wave energy financial viability have stated that current tariff support offered by most countries are inadequate, and need to be at least over €0.30c/kWh, to be financially viable (Dalton, Alcorn et al. 2012, Teillant, Costello et al. 2012).

---

[10]Technology Readiness Level: www.westwave.ie/wp-content/uploads/downloads/2012/10/Wave-Power-Systems-Technology-Readiness-Definition-ESBIoe-WAV-12-091-Rev2.pdf

Ireland, in 2016, completed a second review of the marine energy sector, called "Our Ocean Wealth task force report" (Development Task Force 2015). The report recommended the introduction of an market support scheme, funded from the public service obligation levy, equivalent to €260/MWh and strictly limited to 30 MW for ocean (wave and tidal). This will be allocated by public competition and focused on pre-commercial trials and experiments. A subsequent review will determine the most appropriate form and level of support for projects beyond 30 MW.

Portugal had perhaps the most developed tariff scheme (Figure 4.8), which incorporates the tariff degression method (this scheme has now lapsed). The tariff scheme supported prototype deployments under 4 MW at €0.26/kWh (Brito Melo 2010). Five pre-commercial projects were to be supported of 20 MW each, with FIT of €0.22/kWh. FIT rates for commercial projects would then drop to a range from €0.16/kWh for under 100 MW farms, €0.11/kWh for 100–250 MW and €0.075/kWh for farms over 250 MW.

The UK had the Renewable Obligation, active until the end of 2017, mandating electricity suppliers to deliver a certain proportion of their electricity from renewable sources, evidenced each year through the submission of the appropriate amount of Renewable Obligations Certificates (ROCs). ROCs are distributed to each renewable energy generator for each MWh of electricity sold. This effectively establishes a market for ROCs that is separate to the market for electricity. The price of a ROC in 2008 was approximately £0.047 (Scottish Government 2008). From April 2009, two ROCs was issued for each

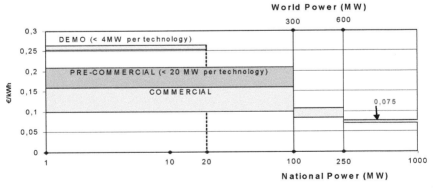

Figure 4.8   The proposed range of FIT offered in Portugal for the various stages of R&D and capacity deployed. (Brito Melo 2010).

MWh of wave generated electricity in England and Wales (equating to a value currently of £0.09/kWh), that is supplementary to the price received for the electricity). In Scotland five ROCs was allocated for each MWh of wave and tidal generated electricity (equating to £0.225/kWh based on current prices), also in addition to the electricity market price.

Post 2017 projects rely on Contract for Difference (CfD) for support in the UK market (Department of Energy and Climate Change 2014). CfD offers a fixed price above the market price for electricity, guaranteed for a period of time. Changing from the ROCs systems to the CfD is a major change for the UK renewable electricity sector. UK Government states that CfD will give Wave and Tidal much benefits and greater certainty[11]. It is argued that CfD will lead to lower finance costs, which will reduce the overall project costs. A potential wave or tidal development would need to bid into the new system and need win a successful bid to get access to the long term contracts. Once this is secured, CfD offers more revenue certainty, relative to the previous ROC regime. Wave and tidal developers will have access to a general pot of £260m which includes other renewable sectors such as advanced conversion, anaerobic digestion, dedicated biomass with CHP, geothermal. This does mean that wave and tidal will be competing with these other technologies to secure funding in a mechanism where the support will go to the cheapest technology. The highest strike price for both wave and tidal with be of 305 £/MWh, this is the Initial administrative (maximum) strike prices (£/MWh in 2012 prices). This change may have an initial settling period, where investors will be uncertain of the new market.

4.2.5.3 Push support scheme

Technology push support mechanisms for wave and tidal include public grants and private equity. Table 4.4 presents push mechanisms implemented by four EU member states to favour the development of ocean energy (Magagna, Monfardini et al. 2016). Push mechanisms tend to provide upfront capital for the deployment of pilot projects.

Examples include €26 million in Ireland to more than about €285 million in the United Kingdom.

The largest push support fund to come from the EU is called NER 300[12]. It is composed of European Commission, European Investment Bank

---

[11] https://www.gov.uk/government/publications/contracts-for-difference/contract-for-difference

[12] http://www.ner300.com/

Table 4.4 Summary of Push schemes for wave and tidal energy. Information from JRC report (Magagna, Monfardini et al. 2016)

Country	Fund	Total Million
France	Two projects	€103
Ireland	SEAI Prototype Development Fund,	€4
	Ocean Energy Development Budget	€26
Portugal	Fundo de Apoio à Inovação (FAI)	€76
UK	Marine Energy Array Demonstrator (MEAD),	£20
	Energy Technologies Institute (ETI),	£32
Scotland	Renewable Energy Investment Fund (REIF) Scotland,	£103
	Marine Renewables Commercialisation Fund (MRCF)	£18
	Saltire Prize, Scotland,	£10
	Wave Energy Scotland funding,	£14.3

and Member States. The NER 300 is a common pot of €300 M EU ETS allowances set aside for supporting 8 CCS and 34 renewable energy projects. The allowances will be sold on the carbon market and the money raised could be as much as €4.5B if each allowance is sold for €15. Up to 50% of "*relevant costs*" are funded under the scheme. Each member state will allocated at least one and a maximum of three projects[13]. The maximum return would be achieved by securing funding for the three largest demonstration projects that are in the public interest. The remaining costs will need to be co-funded by Member State governments and/or the private sector. A total of three ocean energy projects will be funded including wave, tidal and ocean thermal. Wave energy devices of up to 5 MW nominal power are eligible to apply[14].

NER 400[15] will supersede NER 300. Called ETS Innovation Fund, and proposes €2.1 bn EUR awarded for the period 2021–2030 (with some amount possibly made available before 2021). NER 400 will fund 38 innovative renewable energy and one CCS project and will additionally include measures to decarbonise industrial production.

Figure 4.9 provides a visual summary of market push and pull mechanisms for ocean energy, based on developers stage of technology or commercial development stage (Magagna, Monfardini et al. 2016, Vantoch-Wood 2016).

---

[13]http://ec.europa.eu/clima/funding/ner300/docs/faq_en.pdf

[14]http://ec.europa.eu/clima/funding/ner300/00031/index_en.htm

[15]http://ner400.com/

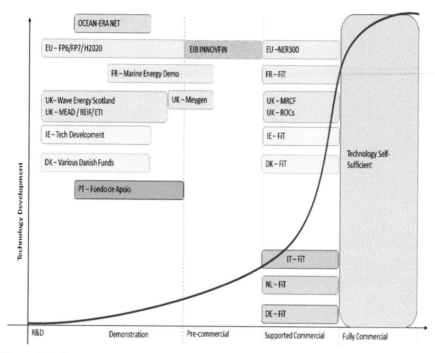

Figure 4.9 Summary of market push and pull mechanisms for ocean energy in the EU based on Carbon Trust deployment scenarios. Image taken form JRC report and Vantoch-Woods (Magagna, Monfardini et al. 2016, Vantoch-Wood 2016).

The OES Annual report (OES 2016) presents an excellent summary, country by country of:

- National strategy
- Market Incentives
- Financing

4.3 Sector Industry Structure and Lifecycle

4.3.1 Wave and Tidal Sectors – Present and Future Centres of Developer Activity

ReNews (ReNews 2014) in 2014 compiled an exhaustive list of stakeholder companies in the Wave and Tidal sectors, viewable in the following reference link: http://renews.biz/wp-content/assets/WTP-Research-Review-Winter-2014.pdf

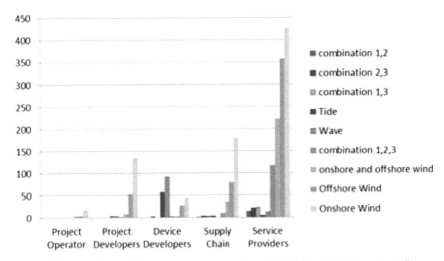

Figure 4.10   Number of Ocean energy companies defined by technology including on-shore wind. Figure provided by Exceedence Ireland[16](1=tide, 2=wave, 3=offshore wind, combination 1,2 = tide and wave).

JRC Ocean report (Magagna, Monfardini et al. 2016) contains a non-exhaustive list of companies currently active in the field of ocean energy, ranging from technology developers to component suppliers. The majority of technology developers are based in countries with significant ocean energy resources, many intermediate components suppliers are based across the EU (Germany, Sweden, Finland, Italy, Austria).

Figure 4.10 presents an analysis of the spread of sectors for the global wave and tidal industry, conducted by Exceedence[16]. The figures shows that service providers are by far the largest category, followed by supply chain. As anticipated, the majority are focused on onshore and offshore wind. These service providers are mostly based in the UK currently (Figure 4.11). It is anticipated that there will be transferable skills and business prospects.

The majority of wave and tidal developer companies are based in the UK and USA, Figure 4.12, with very sizeable annual turnover in USA as presented in Figure 4.13.

A visual representation of the European spread of wave and tidal industry is presented in Figure 4.14, created by SETIS[13], Eurostat for JRC. The map concurs with Exceedence findings, namely that the UK contains the most of the wave and tidal companies in Europe. The image also concurs that that

---

[16]www.exceedence.com

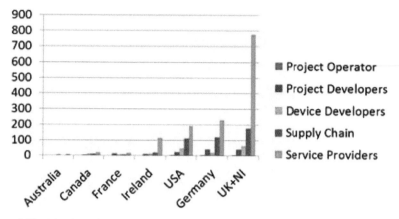

Figure 4.11 Number of companies in sample countries defined by stakeholder type. Figure provided by Exceedence Ireland[11].

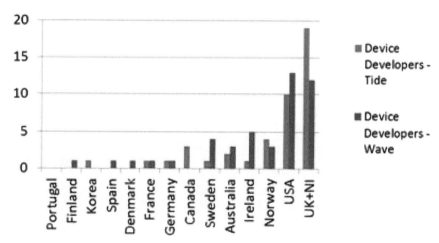

Figure 4.12 Number of wave and tidal developer companies in sample countries. Figure provided by Exceedence Ireland[11].

wave and tidal developers only comprise a small proportion of the overall stakeholder industry representation.

An important recent milestone has been a number of large engineering firms taking controlling stakes in device development companies, primarily in tidal technology companies, indicating that the tidal industry is closer to maturity than wave (Alcorn, Dalton et al. 2014). Companies include Siemens,

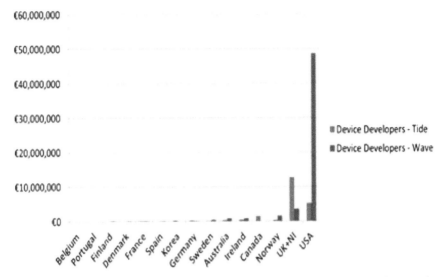

Figure 4.13   Total Annual turnover of all companies in Wave and Tidal in sample countries. Figure provided by Exceedence Ireland[11].

DCNS, Andriz Hydro, Alstom and others. In the last 7 years up to 2014, total private sector investment has been over €600 m in the last 7 years in Europe (EU-OEA 2013).

4.3.2 Supply Chain

Current market conditions and technology status of ocean energy converters have affected the consolidation of the supply and value chain of the sector (Magagna, Monfardini et al. 2016).

Supply chain consolidation is project-driven for technologies that are commercially viable. As witnessed in the wind energy sector, a strong project pipeline ensures that there is sufficient demand for Original Equipment Manufacturers (OEMs), and as a result guarantees demand for the manufacturing of components and subcomponents and for the supply of raw materials. On the other hand, for technologies that are not yet market-ready, such as ocean energy technology, the consolidation of the supply chain is dependent on the ability of reliability of the technology and its progress to higher TRL. Uncertainties in the project-pipeline are amplified throughout the supply chain, with potentially serious implications for the providers of components and raw materials. This can result in both price variation of good and materials, and in limited supply of products.

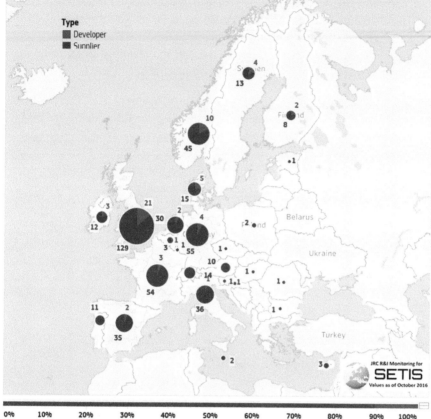

Figure 4.14 Ocean Energy patenting companies in the EU in 2008–2013 Companies identified as wave and tidal energy developers are represented in blue, supply chain and components manufacturers are classified as suppliers and represented in red. Image from SETIC JRC.[17]

One of the critical issues for the ocean energy sector over the past few years has been the lack of engagement of OEMs. Currently, however, as the separation between tidal and wave energy is more marked, it can be seen that, OEMs are either acquiring or investing tidal energy developers with DCNS, Andritz Hydro-Hammerfest, Lock-heed Martin, General Electric all making investments despite the exit of Siemens from the sector. For wave energy, however, since 2012 an exodus of OEMs has been witnessed.

---

[17]https://setis.ec.europa.eu/sites/default/files/report_graphs/patenting_companiese_eu_0.png

The necessity of reducing the cost of ocean energy technology, also through economy of scales, implies that the presence of OEMs with access to large manufacturing facilities could be seen as an indicator of the consolidation of the supply chain.

The Exceedence[11] company compiled a list of main supply chain companies supporting Wave and Tidal, categorised by marine basin, and is presented in Table 4.5.

Table 4.5 Table of major supply chain companies in the Wave and Tidal industry, spanning all the stakeholder categories categorised by marine basin (compiled by Exceedence)[11]

			Wave			
			Atlantic			
PTO & Generator	Electrical & Automation	Bearings	Marine Operations	Hydraulic Components	Coating	Diagnostic
Bosch Rexroth	ABB	Hutchinsons	Mallaig Marine	Mallaig Marine	Hempel	BAE Systems
Siemens	KTR Couplings	Schaeffler	Fugro Seascore	Hunger Hydraulics	Protective & Marine Coatings	Brüel & Kjær Vibro GmbH
Winco/ Dayton	Bailey	SKF	SeaRoc	Hydac	Akzo Nobel Coatings	SKF
Alstom/TGL	Eaton	Bailey	aquamarine power	Bailey	ICI paints	James Fisher Marine Services
Andritz Hydro/ Hammerfest	SKF	NSK	James Fisher Marine Services	Seaproof Solutions	Jotun	
			Baltic			
PTO & generator	Electrical & automation	Bearings	Marine O&M	Hydraulic components	Coating	Diagnostic
Bosch Rexroth	ABB	Schaeffler	A2SEA A/S	Hunger Hydraulics	Hempel	Voith
SKF	Eaton	SKF	EDF	Andritz Hydro/ Hammerfest	Protective & Marine Coatings	SKF
Siemens	Metso	NSK	DNV GL	Hydac	Sherwin-Williams	Brüel & Kjær Vibro GmbH
The Switch	KTR Couplings	NKE		Parker	ICI paints	
Schottel	VEO	Wolfgang Preinfalk			BASF Coating AG	

			Wave			
			Mediterranean			
PTO & Generator	Electrical & Automation	Bearings	Marine Operations	Hydraulic Components	Coating	Diagnostic
Siemens	ABB	Hutchinsons	Oceantec	D&D Ricambi	Protective & Marine Coatings	Metrohm
Bosch Rexroth	Eaton	SKF	Robert Bird	Hydac	Akzo Nobel Coatings	SKF
Alstom/TGL	SKF	NSK		Parker	Hempel	
SKF	Emerson Industrial Automation	NKE			Jotun	
	Leroy-Somer	Bosch Rexroth				

			Caribbean			
PTO & Generator	Electrical & Automation	Bearings	Marine Operations	Hydraulic components	Coating	Diagnostic
Northern Lights	Bailey	Waukesha Bearings		Hydac	Protective & Marine Coatings	C&C Technologies
Winco	Eaton	SKF		Parker	Hempel	SKF
SKF	ABB	Hutchinsons		Prince	Akzo Nobel Coatings	Hoffer Flow Controls Inc.
Marathon generators	SKF	NSK		Bailey	Jotun	
Bosch Rexroth	General Electrics	Bailey				

4.3.3 Lifecycle Stage

Figure 4.15 presents the life cycles stages for ocean renewables (Ecorys 2013). It will be noted that the stages are similar to those of offshore wind.

Table 4.6 presents the Life Cycle Stages for Wave and Tidal technology types.

It can be seen that the tidal industry has two technology types in the Growth phase.

The Wave energy industry has no technology types in the growth phase, all still in the embryonic phase. In addition to this negative picture, is the recent news of four companies liquidating, each company a flagship representative of a wave energy technology type of subsector. Oscillating water

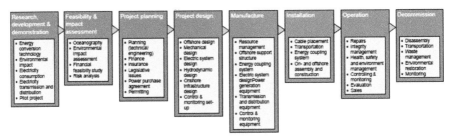

Figure 4.15   Life Cycle stages for Ocean energy. Image taken for Ecorys report (Ecorys 2013).

Table 4.6   Life cycle stages for Wave and Tidal industry, subdivided in technology types

Sector	Sub Sector	Life Cycle Stage
Tidal Energy	Fixed 3 blades	Growth Stage: multiple companies at array testing
	Fixed open centre	Growth Stage: Open Hydro at array testing phase
	Floating Tidal	Embryonic Stage; At prototype development phase
Wave Energy	OWC	Embryonic Stage; At prototype development phase Ocean Energy Buoy and GRS at prototype testing in Hawaii
	Over Topping	Embryonic Stage; At prototype development phase: WaveDragon in Wales and Fred Olsen Bolt
	Small scale devices kW	Embryonic Stage; At prototype development phase: Albatern and Seabased
	Point Absorber	Liquidated: WaveBob Carnegie Australia, OPT USA, SeaTricity UK
	Multiple point absorber	Liquidated: Wavestar
	Attenuator	Liquidated: Pelamis
	Hinge Flap	Liquidated: Aquamarine Wave Roller: Embroyonic

columns and Overtopping are the only technologies types remaining, thus indirectly demonstrating technology convergent through attrition.

See Section 4.5 'Innovation' for more details on wave and tidal companies and their lifecycle stage.

4.4 Working Environment

4.4.1 Job Creation and GVA

The European Commission 2012 report on Blue Economy (European Commission 2012) stated that the EU's blue economy represents 5.4 million jobs and a gross added value of just under €500 billion per year. In all, 75% of Europe's external trade and 37% of trade within the EU is seaborne. Much of this activity is concentrated around Europe's coasts, but not all. Some land-locked countries host very successful manufacturers of marine equipment.

Figure 4.16 shows that Ocean Energy comprises a small proportion of the Blue growth Jobs and GVA total percentages (European Commision 2012). However, Ocean energy is well positioned to contribute to regional development in Europe, especially in remote and coastal areas. Parallels can be drawn with the growth of the wind industry.

Based on the projections for installed capacity for ocean energy, the following reports quote a wide range of job creation potential for ocean energy and summarised in Figure 4.17:

- Ecorys (2010) (Ecorys 2013) In 2010 about 1000 people were esti-mated to be employed in the ocean renewable energy sector and about

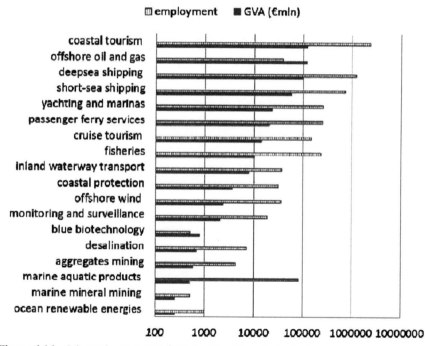

Figure 4.16 Job employment and GVA for Blue Growth, including ocean energy wave and tidal. Image taken form European Commission Blue Growth Opportunities COM(2012) 494 final (European Commision 2012).

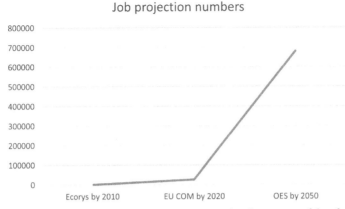

Figure 4.17 Job projection numbers for ocean energy – visual summary of data from reports.

€250 million of GVA was created in the EU. The great majority was depending on the developments in the Atlantic Arc.

- EU-OES (2010): by 2020 the ocean energy sector will generate over 26,000 direct and 13,000 indirect jobs, for a total of close to 40,000 (EU-OEA 2010). By 2050 these numbers would increase to 314,213, 157,107 and 471,320 respectively. The EU-OEA report further states that if 3,6 GW was installed in Europe by 2020 it would result in an investment of around €8,544 M, generating 40 thousand jobs. By 2050, achieving 188 GW could lead to an investment of €451B and the creation of around 471 thousand jobs.

- European Commission (2014) (European Commision 2014) indicates that indicative job estimates from the impact assessment show that 10,500–26,500 permanent jobs and up to 14,000 temporary jobs could be created by 2035. Other, more optimistic sources estimate 20,000 jobs by 2035 in UK alone (RenewableUK 2013) and 18,000 in France by 2020[18]. A substantial proportion of these employment opportunities will arise in the Atlantic coastal areas, which currently suffer from high unemployment.

- By 2050, the OES (OES 2016) has updated its international vision for ocean energy stating that by 2050 ocean energy has the potential to have deployed over 300 GW economic growth and job creation, estimated by the OES in 680,000 direct jobs.

---

[18]French Senate (2012), Report on Maritime Affairs at: http://www.senat.fr/rap/r11-674/r11-6741.pdf

- Other job predictions:
    - UK based (RenewableUK 2011, Energy and Climate Change Committee of the House of Commons 2012): 70 GW creating 68,000 jobs
    - US Based (Ocean Renewable Energy Coalition (OREC) 2011): 15 GW creating 36,000 jobs

4.4.1.1 Jobs/MW for wave and tidal in comparison to wind

Dalton et al. published a detailed paper analysing the metric of Job/MW relating to wind, wave and tide (Dalton and Lewis 2011). The paper stated that the onshore wind industry in Europe reported a total of 13 jobs/MW (direct jobs) were created on average for wind capacity installed in one year only (2007 in the study), or 1.9 jobs/MW (direct jobs) if using cumulative MW was used in the estimation. Installation job rates for many renewable energy technologies can be as labour intensive as fabrication. The European Photovoltaic Industry Association (EPIA) (EPIA 2004) states that more jobs could be created in the installation and servicing of PV systems than in their manufacture (30 jobs/MW). However, this figure contrasts dramatically to the wind energy installation job/MW figure quoted by the EWEA; 9 jobs/MW in their 2004 report (EWEA 2004), and 1.2 jobs/MW in their 2008 report (EWEA, Blanco et al. 2008) (perhaps because they used cumulative MW in estimations).

Wave and tidal studies on jobs/MW are very few as there is no real data to model.

Batten et al., (Batten and Bahaj 2006) in 2006 produced a comprehensive prediction of job creation for wave and tidal, based on each stage of the development of an ocean energy project, as well as direct and indirect jobs (Figure 4.18). This data was used in the report European Ocean Energy Association 2010 report, "Waves of Opportunity" (European Ocean Energy Association 2010). The analysis predicts the job/MW rate for both wave and tidal, direct and indirect, to be very similar, with wave having on average 1 job/MW more than tidal for each category. The greatest job intensities in device construction supply and foundation constructing (4-5jobs/MW for wave, 3jobs/Mw for tidal), followed by installation 1 job/MW. Batten's report predicts that by 2015, 19 direct and indirect jobs/MW at the start, falling to 7 jobs/MW by 2020. Direct jobs in device and foundation supply are quoted at around 10 jobs/MW falling to 3.5 jobs/MW.

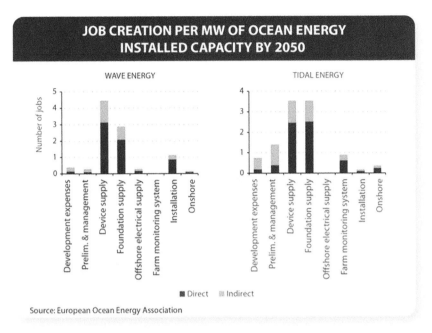

Figure 4.18   Job creation per MW of Ocean Energy. Image taken from SETIS Ocean Energy Association 2010 (European Ocean Energy Association 2010).

Further reports predicting jobs/MW figures for 2050 based on at least 10,000+MW installed are:

- Ireland (SEAI 2012): 2.4 Jobs/MW (based on 70,000 jobs created installing 29 GW)
- UK (RenewableUK 2011, Energy and Climate Change Committee of the House of Commons 2012): 1.08 Jobs/MW (based on 68,000 jobs created installing 70 GW)
- USA (Ocean Renewable Energy Coalition (OREC) 2011): 2.4 Jobs/MW (based on 15 GW installed and 36,000 jobs)

4.4.2 Skills Required, Workforce Mobility and Availability/ Competition for Skills

It is expected that workforce characteristics for ocean renewable energy will be similar to offshore wind and other offshore activities (Ecorys 2013). Ocean renewable energy requires a combination of skills from hydropower and offshore skills also needed for offshore wind, but also offshore oil & gas.

For different parts of the value chain, different skills are needed. Furthermore, as the sector is still under development, there are many research and consulting skills required.

4.4.2.1 Population centres versus ocean energy 'Hotspot' centres

Ideally, power production is located as close as possible to population centres to reduce energy loss via cable transmission. In the majority of northern European cases, the premium 'hotspot' sites are in remote locations, far from population centre. Analysis will be necessary to ascertain the economic optimum location taking both these factors into account.

Table 4.7 presents general information on skilled labour trends in 4 maritime basins. The following observations can be observed that are of relevance to wave and tidal development in the Atlantic and Baltic nations:

- Economies of Atlantic and Baltics nations are strong, with positive political stability favouring investment in the high-risk areas of Ocean Energy development.

Table 4.7 Population stats for 4 maritime basins, labour costs and migration trends (information taken for Maribe WP4- Wave and Tide Context report)[19]

	Atlantic	Baltic	Mediterranean	Caribbean
Population Stats	311,871,390	145,911,069	482,217,455	344,520,725
Pop growth or decline [%]	0.27	–0.05	0.81	1.03
Economic climate (growing, static, decline) (GDP) [%]	1.68	1.96	0.18	2.29
Political stability (stable, neutral, unstable) [from –2.8 to 1.5]	0.78	0.94	–0.44	0.19
Skilled labour (workforce with tertiary education) [%]	33.8	33.1	23.3	21.3
Skilled Migration trends	low labour mobility	relatively low labour mobility	relatively high labour mobility	high labour mobility
Annual average wage cost [$]	49,193	35,345	16,851	14,658

---

[19] http://maribe.eu/blue-growth-deliverables/blue-growth-work-packages/

- Third level skilled labour numbers are is high in Atlantic and Baltics nations favouring R&D in the high tech areas required for development of Ocean Energy.
- Negatives for the Atlantic and Baltics nations in developing ocean energy sector:
  - Labour mobility is low, posing a barrier to development of ocean energy in remote locations. Labour might be filled by highly mobile skilled workforce form Mediterranean and Caribbean.
  - Wages are high, posing a financial barrier to device development. Cheaper labour sourced from Mediterranean and Caribbean might be the solution.

4.4.2.2 Construction and fabrication skilled workforce

Manufacturing of turbines and other parts of ocean energy spare parts is mainly done by companies which have experience in related technologies. These bigger companies can easier shift workforce from one sector to the other. For example, Voith[20] used its knowledge from automotive industry, aerospace industry and apply it towards ocean renewable energy. Andritz[21] used its experience and knowledge on hydropower plants and transfers this towards the ocean tidal devices.

4.4.2.2.1 *Shipyards*

WEC devices will more than likely need to be built in shipyards (Previsic 2004), where existing maritime construction expertise and facilities exist. So far, most of the WEC prototypes have been constructed in local shipyards e.g. OE buoy in Cork Dockyards[22], Wavebob in Harland and Wolf, Belfast[23] and the 'Mighty Whale' in the Ishikawajima Harima shipyards in Japan[24]. The steel sections and power conversion modules of Pelamis were constructed in Scotland, but were assembled at the site of deployment: e.g. Peniche shipyards in Portugal[25] and Hunters Bay shipyards in San Francisco (Previsic 2004). The last two decades have witnessed a major contraction in

---

[20]http://voith.com/en/index.html

[21]http://www.andritz.com/

[22]http://www.irishexaminer.com/business/eco-energy-company-rides-on-a-wave-of-success-80844.html

[23]http://www.irishtimes.com/business/wave-generator-damaged-by-storm-1.1018087

[24]http://www.nsf.gov/pubs/1998/int9815/ssr9809.doc

[25]http://www.ain.pt/index.php/178703956051dad39d28963.pdf?mod=articles&action=downloadDocument&article_id=+++++++++++++++++237&document_id=256

Table 4.8 Shipyards for the four marine basins

Shipyards			
Atlantic	Baltic	Mediterranean	Caribbean
Harland & Wolff, Belfast, UK	Riga Shipyard, Riga, Latvia	Hellenic Shipyards, Piraeus, Greece	Grand Bahama Shipyard, Bahamas
Luerssen-Werft, Bremen, Germany	Western Shipyard, Klaipėda, Lithuania	Gibdock, Gibraltar	Ciramar Shipyards, Dom.Rep.
Peniche PT, Peniche, Portugal	Admiralty Shipyards, St. Petersburg, Russia	Tuzla Shipyard, Istanbul, Turkey	CL Marine Limited Caribbean Dockyard, Trinidad & Tobago
Damen Shipyard, Gorinchem, Netherlands	Meyer-Werft, Turku, Finland	Palumbo Shipyard, Messina, Italy	Cotecmar Shipyard, Colombia
Les Nefs Shipyard, Nantes, France			

Europe's shipbuilding capacity (Stopford 1997). Consequently future large-scale production of WEC devices in European shipyards may not be viable. Even if the choice were available, overseas competing shipyards in Poland, Korea and China, could feasibly outbid local contractors even factoring in shipping costs, due to lower overseas wages and cost of materials (Salonen, Gabrielsson et al. 2006).

Table 4.8 presents a list of shipyards, categorised into four marine basins, that may potentially serve the wave and tidal industry in construction and maintenance.

4.4.2.3 Installation and operations & maintenance (O&M) skilled workforce

Installation and operations & maintenance (O&M) of the ocean energy devices, cables and moorings also requires a skilled workforce and facilities. Specialised tugs companies are required to toe the devices to site, experienced underwater divers are required for deployment and maintenance of WEC and moorings, and specialised cable laying services for the electricity connector cable. A local skilled workforce may not available in the location for construction and deployment, or may be in limited supply due to competing technologies such as offshore wind. An example of this situation was when Seagen's tidal turbine was supposed to have been installed by

Table 4.9   Employment in operations and maintenance on ocean energy b7 2035. Table taken form Ecorys (Ecorys 2013)

Jobs in operation and maintenance of OE in 2035 under the three different scenarios			
	Direct	Indirect	Total
Scenario 1–Baseline	3,000–7,500	1,500–4,000	4,500–11,500
Scenario 2–Intensified Coordination	4,500–11,000	2,000–5,500	6,500–16,500
Scenario 3–Strong Stimulus	7,000–17,500	3,500–9,000	10,500–26,500

a local specialised tug early 2008. A higher offer made by the Thames off-shore wind project for the tug services left Seagen without a boat for installation (ReNews 2008). It took another 3 months for another contractor to be sourced, at a far higher cost, for the single installation.

Ecorys (Ecorys 2013) predicts that in 2035 total employment in operations and maintenance on ocean energy ranges from 4,500–26,500, depending on the scenario chosen.

4.4.3 Availability/Competition for Skills

As in other related sectors, shortages in engineering skills might occur and ocean energy may have to compete with the main competing sector; offshore wind. In offshore wind in the UK from 2013 onwards bottlenecks are expected as energy sectors are expected to grow at the same time (Scott Dickinson, Jonathan Cook et al. 2011). This affects ocean renewable energy. In the short-run employment will need to come from other sectors (e.g. offshore wind, offshore oil & gas) (Scott Dickinson, Jonathan Cook et al. 2011).

Ecorys (Ecorys 2013) predict that SMEs may struggle to attract skilled people from related sectors to fill skill requirements. Big companies will not be exposed to this risk due to the fact that they should be able to shift employment within their organisation, as per example of Voith and Adritz detailed above.

4.4.4 Infrastructure and Support Service Requirements

The necessary infrastructures such as reinforcing electrical grid networks and deepening of ports required for the roll-out of large-scale ocean renewables are still many years from materialising (Intelligent Energy Europe (IEA) 2010). Investors see that most sites of high ocean renewable potential are very remote from population centres, with inadequate current plans for upgrading facilities to the scale of development planned. Investor confidence will be significantly boosted if it sees major government funding to upgrade infrastructures at this current time, providing the ingredients for a successful future technology development roll-out.

4.5 Wave Technology Innovation

4.5.1 Wave Technology Innovation

Technology Categories	Company Examples	Technology Innovation and Future Development	Future Prospects
Attenuator	Pelamis	The Scottish based company Pelamis Wave Power went into administration in November 2014. The company was after being unable to secure the level of additional funding required for the further development of their technology[26]. Development agency Highlands and Islands Enterprise (HIE) has acquired th intellectual property and a range of physical assets previously owned by Pelamis. HIE has obtained the assets on behalf of Wave Energy Scotland (WES)	Liquidation (Assets are owned by WES)
	Dexa-Wave	Danish company, Blue Ocean Energy (BOE) project aims to adapt and test the feasibility of the DEXA WAVE. The company participated in €6 million EU funded research, H2Ocean, on wind-wave power open-sea platform equipped for hydrogen generation with support for multiple users of energy. No news since 2012	No news
	AlbaTERN	Scotland's Albatern WaveNET device is a scalable array of floating "Squid" generator units that harvest wave energy as their buoyant arms rise and fall with the motion of the waves. Each Squid can link up to as many as three others, effectively	Progressing

(Continued)

---

[26] http://tidalenergytoday.com/2015/01/19/wave-energy-scotland-bags-pelamis-assets/

Table: Continued

Technology Categories	Company Examples	Technology Innovation and Future Development	Future Prospects
		creating a large, floating grid that is flexible in every direction. The bigger this grid gets, the more efficient it becomes at harvesting energy, and the more different wave movements it can extract energy from. Albatern's 10-year target is to have 1.25 kilometre-long floating energy farms pumping out as much as 100 megawatts by 2024	
Flap	Aquamarine Power	Aquamarine the company which developed the Oyster 800 device is now in liquidation. Emerging from the group was the WavePOD consortium which aimed at developing a sealed sub-sea generating unit that can be used by many different WECs. The WavePOD is a standardised self contained generator, at tenth scale testing for the moment. In November 2015, there were no offers made for Aquamarine Power as a going concern, and Aquamarine ceased trading.	Liquidation
	AW Energy	2016–19, 5.6 MW nominal capacity, Installation in Peniche. 11–12 GWh targeted annual output, Project funding: EUR 9 million EU NER300 grant, EUR 13.5 million private investments, EUR 1.5 million Carbon Fund grant AW Energy has commissioned a PTO testing centre to test real scale PTO units. WaveRoller has got the second endorsement from Lloyd's Register Energy (LRE).	Progressing (Not Static)
	Bio Power Systems Australia	The Bio Power Systems device, the BioWAVE, will soon be at ocean-testing phase. The data collected through this final test phase will enable the development of a larger 1 MW device commercial scale BioWAVE unit.	Progressing

Technology Categories	Company Examples	Technology Innovation and Future Development	Future Prospects
Single point absorber	Carnegie Australia	Carnegie is developing the new CETO 6 device. Size, efficiency and power generation capacity are increased (compared to CETO 5). The aim is to be able to harvest wave energy further offshore, in higher sea states, and at lower cost. The innovation lies in the fact that the buoy will integrate the power generation. Thus power will be generated offshore and then transferred onshore with cables. 2016, $7.5 million microgrid project, a 2 megawatt solar photovoltaic array, a 2MW/0.5 megawatt hour battery energy storage system and a "sophisticated" control system integrated with Carnegie's CETO 6 wave technology and existing desalination plant	Progressing
	Ocean Power Technologies (OPT) USA	OPT is currently working on its PTO technology. This new technology will be integrated in the new device APB 350 (A1), followed by the APB 350 (A2) which geometry will be improved for a better operational stability and so that it can fit into a standard 40-foot container (to reduce transportation and deployment costs). In 2016, OPT announced the deployment of its commercial design of the PB3 PowerBuoy approximately four miles off of the coast of New Jersey	Progressing
	Seatricity UK	Future improvements are two-fold: research optimisation options for predicted device outputs used to compare the results with the full scale Oceanus 2 testing, and examine the tether loadings in storm conditions to improve the mooring system.	Progressing

(Continued)

Table: Continued

Technology Categories	Company Examples	Technology Innovation and Future Development	Future Prospects
Multipoint absorber	Wavestar	Wavestar was one of the longest surviving wave energy companies. Private investment of approx. €80 M over 18 years led to 1/4 scale testing of its device at Hanstholm. Wavestar succeeded in H2020 LCE3 funding, total €30 M. Unfortunately key partner financing withdrawal, and uncertainty if deployment location, led to the H2020 fund cancelation, ultimately leading the liquidation of Wavestar.	Liquidation
	Global Renewable Solutions (GRS) Australia	GRS is currently in the process of project planning for a 1/4 scale deployment. GRS is working closely with The SEA Ireland to develop the Atlantic Marine Energy Test Site which will enable GRS to test the performance of their pre commercial Power Platform.	Progressing
Oscillating Water Column	Oceanlinx	Oceanlinx wave energy device 'greenWAVE' sank during the transportation from Port Adelaide to Port MacDonnell. The company then went into liquidation.	Liquidation
	OE Bouy Ireland and USA	The longest surviving OWC technology company. Received funding from US DOE in 2016 for deploying 4/5 scale device at *US* Navy's Wave Energy Test Site, Kaneohe, *Hawaii* in Hawaii at 4/5 scale. https://energy.gov/sites/prod/files/2016/04/f30/100590.pdf	Progressing
	Voith Hydro WaveGen	In March 2013 Voith Hydro decided to close down Wavegen choosing to concentrate on tidal power projects.	Liquidation
Overtopping	Wave Dragon Denmark and UK	Applying for Wales/Ireland funding, deploy 4 MW full scale device in Wales for 2019.	Static
	Fred Olsen Bolt Norway	Sound & Sea Technology (SST) has completed the assembly of Fred. Olsen's Lifesaver wave energy converter ahead of its planned deployment at Navy's Kaneohe Bay Wave Energy Test Site (WETS) in Hawaii.	Progressing

4.5.2 Tidal Technology Innovation

Technology Categories	Company Examples	Technology Innovation and Future Development	Future Prospects
Horizontal Axis 3 blade Fixed	Atlantis Resources Corp UK	Atlantis Resources Limited has almost completed construction of the first phase of the MeyGen project – the world's largest planned tidal stream array; in Scotland's Pentland Firth. 2017 is due to be spent expanding the array to a capacity of 6 MW, thus completing phase 1A of the project. Full capacity across all phases is to be up to 398 MW.	Progressing
	Andritz Hydro Hammerfest Norway	ANDRITZ HYDRO delivered three turbines to MeyGen project; The Project "Development and Optimization of a Drive Train for Tidal Current Turbines" was successfully completed in 2015 after running for more than two and a half years.	Progressing
	Sustainable Marine Energy UK	successfully installed four subsea drilled rock anchors at its Fall of Warness for their first PLAT-O system, which hosts two SCHOTTEL Instream Turbines (SIT).	Progressing
	Nova Innovation Scotland	Nova Innovation are currently exporting power from two turbines installed off the coast of Shetland in Scotland, with a third turbine due to go live in early 2017.	Progressing
Horizontal Axis 3 blade Floating	Nautricity Ltd UK	Nautricity) are due to run test and demonstration projects at EMEC in the course of 2017	Progressing
	Scotre newables UK	Construction of first phase (10 MW) expected to start in 2017 550-tonne 2 MW tidal turbine arrived at EMEC in 2016,	Progressing
	TidalStream Limited UK	The TRITON, developed by SCHOTTEL HYDRO subsidiary TidalStream Ltd., carries. 40 SCHOTTEL Instream Turbines, reaching a total nominal	Progressing

(Continued)

Table: Continued

Technology Categories	Company Examples	Technology Innovation and Future Development	Future Prospects
		power output of 2.5 MW. Deployment at FORCE, Bay of Fundy, Canada, is scheduled for 2017.	
Venturi	Open Hydro/ DCNS Ireland/ France	Openhydro installing a turbine in the Bay of Fundy (a scaled-up version of the 6m turbines. They have been testing at EMEC since 2007)	Progressing
Kite	SeaCurrent NL	SeaQurrent has conducted the first tests on its 'multi wing' tidal kite technology at the MARIN research institute in the Netherlands.	
	Minesto Sweden	In 2017, Minesto plans to build and commission the first demonstrator of the Deep Green technology at commercial scale. The device will be installed at Minesto's site in Holyhead Deep, some 8 km outside the coast of northern Wales. In Holyhead Deep, for which Minesto holds an Agreement for Lease from the Crown Estate, the company will gradually expand installed capacity to a 10 MW commercial array (20 Deep Green units). Minesto has received funding from KIC Innoenergy and European Regional Development Fund through the Welsh Government.	Progressing

4.6 Concluding Remarks

Ocean energy research and development started in earnest in the early 1970's, in the wake of the oil crisis (Cruz 2008). In 2006, the Carbon Trust stated that the value of worldwide electricity revenues from wave and tidal stream projects could potentially be substantial, with predictions of electricity revenues between €75 billion/year and €237 billion/year, requiring Investments of over £500 bn (€600 bn) contributing 2000 TWh/year worldwide (Carbon Trust and Callaghan 2006).

With such commercial potential, a question that must be asked in 2017 is *"Why has the wave and tidal industry in 2017 not established itself as a competing renewable technology"* (Dalton 2010, Dalton 2014)? The contributing ingredients to the delay in consolidation of the sector are multidimensional. However significant progress has been made particularly in the tidal sector.

The primary issue for the majority of investors is lack of confidence. Stated simply, there are no fully commercial arrays of wave or tidal devices in the water (Meygen may be considered a commercial array depending on definition), demonstrating that neither technology currently have the technical capacity to generate reliably.

On the positive side, tidal technology development is moving to the final stages of pre-commercial demonstration (eg Meygen), raising the confidence levels in that sector substantially. In many respects tidal technologies are an extension of well-proven wind technologies. Tidal technologies are now being tested at pre-commercial phase via private and public (FP7/NER300) project funding, with relatively few technical setbacks. Tidal energy seems certain to be technically viable, and in time should become economically and commercially viable. However, the market is niche, due to the limited global tidal energy resource.

Wave energy development, on the other hand, has been hampered by a lack of confidence in current existing technology concepts. It has been questioned how so many wave energy companies move all the way through the TRL levels, reaching pre-commercial scale, and fail. The current lack of confidence in wave energy technology development is reflected by the recent closures of some longstanding wave development companies e.g. Pelamis, Aquamarine, and, Wavestar. Moreover, two major NER300 projects for wave demonstration have also been withdrawn or postponed: Waveroller, as well as the Westwave project.

Wave projects have failed to achieve, what may be overambitious TRL, design and testing targets, set by funders. Consequently a lack of investor confidence has dried up funding added to this, additional pressures from government support mechanisms which rewards energy production rather than robust designs (Alcorn, Dalton et al. 2014).

More stringent concept evaluation, driven centrally, by government funding bodies, at early stage development would eliminate the weakest design concepts. Stringent adherence to stage testing along the TRL scale should help ensure positive technical results. Investors increasingly require evidence that this standardised technology development approach is implemented. Finally, strong and consistent national government driven policy (Dalton and Gallachóir 2010), combining best practice pull and push market mechanisms based on successful innovation development is crucial to bring pre-commercial ocean energy companies to commercial ready stage.

References

Alcorn, R., G. J. Dalton, M. Healy and M. O'Connor (2014). De-Risking Ocean Energy. Clean Tech, Clean Profits. A. Jolly.

Altantic Action Plan (2013). Blue Growth, opportunities for marine and mari- time sustainable growth, European Commission. https://bookshop. europa.eu/en/blue-growth-pbKL3212321/

Atlantic Action Plan (2013). Action plan for a maritime strategy in the Atlantic area, European Commission. https://bookshop.europa.eu/en/ action-plan-for-a-maritime-strategy-in-the-atlantic-area-pbKL0313320/

Batten, W. M. J. and A. B. Bahaj (2006). An assessment of growth scenarios and implications for ocean energy industries in Europe, Sustainable Energy Research Group, School of Civil Engineering and the Environment, University of Southampton, Report for CA-OE, Project no. 502701, WP5. http://eprints.soton.ac.uk/53003/

Brito Melo, A. (2010). Marine renewable energy in Portugal. Energy of the Sea, Madrid, Spain. http://www.wavec.org/client/files/10_02_02_ Madrid_Ana_Brito_Melo.pdf

Carbon Trust and J. Callaghan (2006). Future Marine Energy – results of the Marine Energy Challenge: cost competitiveness and growth of wave and tidal stream energy Carbon Trust – CTC601. http://www.carbontrust. co.uk/Publications/publicationdetail.htm?productid=CTC601&metaNo Cache=1

Chozas, J. F., Wavec, Ramboll and ReVision (2015). International Levelised Cost Of Energy for Ocean Energy Technologies, OES IEA Ocean Energy Europe. https://www.ocean-energy-systems.org/documents/ 65931-cost-of-energy-for-ocean-energy-technologies-may-2015-final. pdf/

Cruz, J. (2008). Ocean Wave Energy: Current Status and Future Prespectives, Springer Book: Green Energy and Technology. http://www.springer. com/engineering/energy+technology/book/978-3-540-74894-6

Dalton, G., R. Alcorn and T. Lewis (2009). Economic assessment of Pelamis wave energy converter: a case study off the Irish Atlantic coast. Renewable Energy World, Cologne, Germany. http://pennwell.websds.net/ 2009/cologne/_REWE/results.php?Track=1;Session=4

Dalton, G., G. Allan, N. Beaumont, A. Georgakaki, N. Hacking, T. Hooper, S. Kerr, A. M. O'Hagan, K. Reilly, P. Ricci, W. Sheng and T. Stallard (2016). "Economic and socio-economic assessment methods for ocean renewable energy: Public and private perspectives." Renewable and Sustainable Energy Reviews **45**(0): 850–878.

Dalton, G., G. Allan, N. Beaumont, A. Georgakaki, N. Hacking, T. Hooper, S. Kerr, A. M. O'Hagan, K. Reilly, P. Ricci, W. Sheng and T. Stallard (2016). "Integrated methodologies of economics and socio-economics assessments in ocean renewable energy: Private and public perspectives." International Journal of Marine Energy **15**: 191–200.

Dalton, G. and B. P. ó. Gallachóir (2010). "Building a wave energy policy focusing on innovation, manufacturing and deployment." Renewable and Sustainable Energy Reviews **14**(8): 2339–2358.

Dalton, G. J. (2010). Why wave energy – Market driver analysis for investors and policy makers. Annual Report 2010 – Implementing Agreement on Ocean Energy Systems, OES-IA. http://www.ocean-energy-systems.org/library/oes-reports/annual-reports/document/oes-annual-report-2010/

Dalton, G. J. (2011). Annual Report 2010 – Implementing agreement on ocean energy systems, OES IA and IEA Energy Technology Network: 93–110. http://www.iea-oceans.org/_fich/6/2010_Annual_Report.pdf

Dalton, G. J. (2014). Investments in Ocean Energy – A review of the global market. ICOE, Halifax, Nova Scotia. http://www.icoe2014canada.org/wp-content/uploads/2014/11/1-ICOE2014-Gordon-Dalton-53e23ba6c76748.10106971–8.pdf

Dalton, G. J., R. Alcorn and T. Lewis (2010). "Case study feasibility analysis of the Pelamis wave energy convertor in Ireland, Portugal and North America." Renewable Energy **35**(2): 443–455.

Dalton, G. J., R. Alcorn and T. Lewis (2010). Operational expenditure costs for wave energy projects; O/M, insurance and site rent, International Conference on Ocean Energy (ICOE), Bilbao, Spain.

Dalton, G. J., R. Alcorn and T. Lewis (2012). "A 10 year installation program for wave energy in Ireland, a sensitivity analysis on financial returns." Renewable Energy **40**(1): 80–89.

Dalton, G. J. and T. Lewis (2011). "Metrics for measuring job creation by renewable energy technologies, using Ireland as a case study" Renewable and Sustainable Energy Reviews **15**(4): 2123–2133

Dalton, G. J. and T. Lewis (2011). Performance and economic feasibility analysis of 5 wave energy devices off the west coast of Ireland, EWTEC, Southampton, UK.

Department of Energy and Climate Change, D. (2014). Contracts for Difference (Standard Terms) https://www.gov.uk/government/uploads/system/uploads/attachment_data/file/348202/The_Contracts_for_Difference_Standard_Terms_Regulations_2014-_CFD_Standard_Terms_Notice_29_August_2014_.pdf

Development Task Force (2015). Our Ocean Wealth Development Task Force Report. https://www.ouroceanwealth.ie/sites/default/files/sites/default/files/Documents/Final%20DTF%20Consolidated%20Input%20Material.pdf

Ecorys (2013). Study in support of Impact Assessment work for Ocean Energy, EC DG Maritime Affairs and Fisheries FWC MARE/2012/06 – SC C1/2012/01. https://webgate.ec.europa.eu/maritimeforum/sites/maritimeforum/files/Final%20Report%20Ocean%20Energy_0.pdf

Ecorys (2013). Study on Deepening Understanding of Potential Blue Growth in the EU Member States on Europe's Atlantic Arc, DG Maritime Affairs and Fisheries FWC MARE/2012/06 – SC C1/2013/02. https://webgate.ec.europa.eu/maritimeforum/sites/maritimeforum/files/Blue%20Growth%20Atlantic_Seabasin%20report%20FINAL%2007Mar14.pdf

Energy and Climate Change Committee of the House of Commons (2012). The Future of Marine Renewables in the UK. Eleventh Report of Session 2010–12 Volume II. https://www.publications.parliament.uk/pa/cm201012/cmselect/cmenergy/1624/1624.pdf

EPIA (2004) "Solar generation." http://www2.epia.org/documents/Solar_Generation_report.pdf

EU-OEA (2010). Oceans of Energy – European Ocean Energy Roadmap 2010–2050, European Ocean Energy Association (EU-OEA). http://www.waveplam.eu/files/newsletter7/European%20Ocean%20Energy%20Roadmap.pdf

EU-OEA (2013). "European Ocean Energy – Industry Vision Paper."

Europeaan Commision (2015). Towards an Integrated Strategic Energy Technology (SET) Plan, C(2015) 6317 final. https://ec.europa.eu/energy/sites/ener/files/documents/1_EN_ACT_part1_v8_0.pdf

European Commision (2007). Conclusions from the Consultation on a European Maritime Policy COM(2007) 575, 10.10.2007. http://www.programmemed.eu/fileadmin/PROG_MED/Reglements/COM_2007–575_final_EN.pdf

European Commision (2007). "Promotion of the use of energy from re-newable sources-repealing Directives 2001/77/EC and 2003/30/EC." Official Journal of the European Union **L140**: 16.

European Commision (2008). Offshore Wind Energy: Action needed to deliver on the Energy Policy Objectives for 2020 and beyond COM(2008) 768 final 13.11.2008.

European Commision (2010). National Renewable Energy Action Plans (NREAPs), Joint Reserach centre (JRC). http://iet.jrc.ec.europa.eu/reme a/national-renewable-energy-action-plans-nreaps

European Commision (2010). A strategy for smart, sustainable and in-clusive growth COM(2010) 2020, 3.3.2010. http://eur-lex.europa.eu/legal-content/EN/TXT/PDF/?uri=CELEX:52010DC2020&from=en

European Commision (2011). A Roadmap for moving to a competitive low carbon economy in 2050 COM(2011) 112 final. http://eur-lex.europa.eu/legal-content/EN/TXT/PDF/?uri=CELEX:52011DC0112&from=EN

European Commision (2012). Blue Growth opportunities for marine and maritime sustainable growth COM(2012) 494 final. https://ec.europa.eu/maritimeaffairs/sites/maritimeaffairs/files/docs/body/com_2012_494_en.pdf

European Commision (2013). Action Plan for the Atlantic Ocean area-Delivering smart, sustainable and inclusive growth, COM(2013) 279, 13.5.2013. http://eur-lex.europa.eu/legal-content/EN/TXT/PDF/?uri=CELEX:52013DC0279&from=EN

European Commision (2013). Communication on Energy Technologies and Innovation. COM(2013) 253, 2.5.2013. https://ec.europa.eu/energy/sites/ener/files/comm_2013_0253_en.pdf

European Commision (2014). Action needed to deliver on the potential of ocean energy in European seas and oceans by 2020 and beyond COM (2014) 8 final. http://eur-lex.europa.eu/legal-content/EN/TXT/PDF/?uri=CELEX:52014DC0008&from=en

European Commision (2014). Blue Energy Action needed to deliver on the potential of ocean energy in European seas and oceans by 2020 and beyond, COM/2014/08 final http://eur-lex.europa.eu/legal-content/EN/TXT/PDF/?uri=CELEX:52014DC0008&from=en

European Ocean Energy Association (2010). Waves of Opportunity – Road Map for Ocean Energy in Europe, SETIS TREN/07/FP6EN/S07.75308/038571 EU-OEA. https://webgate.ec.europa.eu/maritimeforum/sites/maritimeforum/files/Ocean%20Energy%20Roadmap%202010.pdf

Eurostat (2016). Renewable energy statistics. http://ec.europa.eu/eurostat/ statistics-explained/index.php/Renewable_energy_statistics

EWEA (2004). Wind energy – the facts. Industry and employment. http:// www.ewea.org/fileadmin/ewea_documents/documents/publications/WE TF/Facts_Volume_3.pdf

EWEA, G. Blanco and I. Kjaer (2008). Wind at work: Wind energy and job creation in the EU, European Wind Energy Association. http://www. ewea.org/fileadmin/ewea_documents/documents/publications/Wind_at_ work_FINAL.pdf

Global CCS Institute (2013). Science for energy: JRC thematic report, JCR European Commission. https://hub.globalccsinstitute.com/publications/ science-energy-jrc-thematic-report/63-other-scientific-research-support-european-energy-and-climate-policy

IEA-OES (2011). An International Vision for Ocean Energy.

IEA-OES., J. Khan and M. M. H. Bhuiyan (2008). Annual report 2007 of the International Energy Agency Iimplementing agreement on ocean energy systems IEA-OES. http://www.vliz.be/imisdocs/publications/213 647.pdf

Intelligent Energy Europe (IEA) (2010) "Waveplam." Wave Energy: A guide for investors and policy makers. http://www.waveplam.eu/files/ downloads/D.3.2.Guidelienes_FINAL.pdf

Lewis, A., S. Estefen, J. Huckerby, W. Musial, T. Pontes, J. Torres-Martinez (2011). Ocean Energy. IPCC Special

Report on Renewable Energy Sources and Climate Change Mitigation.

Magagna, D., R. Monfardini and A. Uihlein (2016). JRC Ocean Energy Status Report 2016 Edition, Joint Research Centre, Directorate for Energy, Transport and Climate Change. http://publications.jrc.ec.europa.eu/ repository/bitstream/JRC104799/kj1a28407enn.pdf

Magagna, D. and A. Uihlein (2015). "Ocean energy development in Europe: Current status and future perspectives." International Journal of Marine Energy **11**: 84–104.

Mork, G., S. Barstow, A. Kabuth and M. T. Pontes (2010). Assessing the global wave energy potential. Proc. of 29th International Conference on Ocean, Offshore and Arctic Engineering, ASME, paper.

Ocean Energy Europe (2014). The Technology and Innovation Platform for Ocean Energy Terms of Reference. https://www.oceanenergy-europe. eu/images/Documents/TPOcean/TPOcean-Terms_of_Reference.pdf

Ocean Energy Forum (2016). Ocean Energy Strategic Roadmap, Ocean Energy Forum. https://webgate.ec.europa.eu/maritimeforum/sites/mari

timeforum/files/OceanEnergyForum_Roadmap_Online_Version_08Nov 2016.pdf

Ocean Renewable Energy Coalition (OREC) (2011). U.S. Marine and Hydrokinetic Renewable Energy Roadmap for 2050. http://www.policy andinnovationedinburgh.org/uploads/3/1/4/1/31417803/mhk-roadmap-executive-summary-final-november-2011.pdf

OES (2015). OES 2015 Annual Report. A. B. e. M. a. J. L. Villate, International Energy Agency (IEA). https://report2015.ocean-energy-systems.org/

OES (2016). OES 2016 Annual Report. A. B. e. M. a. J. L. Villate, International Energy Agency (IEA). https://report2016.ocean-energy-systems.org/

Previsic, M. (2004). Offshore wave energy conversion devices, EPRI. http://oceanenergy.epri.com/attachments/wave/reports/004_WEC_Device_Ass ess_Report_Rev1_MP_6-16-04.pdf

Previsic, M. (2004). System level design, performance, and costs of California Pelamis wave power plant, EPRI. http://oceanenergy.epri.com/attachments/wave/reports/006_San_Francisco_Pelamis_Conc eptual_Design_12-11-04.pdf

RenewableUK (2011). SeaPower funding the marine energy industry 2011–2015, BWEA. http://www.bwea.com/pdf/marine/SeaPower_Fund _Paper.pdf

RenewableUK (2013). Wave and Tidal Energy in the UK. www.renewableuk. com/en/publications/reports.cfm/wave-and-tidal-energy-in-the-uk-2013

ReNews (2008) "Seagen delay." http://www.renews.biz/

ReNews (2014). Wave and Tidal Power Resreach Review. http://renews.biz/ wp-content/assets/WTP-Research-Review-Winter-2014.pdf

Salonen, A., M. Gabrielsson and Z. Al-Obaidi (2006). "Systems sales as a competitive response to the Asian challenge: Case of a global ship power supplier." Industrial Marketing Management **35**(6): 740–750.

Scott Dickinson, Jonathan Cook, Jean Welstead, Grendon Thompson, A. Yuille and K. Chapman (2011). Maximising employment and skills in the offshore wind supply chain, SQW for UK Commission for Employment and Skills (UKCES). http://www.amtu.fi/download/141097_ evidence-report-34-maximising-employment-offshore-wind-vol1.pdf

Scottish Government (2008) "Support for wave and tidal energy." http://www.scotland.gov.uk/News/Releases/2008/09/19111827

SEAI (2012). Ocean Energy roadmap, SEAI. http://www.seai.ie/Renewables/ Ocean_Energy_Roadmap.pdf

SI Ocean (2014). Wave and Tidal Energy Market Deployment Strategy for Europe, Intelligent Energy Europe. https://www.oceanenergy-europe.eu/images/OEF/140037-SI_Ocean_Market_Deployment_Strategy.pdf

Stopford, M. (1997). Maritime economics. New York, Routledge.

Teillant, B., R. Costello, J. Weber and J. Ringwood (2012). "Productivity and economic assessment of wave energy projects through operational simulations." Renewable Energy **48**(0): 220–230.

TP Ocean (2016). Strategic Research Agenda for Ocean Energy Ocean Energy Europe: European Technology and Innovation Platform for Ocean Energy http://oceanenergy-europe.eu/images/Publications/TPOcean-Strategic_Research_Agenda_Nov2016.pdf

Vantoch-Wood, A. (2016). Best Practice Innovation Support. Insights from the Carbon Trust's innovation activity. ICOE Nantes.

World Energy Council. (2007). "Survey of energy sources." from http://www.worldenergy.org/wec-geis/publications/reports/ser04/overview.asp

5

Offshore Wind Energy

Mike Blanch*, Clare Davies and Alun Roberts

BVG Associates, UK
*Corresponding Author

5.1 Introduction

Offshore wind is the world's most commercially and technologically developed marine renewable energy subsector and is changing fast from being a niche technology into a mainstream supplier of electricity. At the end of 2016, global offshore wind capacity reached over 14.8GW with 12.9GW in Europe and 5.3GW (41%) of this in the UK.[1,2] At the end of 2016, there were 81 operational offshore wind farms spread across the waters of 10 European countries with 11 more projects in construction, totalling an additional 4.8GW. Offshore wind market activity is currently focused in the Atlantic, and Baltic and North Sea basins, which function as a single market. The UK, Germany and Denmark have principally driven the market in the North Sea, Irish Sea and Baltic Sea to date. Future activity in Europe focuses on these areas as well as expanding to the English Channel and the Bay of Biscay.

The countries around these basins have relatively low electricity prices so output from offshore wind farms has been explicitly subsidised. Some subsidy is necessary for all new electrical generating plant, but the cost of energy of offshore wind and subsidies needed continue to fall. Table 5.1 shows the recent winning level of price support from auctions.

---

[1]Global Wind Energy Council (2017), Global Wind Report 2016, 76 pp. Last accessed August 2017 http://www.gwec.net/publications/global-wind-report-2/global-wind-report-2016/

[2]WindEurope (2017). The European offshore wind industry – key trends and statistics 2016. A report by WindEurope (formally European Wind Energy Association). 25 pp. Last accessed August 2017. https://windeurope.org/wp-content/uploads/files/about-wind/statistics/WindEurope-Annual-Offshore-Statistics-2016.pdf.

Table 5.1   Recent price support for offshore wind farms

Country	Owner	Project	Unit	Price[3]	Date of Auction Win	First Operation Expected at Time of Bid
UK	Mainstream	Neart na Gaoithe	£/MWh	114.39	Feb-15	2018
UK	Scottish Power Renewables	East Anglia 1	£/MWh	119.89	Feb-15	2019
DK	Vattenfall	Horns Rev 3	€/MWh	103.1	Feb-15	2020
NL	DONG Energy	Borssele 1&2	€/MWh	72.7	Jul-16	2020
DK	Vattenfall	Vesterhav (Nord & Syd)	€/MWh	64.0	Sep-16	2020
SE	EnBW/Macquarie Capital	Kriegers Flak (Baltic 2a&2b)	€/MWh	49.9	Nov-16	2021
NL	Shell/Eneco/Van Oord/Mitsubishi DNG	Borssele 3&4	€/MWh	54.5	Dec-16	2022
UK	Innogy	Triton Knoll	€/MWh	83.6	Oct-17	2021/22
UK	DONG	Hornsea Project 2	€/MWh	64.3	Oct-17	2022/23
UK	EDP	Renewables Moray (East)	€/MWh	64.3	Oct-17	2022/23

The Mediterranean and Caribbean basins do not currently have any commercial offshore wind installations. The Mediterranean could capitalise on its proximity to the established markets in the Atlantic and Baltic to develop a commercial market. This is not likely by 2020 due to the limited number of projects currently in development. The Caribbean could capitalise on synergies in the established oil and gas industry. Despite relatively high costs of electricity from new electricity generating plant, low annual electricity demand may limit the ability to establish a cost competitive market beyond a few projects. Combining with another sector such as desalination or with a floating deeper-water shipping terminal might help enable a bigger market.

A report prepared for WindEurope by BVG Associates and Geospatial Enterprises, highlighted the economically attractive offshore wind resource that is potentially available to Europe in the Baltic, North Sea and Atlantic from France to the north of the UK in 2030. Offshore wind could, in theory, reach a capacity of between 600 GW and 1,350 GW in the modelled baseline and upside scenarios respectively. This would generate between 2,600 TWh

---

[3]Note that the figures are not directly comparable. The duration of the support is different and not all include the cost of transmission.

and 6,000 TWh per year at a competitive cost of €65/MWh or below, including grid connection and using the technologies that will have developed by 2030. This economically attractive resource potential would represent between 80% and 180% of the EU's total electricity demand.[4]

5.2 Market

5.2.1 Atlantic and Baltic Basins

European activity is primarily in the Atlantic and Baltic basins, which function as a single market. Wind farms are developed either by large utilities, which they subsequently operate, or by independent developers. Often, they are developed by joint ventures to spread the risk. In the past, utilities tended to finance projects from their balance sheets and contracted up to 10 large packages (multi-contracting). With the increasing scale and complexity of projects, there are an increasing number of project-financed wind farms. Here investors typically prefer a small number of contracted packages to push the project risk down the supply chain and minimise the project's interface risks. There is a decreasing role for independent developers as they can rarely support project development teams for several years.

The key players are the turbine manufacturers. Turbines typically cost 30–40% of CAPEX (capital expenditure) and manufacturers have a major role in driving innovation and reducing costs. In an engineer, procure, construct (EPC) contracting environment, the turbines are usually procured separately as this contract has to be awarded before detailed engineering can begin. EPC contractors are usually offshore construction companies active in a range of sectors. The turbine contract generally includes a service agreement. Historically, this has been five years but this is increasingly variable as owners seek either to reduce cost by breaking the tie with the turbine manufacturer early or to increase the project's attractiveness to investors by negotiating a longer service agreement.

5.2.2 Mediterranean Basin

The Mediterranean is an emerging market and a small number of projects may be built before 2020. The characteristics of the market have not yet

---

[4]BVG Associates for Wind Europe (2017), Unleashing Europe's offshore wind potential, A new resource assessment, available online at: https://bvgassociates.com/publications/, last accessed August 2017.

emerged but the size of projects means that independent developers can play a significant role. In time, the market is likely to evolve in line with the Atlantic and Baltic basin markets. This is because the turbine manufacturers will be the same and they, largely, shape the market.

5.2.3 Caribbean Basin and Rest of World

Outside of Europe, there is an establishing market in China, and the first commercial sites are being developed in Japan, South Korea, Taiwan and the US. There is no commercial market in the Caribbean.

Table 5.2 presents key data for the offshore wind market of each basin.

5.3 Sector Industry Structure and Lifecycle

Offshore wind market activity is currently focused in the Atlantic and Baltic basins as Northern European countries such as UK, Germany and Denmark have driven the market in the North Sea, Irish Sea and Baltic Sea. Future activity in Europe will continue to focus on these areas. There is a good level of confidence that the geographic spread will expand to the English Channel and Bay of Biscay as there are a number of UK and French projects under development that are expected to be operating, under construction, or have reached final investment decision (FID) by 2020. To date, there is no commercial activity within the Mediterranean or Caribbean basins.

At the end of 2016, the Mediterranean had 1.1% of Europe's total consented offshore wind capacity.[5] The basin is therefore not expected to have any significant commercial activity before 2020. The Caribbean basin is currently dominated by oil and gas activity. This could provide synergies with existing infrastructure and supply chain capability if offshore wind was ever developed and deployed at a commercial scale. Offshore wind commercially leased areas and demonstration sites in the US are currently located elsewhere along the East Atlantic coast, with some early activity also in the Pacific Northwest, California, Hawaii and the Great Lakes.

The Atlantic and Baltic basins together can primarily be classified into lifecycle stage two (growth stage) and three (mature stage). The Mediterranean and Caribbean basins together can primarily be classified into lifecycle stage nought (development) and one (embryonic). To date, offshore wind

---

[5]The European offshore wind industry – key trends and statistics 2016, WindEurope, 2017, available online at https://windeurope.org/wp-content/uploads/files/about-wind/statistics/WindEurope-Annual-Offshore-Statistics-2016.pdf, last accessed August 2017.

Table 5.2 Offshore wind market by basin

Basin	Country	Installed Capacity[6] (GW)	Consented Capacity[7] (GW)	Cost (CAPEX)	Price Support Last Awarded[8]	Targets[9]
Atlantic	UK	5.3	14.6 (57%)	CAPEX ~€4 million per MW	£57.5–£74.75/MWh[10]	No formal targets, but range of 8GW to 13GW stated by Government by 2020. Provisional budget for about 10GW by 2020
	Germany	3.8	5.3 (21%)		€60/MWh[11]	6.5GW by 2020 (covering both Atlantic and Baltic basins)
	Netherlands	1.1	2.1 (8%)		€54.5/MWh[12]	5.2GW by 2020
	Denmark	0.4	0.0 (0%)		€64.0/MWh[13]	1.4GW by 2020 (covering both Atlantic and Baltic basins)
	France	0.0	0.0 (0%)		FIT of €130/MWh, or competitive tender[14]	6GW by 2020 (covering both Atlantic and Baltic basins)

(Continued)

6 Installed capacity to the end of 2016.

7 Share of consented offshore wind farms (totalling 25.8GW) in Europe at the end of 2016. Some countries with consented capacity are not included in Table 5.2.

8 Note that the figures for different countries are not directly comparable. The duration of the support is different and not all include the cost of transmission.

9 Targets taken from National action plans available online at: https://ec.europa.eu/energy/en/topics/renewable-energy/national-action-plans, last accessed August 2017.

10 Competitive contract for difference. Over 15 years for projects reaching final investment decision in 2017. Includes transmission.

11 Auction April 2017 price for Gode Wind 3 offshore wind farm. There were three other wind farms that were bid at zero whose effective price is unclear.

12 Netherland price support awarded through tender process. Price support awarded to Shell/Eneco/Van Oord/Mitsubishi DNG for Borssele 3 and 4 in December 2016. Transmission is excluded.

13 Price support awarded to Vattenfall for Vesterhav (Nord & Syd) in September 2016.

14 A FIT is available for offshore wind projects in France at €130/MWh, however a competitive tender was run in 2012 for 3GW of projects. Transmission costs are financed by developer in a competitive tender, but compensated at-cost as an incremental part of the tariff in €/MWh.

Table 5.2   Continued

Basin	Country	Installed Capacity[6] (GW)	Consented Capacity[7] (GW)	Cost (CAPEX)	Price Support Last Awarded[8]	Targets[9]
Baltic	Denmark	0.9	0.9 (3%)	Lower costs than Atlantic basin, as	€103.1/MWh[11]	1.4GW by 2020 (covering both Atlantic and Baltic basins)
	Germany	0.35	0.3 (1%)	sites are less challenging	€150–180/MWh[10]	10GW by 2020 (covering both Atlantic and Baltic basins)
	Sweden	0.2	2.6 (10%)		Tradable Green Certificates	0.2GW by 2020
	Finland	0.03	0 (0%)	Lower wind	€83.5/MWh[15]	No offshore specific target[16]
	Poland	0.0	0.0 (0%)	speeds lead to lower yields	Tradable Green Certificates	500MW by 2020
	Estonia	0.0	0.0 (0%)		System in the process of reform	Overall 2020 renewables targets are expect to be exceeded.
Mediterranean	Italy	~0	0.03 (0.1%)	Likely to be similar to the Baltic basin	Tradable Green Certificates moving to a FIT in 2016	0.7GW by 2020
	Greece	~0	0.0 (0%)		€108.30/MWh[17]	No offshore specific target
	France	~0	0.0 (0%)		FIT of €130/MWh, or competitive tender[13]	6GW by 2020 (covering both Atlantic and Baltic basins)
Caribbean		~0	n/a	n/a		

[15] The FIT price set for wind power (onshore and offshore). There is no offshore wind specific subsidy.

[16] Target of 2.5GW for onshore and offshore wind by the end of 2020.

[17] An increase of up to 30% on the base FIT for offshore wind is available on a project-by-project basis if the Regulatory Authority for Energy deems it required for the level of investment required.

farms have primarily been developed as stand-alone projects. In the next five years this is unlikely to change. As markets and technology matures, especially where the state develops the site and then auctions it, leading developers such as DONG and Vattenfall are already adopting a 'pipeline' approach to site selection, progressive technology and procurement decisions.

5.3.1 Lifecycle

The offshore wind lifecycle can be classified into five main stages:

1. Development and consenting,
2. Final investment decision (FID),
3. Supply, installation and commissioning,
4. Operations, maintenance and service (OMS), and
5. Decommissioning.

Development and Consenting

Wind farm development and consenting covers work on the offshore wind farm from the point of site identification, to FID. Processes for completing activities vary widely between countries and basins. Here they are described typical to the UK and German markets. The main activities undertaken include:

- Site identification to establish areas of seabed suitable for wind farm development. This is typically undertaken by a leasing body, such as UK's The Crown Estate and Germany's BSH (Federal Maritime and Hydrographic Agency).
- Front-end engineering and design (Pre-FEED and FEED) studies to identify and address areas of technical uncertainty and develop the concept design of the wind farm in advance of contracting. The developer will use specialist subcontractors for specific activities like preliminary foundation design. Key parameters such as turbine rating, foundation type, wind farm layout, and grid connection method are considered to optimise economic viability. Onshore and offshore operation strategy is formed and procedures are planned, contracting methodologies determined, and key risk management and health and safety policies developed. Construction management teams use the studies to implement the wind farm.
- Wind farm design, which includes input from the FEED studies, wind modelling and turbine wake analysis, array optimisation, and wind

resource assessment. The developer typically completes most wind farm design in-house but places contracts with specialist engineering firms for key component design. Meteorological stations are often installed at wind farm sites at an early stage of development to monitor meteorological and oceanographic conditions.

- Surveys are typically contracted by the developer to specialist data acquisition companies. Surveys include environmental surveys (benthic and pelagic, marine mammal and ornithological), coastal process (sedimentation and erosion impact) and geotechnical and geophysical surveys.
- Stakeholder engagement is undertaken by the developer in parallel with the wind farm design and surveys. Stakeholders engaged included statutory bodies, non-statutory bodies, businesses and members of the public.
- Consenting is the process of regulatory approval for offshore works and grid connection. This is a process that varies between countries and basins.
- Procurement, which is the process of the developer contracting work packages for the supply and installation of components. Potential suppliers are qualified and progress through a bespoke selection process.

Final Investment Decision
FID is defined as the point of a project life cycle at which all consents, agreements and contracts required to commence project construction have been signed (or are at or near execution form) and there is a firm commitment by equity holders and debt funders to provide funding to cover the majority of construction costs.

Supply, Installation and Commissioning
Supply, installation and commissioning is the period where the offshore wind farm components are manufactured, installed, fully grid connected and brought into operation. Installation and commissioning covers work on all balance of plant as well as turbines. It can be broken down into the following areas: transport of completed sub-assemblies from manufacturing facilities; installation port facilities preparation and marshalling of sub-assemblies; foundation installation; array and export cable installation; offshore sub-station installation; turbine installation and commissioning; and sea-based support.

Operations, Maintenance and Service

A typical offshore wind farm is expected to have an operating lifetime of around 20 to 25 years, during which time maintenance and minor service and major service activities will take place including:

- Operation relating to the day-to-day control of the wind farm, including minor spares and consumables;
- Condition monitoring;
- Rental of the operations base, port facility, mother ship and crew transfer vessels;
- The repair or replacement of minor components using the wind farm's normal staff and equipment;
- The repair or replacement of major components that cannot be undertaken using the wind farm's normal staff and equipment;
- The use of any additional vessels required to repair faults;
- The implementation of improvements to equipment.

Decommissioning

No commercial scale offshore wind farm has yet been decommissioned, however some single turbines and small projects have been decommissioned. There is a lot of uncertainty about the process. Generally it is assumed that turbines and transition pieces will be removed with foundations cut off at a depth below seabed which is unlikely to lead to uncovering. Cables are likely to be pulled up, due to the recycling value. Environmental monitoring will be conducted after the decommissioning process. It may be that some wind farms will be repowered using new foundations, array cables and turbines, re-using most transmission and grid infrastructure.

5.3.2 Economics

The cost breakdown of a typical offshore wind farm can be classified into five main areas:

1. Development and project management up to the end of commissioning (2%)
2. Wind turbine supply (26%)
3. Balance of plant (19%)
4. Installation and commissioning (14%)
5. Operation, maintenance and service (39%)

These costs represent a breakdown of undiscounted capital and operational costs of a typical 500MW wind farm using 6MW turbines on jacket foundations, with a 20 year operating life.[18]

5.3.3 Supply Chain

The offshore wind supply chain is increasingly formed from companies also active in other sectors. The participation of North Sea oil and gas companies is lower than many anticipated although many offshore wind personnel will have had experience in oil and gas. The offshore wind farms have multiple units spread over a wide area (each turbine is 1 km to 2 km apart) and this better suits other onshore and offshore construction sectors, such as dredging, aggregates and harbour construction. The characteristics of companies in each element of the supply chain are shown in Table 5.3.

Table 5.3   Leading companies in the offshore wind market and their characteristics

Element of Supply	Leading Companies	Characteristics
Developers	DONG, EnBW, E.ON, Iberdrola, Innogy, Vattenfall, WPD	Large multinational utilities with a strategic focus on renewables
EPC contractors	Boskalis, DEME, Van Oord, VolkerWessels	Dominated by Dutch and Belgian dredging companies; companies typically have their own vessels; oil and gas contractors not successful
Turbine nacelles	Adwen, GE, MHI-Vestas, Senvion, Siemens Gamesa	Offshore wind joint ventures (Adwen, MHI-Vestas) and engineering conglomerates. Specialist wind companies such as Senvion struggling to compete
Turbine blades	Adwen, LM Wind Power, MHI-Vestas, Senvion, Siemens Gamesa	Mostly produced in house by turbine suppliers; potential role for an independent
Turbine towers	Ambau, Titan, Valmont	Suppliers to offshore and onshore construction industry
Foundations	Ambau, Bladt, EEW, Navantia, Sif, Smulders, Steelwind Nordenham, ST^3	Suppliers to offshore and onshore construction industry. Market mostly monopiles and split between steel rollers (EEW, Sif) and fabricators (Bladt, Smulders)

---

[18]The costs are a combination of real project and modelled data. The operations, maintenance and service includes the maintenance of transmission assets. The cost of building the transmission assets is included in balance of plant.

Substation electrical	ABB, Alstom, CG Power, Siemens	Large multinational high voltage electrical suppliers. May include the offshore substation structure supply and installation within their scope
Offshore substation platforms	Bladt, Fabricom, Harland & Wolff, STX	Large fabricators usually involved in oil and gas and shipbuilding industries
Cables	ABB, JDR, Nexans, Prysmian, NKT	At medium voltage (array cable) companies supplying wind and oil and gas. At high voltage (export cable) companies also supplying subsea interconnector market
Turbine installation	A2SEA, MPI Offshore, DEME, Fred Olsen, Swire Blue Ocean, Van Oord	Mostly specialist wind companies with some having the capability to work in oil and gas. Includes EPC contractors which are active in inshore construction and dredging
Foundation installation	A2SEA, MPI Offshore, DEME, Fred Olsen, Swire Blue Ocean, Van Oord, Seaway Heavy Lifting	As for turbine installation but market leaders are different with the fleets of some operators better suited to foundation installation (typically larger cranes are needed for foundation installation)
Cable installation	Jan de Nul, Siem, VolkerWessels-Boskalis, Van Oord	Usually contractors are different from turbine and foundation installers. Cable vessels also used in oil and gas umbilical laying and in interconnector installation
O&M vessels	Njord Offshore, Seacat, Windcat Workboats	Large number (>30) of specialist operators of crew transfer vessels (up to 26m). Many turbine installation contractors also operate such vessels. Increasing use of service operations vessels (about 90m) permanently stationed offshore, often taken on long-term charter from offshore fleet owners that supply other offshore sectors

5.4 Working Environment

The European Wind Energy Association (EWEA) published a skills gap analysis for the onshore and offshore industries in 2013.[19] It identified skills shortages are likely to be greatest in operations and maintenance roles, though due to the long gestation time for projects, such needs will be known at a project level 2 years before the jobs are needed. In 2010, The Crown Estate

---

[19]EWEA (2013), Workers wanted: The EU wind energy sector skills gap available online at http://www.ewea.org/fileadmin/files/library/publications/reports/Workers_Wanted_TPwind. pdf, last accessed August 2017.

commissioned a careers guide in the offshore wind industry (the guide covers the UK industry, but is relevant to any offshore wind industry).[20] Wage costs generally vary given the vast range of roles in the industry and between basins. The operations, maintenance and service of a 500MW wind farm supports about 400 to 500 FTE jobs.[21]

5.5 Innovation

5.5.1 Atlantic, Baltic and Mediterranean Basins

Innovations are relevant to both Atlantic and Baltic basins where offshore wind is commercially deployed and most mature. Such innovations will be used in the Mediterranean in due course but because there are limited shallow water sites, innovation in floating foundations is likely to have the biggest impact. Offshore wind is still considered as a high technical innovation sector and funding into technological advancement is still significant. The strong focus is to reduce cost of energy for the sector to become more cost-comparative with other renewable and fossil fuel energy generation. Key technology and trends includes:

- Innovations in larger rated turbines. Turbines installed to date have typically had rated capacity of 6MW or below. There are now 8MW turbines installed in a commercial wind farm. Technical innovations are being made for the different components of a turbine. For example, SSP and LM Wind Power, have both received government funding to develop blades in excess of 88m long for offshore wind turbines.
- Innovations in foundation design. Steel tubular monopiles have been the standard foundation choice for projects using 4MW turbines in water depths of up to 25m but industry expected that they would become less cost effective than other foundation types (such as jackets) with larger turbines and deeper water depths. This was a problem because

---

[20]BVG Associates (2010), Your career in offshore wind energy, on behalf of The Crown Estate (with RenewableUK) 32 pp. available online at https://bvgassociates.com/publications/, last accessed August 2017.

[21]Value breakdown for the offshore wind sector, A report commissioned by the Renewables Advisory Board, [RAB (2010) 0365] (2010), BVG Associates for Renewables Advisory Board, available online at https://www.gov.uk/government/uploads/system/uploads/attachment_data/file/48171/2806-value-breakdown-offshore-wind-sector.pdf, last accessed August 2017.

there was a proven supply chain for monopile production and installation and developers were relatively unfamiliar with the new foundation designs. In recent years, however, strong industry collaboration between developers, designers, suppliers and installers has meant monopiles have remained the most cost effective option in much more challenging conditions than previously expected. So-called "XL" monopiles have now been used with 6MW turbines in water depths of up to 35m deep water and for Burbo Bank Extension, with 8MW turbines in 10m water depth. Monopile foundations are planned to be used further with 8MW turbines in deeper water. These monopiles may be up to 10m in diameter, 120m long with a plate thickness of up to 112mm and a mass of more than 1,500 tonnes. This innovation has been based on detailed performance data gathered from existing projects that has allowed designers to stretch the design envelope of the structures.

- Alternative foundation. Where the combination of turbine size, water depth and soil conditions mean monopiles are not the most cost effective solution, jackets are currently the preferred alternative foundation in waters up to about 50m depth, though concrete solutions have been used in some cases. A BVG Associates study looking at how technology innovation is anticipated to reduce the cost of energy from offshore wind farms stated that most innovations in balance of plant are centred on improvements in the manufacture and design of jacket foundations.[22] To date, jacket production has been influenced by one-off or low volume production practices from the oil and gas sector. As the growth of the offshore wind market continues, new fabrication facilities are being developed for example by ST^3 that are more optimised for serial fabrication. The report also identifies that an introduction of commercial scale suction-bucket foundation technology could reduce LCOE. To date, it has typically been demonstrated on small, close to shore turbines, but could be used on up to 25% of projects with FIDs in 2025.
- Floating foundations are also being proposed for deep water offshore wind farms, which generally are likely to be close to shore to take benefit of lower transmission costs. Wind turbines with floating foundations are at the demonstration stage with the Japanese Fukushima floating demonstrator phase two project the first to install a 5MW and 7MW class

---

[22] KIC InnoEnergy, (2014). Future renewable energy costs: offshore wind – How technology innovation is anticipated to reduce the cost of energy from European offshore wind farms. A report by BVG Associates. 80 pp., available online at: https://bvgassociates.com/publications/, last accessed August 2017.

turbine on a floating foundation beginning in 2015. In August 2017, the Hywind demonstrator of 5 turbines each rated at 6MW with Spar Buoy foundations was installed by Statoil in the Buchan Deep off Scotland. In the Atlantic basin, Norway, Portugal, Wales, and Scotland have waters suitable for wind farms with floating foundations but the Atlantic basin has plenty of cheaper shallower sites likely to be developed first with fixed foundations. The Mediterranean basin has fewer potential shallow sites so could most benefit from wind farms with floating foundations. Four demonstration sites off the French Atlantic and Mediterranean coast are in development.

- Innovations in high voltage alternating current (HVAC) subsea cables. HVAC technology is being used for wind farms located further from shore due to innovations in the technology. It was anticipated HVDC cables would be used as export cables as the most cost efficient means to transport electricity back to shore; although the cables are more expensive than HVAC per km, they have fewer electrical losses over longer distances. However innovations in HVAC cable have reduced the amount of electrical losses when using this type of cable over longer distances, making it a more cost effective technology compared to HVDC over greater distances than previously anticipated, though HVDC technology is also progressing quickly.

Cross-sectoral innovation may include the combination of offshore wind and aquaculture. There have been several studies into the applicability of this cross-sector growth.[23,24,25]

5.5.2 Caribbean Basin

Unlike the Atlantic, Baltic and Mediterranean basins, the Caribbean basin is subject to hurricanes. This may mean that standard turbine technology may

---

[23]M. Syvret, (2013), Shellfish Aquaculture in Welsh Offshore Wind Farms The Potential for Co-location (2014), available online at: http://www.thefishsite.com/articles/1918/shellfish-aquaculture-in-welsh-offshore-wind-farms-the-potential-for-colocation/, last accessed August 2017.

[24]L. Mee (2006), Complementary Benefits of Alternative Energy: Suitability of Offshore Wind Farms as Aquaculture Sites Inshore Fisheries and Aquaculture Technology Innovation and Development, SEAFISH – Project Ref: 10517, available online at: http://www.seafish.org/media/Publications/10517_Seafish_aquaculture_windfarms.pdf, last accessed August 2017.

[25]J. Allard (2009), Symbiotic relationship: aquaculture and wind energy? available online at: http://ecologicalaquaculture.org/Allard%282009%29.pdf, last accessed August 2017.

be limited to areas with low hurricane risk in the Caribbean basin. Elsewhere, it will either need to be adapted, perhaps allowing turbines to be lowered, or an alternative technology used such as kites which can be stowed away. Kite technology is being developed but it is at an early stage of readiness with KPS due to operate a 500kW kite in 2018. There have been several reports attempting to quantify the hurricane risk to offshore wind turbines. One simulated around 50% of offshore turbines being damaged by hurricanes during a 20-year operational life.[26] In 2014, the US National Oceanic and Atmospheric Administration (NOAA) has collected hurricane data with the aim of improving offshore wind turbine designs.[27]

5.6 Investment

Investments into offshore wind industry can be generally categorised into three profiles:

1. **Project acquisition and capital ventures:** If a wind farm is owned by several owners in a subsidiary joint venture (JV) company, one usually assumes a lead developer role on the project. The owners making up the JV may have equal or different shares in the project. Acquisition of these shares can take place at any stage of a project lifecycle. Typically, if a project share is acquired at the pre-construction stage, it is by another developer. For example, Dong acquired the remaining 66.6% of 1.2GW Hornsea 1 from its JV partner SMart Wind (a 50/50 joint venture between Mainstream Renewable Power and Siemens Financial Services). Acquisitions into operational wind farms are more likely to be from a wider variety of investors such as pension funds and private investment firms with the original developer, usually a utility, retaining a significant share of the project. For example, in January 2014, La Caisse de Dépôt et Placement du Québec (financial institution managing funds primarily for Québec's public) acquired a 25% share in the operational London Array 1 for £644 million. The project's other partners are E.ON (30%), Dong Energy (25%) and Masdar (20%).

---

[26]S. Rose, P. Jaramillo, M. J Small, I. Goodman and J. Apt (2012), Quantifying the hurricane risk to offshore wind turbines, available online at: http://www.pnas.org/content/109/9/3247.full.pdf, last accessed August 2017.

[27]US storm chasers on a mission (2014), available online at: http://renews.biz/68177/us-storm-chasers-on-a-mission/ last accessed August 2017.

2. **Company mergers and acquisitions:** Offshore wind is a dynamic market where attrition within the supply chain is expected due to the high-risk work and investment required. As a result, many company mergers and acquisitions are seen. For example, in 2017 Siemens and Gamesa merged to form Siemens Gamesa, and in 2014 Mitsubishi Heavy Industries (MHI) and Vestas merged to form MHI Vestas Offshore Wind (MVOW). Within lower tiers of the supply chain, we are seeing company acquisitions. This is shaping a market with fewer companies within the supply chain, but each with greater capability, and greater commitment to the industry and appetite to take on the associated costs and risks.

3. **Technology funding:** Offshore wind is considered as a high technical innovation sector and funding into technology advancement is still significant. Most research is funded by the wind turbine manufacturers (WTM's) for in-house technology developments but funding can be provided by a range of organisations, such as the European Commission and UK's Department of Energy and Climate Change (DECC) to other supply chain companies. R&D funding programmes have a wide range of fund totals. For example The Department of Business, Innovation and Skills 2012–2013 Regional Growth Fund Round 3, had a total fund of £1050 million. In comparison, DECC's 2011–2014 Offshore Wind Component Technologies Development and Demonstration Scheme 1 had a total fund of £5 million. Public funding can be provided to a single company (for example, DECC provided funding to Blade Dynamics' Composite Hub Technology Demonstration project under the Offshore Wind Component Technologies Development and Demonstration Scheme), or to a collaborative project comprised of several members from industry (for example, the European Commission provided funding to Gamesa Innovation and Technology and nine other project partners to undertake 'FLOATGEN: Demonstration of two floating wind turbine systems for power generation in Mediterranean deep waters' under the European Union Seventh Framework Programme).

Financial consultants play a key role in these three profiles. Merger and acquisition advisory services offered by consultants include project modelling, valuation, transaction services, due diligence and post-acquisition integration services.

5.7 Uncertainties and Concluding Remarks

Offshore wind is a significant industry within the Atlantic and Baltic basins, and there is high confidence these basins will continue to be a focus for future activity. These markets have relatively low electricity prices so output from offshore wind farms are currently explicitly subsidised. This level of price support has substantially reduced in 2016 and 2017 and grid parity with combined-cycle gas turbines for the better wind projects is likely to occur at some point in 2023 to 2025 assuming current views of likely future carbon pricing.

The Mediterranean and Caribbean basins do not currently have any commercial offshore wind installations. The Mediterranean could capitalise on its proximity to the established markets in the Atlantic and Baltic to develop a commercial market, but this is not likely by 2020 due to the limited number of projects currently in development. The Caribbean could capitalise on synergies in the established oil and gas industry. Despite relatively high cost of electricity from comparable new electricity generating plant, limited annual electricity demand may limit the ability to establish a cost competitive market beyond a few projects. Combining with another sector might help enable a bigger market.

Table 5.4 summarises each basin's current offshore wind activity and future opportunities.

Table 5.4   Offshore wind subsector summary

Basin	Summary	Opportunities
Atlantic	Focus of current activity. High confidence that basin will be a focus for future activity. Commitment to the sector by European governments. **Recommended basin for future offshore wind.**	Commercial activity at a scale that innovation in technology and competition in supply chain can reduce lifetime cost of energy. Expansion of installations within basin to English Channel and Bay of Biscay.
Baltic	Focus of current activity. High confidence basin will be a focus for future activity. Commitment to the sector by European governments. **Recommended basin for future offshore wind.**	Commercial activity at a scale that innovation in technology and competition in supply chain can reduce lifetime cost of energy.

(*Continued*)

Table 5.4   Continued

Basin	Summary	Opportunities
Mediterranean	No current activity beyond early stage development of test sites. Potential sites tend to be in deeper water than the Atlantic and Baltic so technology will need to be further developed to suit. **Limited evidence to show a significant market will be established due to current higher price of floating foundations and limited market support available from governments.**	Close to existing Atlantic and Baltic supply chains which could support future activity.
Caribbean	No current activity. Limited evidence to show a significant market will be established in the future. There is a significant risk of damage by hurricanes. Cost of offshore wind energy is tied to the scale of wind farm. Trinidad and Tobago has a population of 1.3 million whose total annual electricity consumption is equivalent to the output of a 400MW offshore wind farm. Only Cuba, Haiti, Dominican Republic, Puerto Rico (US) and Jamaica have larger populations.	Potential synergies with oil and gas supply chain/existing infrastructure in Gulf of Mexico. Quantification of the impact of hurricanes is needed across the basin to see if existing technology can be deployed. Innovations in turbine design are probably needed to reduce exposure to damage to acceptable levels.

PART II

The Blue Economy Sectors

6

Fisheries

Kate Johnson

Heriot-Watt University, Scotland

6.1 Introduction

The catching of wild fish is one of the two oldest maritime industries, the other being shipping. Its roots are lost in pre-history and it remains pre-eminent today in its spatial and social impacts. For centuries fishing has been a cornerstone of the Blue Economy. In recent times it has been of less significance in monetary value when compared to other maritime industries such as offshore oil and tourism. However, catch fisheries remain enormously important in terms of employment and subsistence to coastal communities everywhere. They are the largest maritime employer by an order of magnitude, over five times their nearest rival [OECD 2016]. Arguably, of all maritime human activity, they have the greatest impact on the environment. The ancient rights of individuals to navigate and fish the oceans and seas are under pressure from overfishing, illegal fishing, market competition from farmed fish and spatial exclusion due to new industries and conservation areas.

In helping to feed the world, fish are an important source of animal protein. Consumption of fish exceeded a global average of 20 kg per person per year for the first time in 2014 although recent growth in consumption has been supported entirely from aquaculture. Wild fish consumption has flat lined at an annual rate of about 10 kg per person for some time. Farmed fish consumption, on the other hand, has soared from next to nothing in 1974 to around 10 kg or just over 50% of total consumption in 2014. In nearly half the countries of the world, fish contribute more than 20% of protein in the diet [FAO 2016]. As the largest maritime employer, the social importance of

205

catch fisheries exceeds its commercial importance by quite a wide margin. The artisanal and subsistence fishing sector is very large. The OECD [2016] estimate global catch fisheries employment in the commercial sector at more than 11 million but, when the artisanal sector is added, total employment in all wild fisheries is estimated by the FAO [2016] at nearly 40 million. More are employed in processing and support industries. The artisanal sector is catching for subsistence and for small scale commercial sales where possible.

Global fisheries policies are aimed at food security and a sustainable level of stocks for the future and farmed fish are seen as one way to reduce pressure on the wild stocks. However, levels of illegal, unregulated and unreported (IUU) fishing are high, including elements from both commercial and artisanal sectors [FAO 2016]. Fish catching is a free roaming activity where access is of critical importance but environmental damage is caused, from trawls for example. Public rights to navigate and fish the ocean commons are enshrined in international and national law but pressures on stocks create ever more stringent management measures in response. At the same time, new maritime industrial sectors and platforms for energy and aquaculture require exclusive use of marine space with inevitable consequences of displacement for fishers. The sector is therefore under scrutiny and facing increased restriction. However, its social importance attracts high levels of political support often combined with strong national or regional feelings about boundaries and rights.

This chapter describes the fishing industry and its role in the Blue Economy. As a start, Section 6.2 examines the market for fish including the policy ambition of food security and the consumption of fish as an important part of diet. Section 6.3 reviews the structure and lifecycle of the industry highlighting the differences in the business models of the various sectors. It also reviews regulatory and management measures as drivers in strategy. Section 6.4 describes the working environment of fisheries, including safety, because fishing is one of the most dangerous jobs in the world, the nature of fishing communities and the pressure on labour supply. Section 6.5 is about innovation, reducing costs and boosting productivity while at the same time coming to terms with extensive and complex regulatory frameworks aimed at sustainability. Section 6.6 looks at investors and investment, private and public. Some sectors of the industry continue to attract substantial private investment into large and technologically advanced vessels and methods. Extensive public investment is aimed at monitoring and control which supports the industry in a sustainable future for fishers and the world at large. Section 6.7 summarises the uncertainties and makes concluding remarks.

The global fishing industry is huge in numbers and complexity. This chapter has a focus on Europe but set in an international context. Specific area examples are used to exhibit generic points. The chapter tries to capture the essence of fishing as a business within the expanding maritime economy.

6.2 Market

6.2.1 The Demand for Fish

The market for catch fisheries is driven by the demand for food and is a valuable source of animal protein as shown in Figure 6.1. World fish production, catch and farmed, contributes about 180m tonnes per year while meat production (chicken, cattle, pig, and sheep) totals around 265m tonnes per year. It is supported by the availability of catch and the accessibility of catching areas in the marine commons. The pre-eminence of aquaculture in recent fish consumption records reflects the very high figures from Asia which

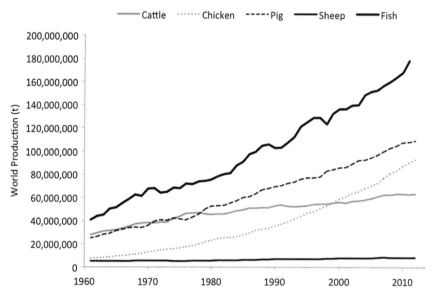

Figure 6.1 World Production of the main sources of animal protein over the period 1960–2010.

Source: FAO Statistics.

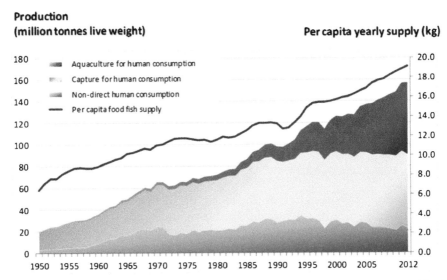

**Production
(million tonnes live weight)**   **Per capita yearly supply (kg)**

Figure 6.2   Relative contribution of aquaculture and capture fisheries to production and food fish supply.

Source: FAO Statistics.

is the leading fish producing and consuming area in the world Figure 6.2. Aquaculture in Asia supplies nearly 60% of consumption there, while for the rest of the world and Europe it is more in the range 12%–20%. Greenland leads per capita fish consumption with more than 60 kg/year for each resident; Norway, France, Spain and China follow in the range 30 kg to 60 kg; North America, Russia, Australia, UK and most of SE Asia consume 20 kg to 30 kg; Central Europe, Latin America, India and most of Africa eat 2 kg to 20 kg while countries like Afghanistan, Sudan and Ethiopia get by on less than 2 kg of fish per person per year.

6.2.2 Pressures on Price

The world population has more than doubled in the period from 1950 to 2014, from three to seven billion, while fish utilisation has more than trebled, from 40 to 130 million tonnes, in the same period. The FAO catch fish price index hovered around the base of 100 for a decade or more before surging to 160 in the period 2004–2014 and falling back to about 140 in 2016 (Figure 6.3). The farm fish index has fluctuated around 120 for three decades but is also currently at about 140. The global market for fish appears strong and has been growing for several decades. Prices are also strong. However, future growth

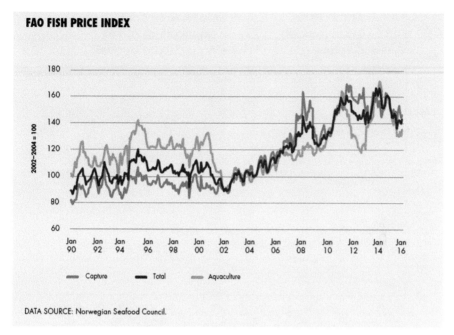

Figure 6.3   FAO Fish Price Index 1990–2016.

Source: FAO 2016.

of the catching industry is uncertain and dependent on several factors. A key question is the extent to which aquaculture and farmed fish will substitute for catch fish. The sectors are inextricably linked as some catch fish are also processed as fish meal for farm fish feed. The proportion of farm fish consumed has grown every year from almost nothing in 1974 to a world average of 50% today and continues to grow. Capture production plateaued in the range 80/85 million tonnes in 1985 and shows no signs of an upturn (Figure 6.4). Pressure on the catch industry also comes from increased sustainability measures and management of wild stocks. In Europe this includes market pressure with major retailers demanding Marine Stewardship Council (MSC) sustainability accreditation, or similar, in the sourcing of the fish they buy and sell [Bell *et al.* 2015]. The MSC is global in its reach [MSC 2016].

6.2.3 Trade across the World

International trade in fish is strong with flows and counter flows of product often reflecting differences in national tastes for certain species. Historic and cultural links to diet are hard to break. Velvet crabs caught in Scotland,

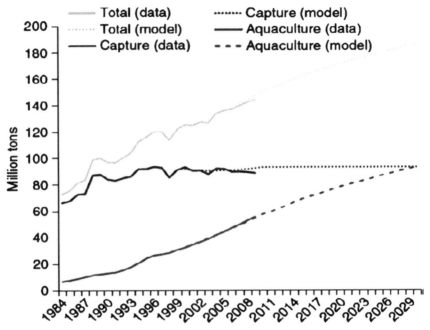

Figure 6.4   Global fish production: Data and projection 1984–2030 from the IMPACT model.

Source: World Bank 2014.

where they are rarely eaten, are exported almost in their entirety to Spain where they are an everyday food. Sea cucumbers from around the world are exported to China and surrounding countries where they are highly valued, the only place in the world where they are. Cod has especial value in UK for fish and chips and salted in Portugal for bacalhau. The strongest exporting region of the world for fish and fish products by value is Europe at over US$50bn. However, it is also the largest importer of fish and fish products at US$60bn. Fish and related products are one of the most traded segments of the world food sector with about 78% of seafood products estimated to be subject to international trade competition. In several countries the fishery is economically essential; in Greenland and Iceland it represents more than 40% of all traded commodities. In overall terms the fishery trade in 2014 was 9% of global exports in the fisheries and agriculture sector and 1% of all world trade [FAO 2016]. The trade values include a wide range of segments essential to the industry including management, harvesting, processing, monitoring, port services, maintenance, crew supply and training, vessel charter, infrastructure and research. Demand for fish and the globalisation of trade, together with improved transport and technology, have all contributed towards a geopolitical role in enforcing these global trends. World trade in fish and related products has increased in the value of exports from about US$72 billion in 1976 to over US$148 billion in 2014 (Figure 6.5).

Catch fish compete in the market with other forms of protein both animal and vegetable. Taste for particular species plays an important role and there will always be a limited demand for high quality high price fish such as tuna and turbot. However, at the general level, the bulk protein

TOP TEN EXPORTERS AND IMPORTERS OF FISH AND FISHERY PRODUCTS

		2004	2014
		(US$ millions)	
EXPORTERS	China	6 637	20 980
	Norway	4 132	10 803
	Viet Nam	2 444	8 029
	Thailand	4 060	6 565
	United States of America	3 851	6 144
	Chile	2 501	5 854
	India	1 409	5 604
	Denmark	3 566	4 765
	Netherlands	2 452	4 555
	Canada	3 487	4 503
	Top ten subtotal	34 539	77 801
	Rest of world total	37 330	70 346
	WORLD TOTAL	71 869	148 147
IMPORTERS	United States of America	11 964	20 317
	Japan	14 560	14 844
	China	3 126	8 501
	Spain	5 222	7 051
	France	4 176	6 670
	Germany	2 805	6 205
	Italy	3 904	6 166
	Sweden	1 301	4 783
	United Kingdom	2 812	4 638
	Republic of Korea	2 250	4 271
	Top ten subtotal	52 119	83 447
	Rest of world total	23 583	57 169
	WORLD TOTAL	75 702	140 616

Figure 6.5 Top ten exporters and importers of fish and fishery products in 2004 and 2014.

food market is working in the range of €15–€25 per kg for catch fish, farmed fish, beef and lamb. Chicken and vegetable proteins can be considerably cheaper in the market with chicken offered for as little as €5 per kg [https://ec.europa.eu/agriculture/index_en]. Public taste for fish is stubbornly fixed on traditional but threatened species such as cod, and reluctant to move to more prolific species such as mackerel and farmed varieties such as Vietnamese tilapia. Education, marketing and innovation in product design, such as supermarket ready meals, are key tools in retaining or increasing market share but price remains a key metric. In relation to farmed fish, catch fish have the advantage of a far greater range of species. Both catch and farm industries face significant problems of sustainability for different reasons – catch because of the problems of overfishing, and farmed because of environmental pollution and disease.

6.3 Structure and Lifecycle

6.3.1 Sectors

The catch fishing sector can be divided and classified into a number of sub-sectors each with its own characteristics and lifecycle. A major division is between commercial, fishing for the market, and artisanal, fishing for subsistence and small scale sales. The commercial fishery, the main subject of this chapter, can be classified in numerous ways including location, target species, catching method and equipment. Particular locations considered in this book are the EU sea basins of the Atlantic (Chapter 10); North and Baltic Seas (Chapter 11); Mediterranean and Black Seas (Chapter 12); and Caribbean Basin (Chapter 13). These broadly correlate to the fishing areas coded by the FAO with the catch data listed in Table 6.1. The European Atlantic area is the most significant of the European areas for catch measured by weight of fish landed. Differences in the businesses related to fishery sub-sectors are illustrated by reference to pelagic, demersal and shellfish species. Each of these groups hosts a fishing industry of different character in respect of impact, value and employment.

  a. **Pelagic Sector (mid-water species).** Typical target species for the pelagic fleets in Europe are herring, mackerel and blue whiting. The pelagic fleets tend to the offshore and employ the largest and most valuable vessels costing in the range £10–20 million each. They are relatively few in number with lengths up to 80 m, displacement of 3000 tonnes and power sometimes in excess of 6000 kw. Taking the

Table 6.1 Wild Fish capture by sea basin

Location	Marine Capture 2014	Major Fishing Countries by Basin (Rounded Figures)
Atlantic, North East (FAO Code 27, includes North and Baltic Seas)	8.7 million × tonnes	Norway (2,300,000t) Iceland (1,100,000t) Spain (1,000,000t) Denmark (700,000t) UK (700,000t) France (500,000t)
Mediterranean and Black Seas (FAO Code 37)	1.1 million x tonnes	Italy (114,000t) Algeria (97,000t) Tunisia (91,000t) Spain (73,000t) Croatia (73,000t) Turkey (58,000t)
Atlantic, Western Central (FAO Code 31, includes Caribbean and around)	1.2 million × tonnes of which about 50% (0.6mt) are sourced from the Caribbean Basin	(Caribbean Basin only) Mexico (190,000t) Venezuela (119,000t) Guyana (48,000t) Suriname (39,000t) Jamaica (25,000t)
WORLD total all species all areas	81.5 million × tonnes	China (14,800,000t) Indonesia (6,000,000t) United States (5,000,000t) Russia (4,000,000t) Japan (3,600,000t) Peru (3,500,000t)

Source: FAO 2016; FAO Statistics 2015.

UK as the example, there are only twenty pelagic vessels, of over six thousand in total registered for all fishery sectors. The twenty vessels are registered in just two Scottish ports – Peterhead and Lerwick. In Scotland, the pelagic catch in 2015 was 291,500 tonnes which represents 66% of the total Scottish catch of 440,000 tonnes. The total Scottish catch is 63% of the whole UK catch of 702,000 tonnes for all species [MMO 2017]. Quota and other management restrictions meant that these twenty pelagic vessels only spent about one month each at sea last year. Not all the Scottish catch is landed in the UK and in addition there is an active import and export business. For example in 2015, 95,000t of mackerel were landed into the UK in addition to which 19,000t were imported. 80,300t of the total were exported with 40% of exports going to the Netherlands. Typical methods and gear include purse seine netting, mid-water trawling or pair trawling. The pelagic fishing industry is the largest by value in Europe, it is big business, and it is concentrated into very few hands.

b. **Demersal or Whitefish Sector (bottom living species).** European targets include cod, haddock and plaice. Fishing methods tend to bottom trawling although hook and line is increasingly promoted for more sustainable fishing. The demersal fishery is working closer to shore and is more likely to be in conflict with static gear fisheries and other activities. It is difficult to draw a distinction between the major commercial demersal vessels and the smaller shellfish vessels which take a significant quantity of finfish in total by virtue of their numbers alone. Taking the example of the part of the Shetland fleet which is defined as demersal, all the vessels are over 20m, and so this delineation is taken here. By this definition about 500 of the 6200 UK registered vessels represent the demersal fleet. These vessels normally fall into the range of 20m–30m in length with registered tonnage of 200t–300t each. Power is in the order of 600 kw. The demersal fishery is the second largest by value in Europe and again is concentrated into few hands in relatively well organised businesses.

c. **The European Inshore Fleet.** The inshore sector of the industry is mainly a shellfish fleet with targets of non-quota species such as lobster, crab, scallops and nephrops. Finfish will also be targets where possible and quotas allow. This is the largest fleet by far numbering about 80,000 across Europe with the great majority of vessels less than 10m in length. Greece has the largest fleet of about 16,000 vessels. In the UK 5700 vessels (of 6200 registered vessels in total) fall into the less than 20m category. Nearly 5000 of these are less than 10m. This is a very different fishery business model in comparison to the demersal and pelagic sectors. It is dominated by owner operators often working by themselves and frequently without crew. The vessels and their equipment are low cost and low tech. Fishing methods include static traps, dredges for molluscs, trawls for nephrops, line fishing for finfish and gill netting. These fishers tend to be poorly organised as a group and less easy to regulate as a result. The majority will not join fisheries associations or federations which they see to be dominated by the 'big fishers' [Noble 2003]. Fishing cooperatives are common whereby fishers in an area band together for purposes of self-management enforced by peer pressure and marketing of product. There are some government supported schemes, such as the 'Regulating Order' (RO) scheme in Scotland, which offers groups of fishers the statutory right to manage their own fishery. The costs of management and rivalry among fishers are an obstacle and this measure has so far only been adopted in Shetland. The inshore sector is often important, and sometimes essential, to vulnerable and peripheral coastal communities feeding into all aspects of way of life and culture. It is also the fishery with least power and participation in regulatory and institutional structures, especially at higher levels of EU governance.

The pelagic, demersal and inshore fisheries sectors exhibit three very different business models and are at different stages in their respective lifecycles, although all may be regarded as mature or even post-mature. The high value pelagic and demersal sectors are big business with a large investment in vessels and equipment. They are capital intensive and are often highly geared with high levels of debt and susceptible to variances in cash flow. They have to keep fishing to service their debts as well as showing significant profits to shareholders and for re-investment. Adequate stocks and regulatory restrictions are their main concerns but they are politically powerful and 'own' most of the available quota for key species. Data about their activities are detailed with programmes of monitoring and evaluation. They range over huge distances in pursuit of stocks which roam freely across national boundaries and into areas beyond national jurisdiction (ABNJ) aka 'the high seas'. They have freedom to move and are the least spatially affected by other maritime activities and 'Blue Growth' ambitions. They employ few people at sea but by virtue of their catch volume but they create substantial shore based employment at the major landing ports in processing and marketing of product – the economies of scale.

Far less is known about the inshore sector and the small vessels which are employed. These tend to focus on non-quota species, only because quota is not available to them for reasons of government distribution policy or cost. As such it is much more open access and vessels below 15m in length are not required to be subject to VMS (vessel management system) tracking. For larger vessels (over 15m), VMS identifies where vessels are steaming and fishing at all times leaving a permanent record. The 'at sea' employment in the inshore sector is large and localised. These fishers do not roam over large distances and tend to have informally specified areas as 'their' area close to home. Their fishing activity in the coastal zone is highly susceptible to other maritime activities which also tend to this area of sea. Available fishing areas can easily be eliminated by developments such as marine energy or the designation of MPAs leaving some fishers with nowhere to go. A lack of knowledge about this very individual and poorly organised sector was clearly exposed in Scottish marine planning in 2010 leading to a concentrated consultation and mapping of inshore fishing areas [Marine Scotland 2016].

6.3.2 Management

The structure of the fishing industry is shaped primarily by the market and the availability of fish for capture. However, there are significant external drivers which are pushing and pulling the industry in strategic directions. The most significant of these is management and regulation aimed at the prevention of overfishing, the mitigation of conflict among users and the promotion of best practice. The seas and their resources are not owned by anyone. The international community and governments are responsible for management of the 'right to fish' and in deciding how the rights, or fishing opportunities, should be allocated [NEF 2016]. There can be three approaches within the overall government ambit although in practice they can overlap to a large degree:

1. Government Management. This is usual for large-scale fisheries because of their over-arching power and relative impartiality among fishers. It also involves intergovernmental organisations called Regional Fisheries Management Organisations (RFMOs) and sometimes supranational governments like the European Union (EU).
2. Common Pool Management. In a small-scale or local context where the same group of fishers target the same stock, the fishers might set and apply the rules, sometimes supported by a statutory mechanism [Dietz *et al.* 2002].
3. Private Management. Privatisation or quasi private mechanisms are often argued as a solution to overfishing because the 'owners' will take care to manage their assets better than if they were in a common pool. In practice this usually means a fishing right or opportunity granted to a private entity and not ownership of the stock itself.

At the international level fisheries regulation is established under several conventions or treaties of which the most comprehensive is the 1982 United Nations Convention on the Law of the Sea (UNCLOSIII). UNCLOS sets zones of sovereign rights to marine resources and stipulates that total allowable catches (TACs) should reflect the best scientific advice available. Wild fish in national waters are considered 'unowned' until captured and then owned by the captors. Additionally, Regional Fisheries Management

Organisations (RFMOs) are intergovernmental institutions composed of member states and play a crucial role in international, deep-sea and migratory stocks which cross borders (straddling stocks).

At the European level, the Common Fisheries Policy (CFP), applies to the waters of all member states and is highly influential under collaboration agreements with neighbouring non-member states such as Norway and Iceland. About two hundred fishing opportunities are set by the EU every year as TACs for the commercial fish stocks in EU waters. Multi-annual management plans (MAPs) specify long-term objectives including rules on effort controls, TAC setting, landing and transport. TAC setting employs the scientific expertise of the International Council for the Exploration of the Seas (ICES) in annual stock assessments. The full annual TAC process comprises:

1. Pooled international dataset made up of sampling landings and research surveys;
2. ICES working groups carrying out annual stock assessments and providing scientific advice;
3. ICES management committee examining annual assessments and providing management advice;
4. European Commission (EC) reviewing ecological, social and economic evidence with additional advice from the Science, Technical and Economic Committee for Fisheries (STECF);
5. EC submits TAC proposals;
6. Annual TAC negotiations with EU member states allows individual TACs to be set by the EU for each member state.

After TACs have been set, it is for Member States to decide on the distribution of national allocations to producer organisations, fishing companies and individual fishers. However, European law also legislates for several technical measures in the form of 'input control' or controls over gears, techniques and other specifications. These input measures are aimed at selectivity in species capture and ecosystem impacts in contrast to the 'output controls' of TACs focusing on what is caught rather than how it is caught.

At the national level the authorities set rules for licensing on who is allowed to enter the industry and the conditions for holding fishing rights. Fishing opportunities are the enforceable restrictions within which fishers can legally fish. They can be grouped into quota management (QM) which are quantitative output controls and effort management (EM) which are input controls. Significant elements of QM and EM fall within the definition of Rights-Based Management (RBM) defined as "*Fishing opportunities that convey secure and exclusive fishing rights to individual fishers or defined groups of fishers*" [NEF 2016].

Quota (output) measures fall into several categories whose use varies by country:

- National Quotas (NQs) applied to the whole fleet;
- Rationed Quotas (RQs) centrally determined often on the principle of equal access;
- Individual Quotas (IQs) made to individual vessels based on quota shares;
- Individual Transferable Quotas (ITQs) similar to IQs but transferable and leasable;
- Community Quotas (CQs) similar to IQs but to a collective unit.

In addition to government reallocation, quotas may be transferred, where allowed, by swaps, leasing and the transfer of quota shares. Quota shares do not normally confer property rights to the owners as the government retains the right to reallocate or reform the system. Some countries have granted rights that guarantee shares for a specified period which gives them a kind of legal status although still ambiguous. Critics of quota management refer to the injustice of the market which has developed in quota sale and purchase. The result has tended to a concentration of fishing opportunities into fewer hands at the expense of the myriad of small and artisanal fishers. Additional tensions are introduced by a trans-national trade in quota with foreign fishers owning quota in other states.

Effort (input) measures include:

- Individual effort quotas (IEQs) granting fishers an allowance for effort usually by gear type (e.g. kilowatt days at sea);
- Territorial Use Rights for Fisheries (TURFs) giving fishers a defined territory with exclusive harvesting rights;
- Limited Licensing (LL) controlling the number of vessels with conditions such as capacity, gears, spatial limits and target species;
- Spatial Management (SM) restricting access to defined areas for reasons of conservation or gear conflict. Marine Protected Areas (MPAs) are one such mechanism;
- Fishing Seasons (FS) determining times of year when fishery may open perhaps to match migratory patterns or to avoid spawning periods;
- Days at Sea (DAS) granting individual vessels the time when they can fish. The catch is therefore limited to the amount possible in their

time allowance. DAS are often linked to the vessel power in kilowatts to make a combined measure.

- Fishery Closures banning all fishing or specified types of fishing.

Taken as a whole the management system has grown piece by piece over decades. As one thing has not worked, another has been tried and so on. Many fishers, from all states see the result as clumsy, unworkable and unfair to their industry. This criticism may be seen clearly in respect to the CFP and the UK referendum decision to leave the EU. 'Fishing for Leave' is an influential organisation with a high public profile and features strongly in the BREXIT negotiation [HoL 2016]. Their argument is not purely nationalist, but is built on a view that a new system of management is needed and that it can be much better [Author interview with the Scottish fishing sector 2017].

6.4 Working Environment

The working environment of the fishing industry is unrelentingly tough. It is a dangerous occupation. Accurate global figures are not available but in the sophisticated United States (US) regime the average fatality rate over the last ten years among commercial fishermen is 1.15 per 1000 per year. This is three times greater than the next most dangerous job (construction) and twenty five times the average across all occupations in the US [Davis 2001]. Severe weather, fatigue and inconsistent use of safety equipment are all contributing factors. An extrapolation across the world is perhaps a step too far, but it is a guide to the sort of figures that might be expected among the 40 million employed in fishing globally, many of whom are working in far less regulated fisheries than those found in North America. Like other hazardous occupations before it, such as coal mining and steel smelting, the danger and the peripheral coastal location of many of the fishing centres breeds strong, tightly bound and self-reliant communities. Outside interference in their way of life and regulation of their livelihoods by distant scientists and politicians can be deeply resented. Fishing permeates every aspect of community life, at sea and onshore. The families and many of those who never go to sea are involved in shore based support. A whole culture of art, music and writing has evolved around fishing communities which these days is highly valued by tourists and urban migrants moving to the coast. So, long established fishing villages like St Ives in Cornwall or Stromness in Orkney are filled with visitors and resident artists.

Take the Shetland Islands and its capital of Lerwick as an example. Shetland is a fishing county with a population of about 23000 and Lerwick is the second most important fishing port in the UK, after Peterhead on the Scottish mainland. It is situated at the junction of the Atlantic Ocean with the North Sea, over 100 miles north of Scotland and half way between Scotland and Norway. Shetland is at the centre of the UK's richest fishing waters and its fleet operates in all three sectors, pelagic, demersal and inshore. It could be seen as remote but sees itself as a North Atlantic hub [Coull 1996] Nearly half the UK pelagic fleet, eight large vessels, is based here. Twenty one demersal vessels are based in Shetland in addition to which Lerwick is the chosen port of landing for vessels from all over the UK and Europe. The inshore fleet numbers about 150 vessels fishing mainly for shellfish. Most of the catch is now exported fresh or processed and exported. In times past it was the main source of food for the islands in common with subsistence fisheries over very large areas of the world today.

The Shetland fleet could not operate without comprehensive and extensive support industries onshore – ports and harbours for shelter, slipways and engineering bases for maintenance, chandlers and fuel merchants to keep the crews and boats going, the fish market itself, shipping and marine transport hubs to export the catch including live crab in vivier trucks to Spain. The 'Shetland Catch' pelagic processing plant in Lerwick is the largest in Europe. To meet a constant demand for officers and crew, the North Atlantic Fisheries College (NAFC) in Scalloway takes trainee deck crew and officer cadets from all over the UK and Europe. In terms of labour, skilled crews can be found and exchanged here. A 'Fishermen's Mission' helps them with their spiritual and welfare needs.

Crew supply has become a controversial matter with recruitment of foreign crews to man European vessels at greatly reduced rates of pay – a much discussed feature of globalisation. The International Transport Workers' Federation (ITF) has reported on 'Migrant Workers in the Scottish and Irish Fishing Industry' [ITF 2017]. They report migrant fishers earning as little as £268 per month at an equivalent rate of £1.29 per hour. The minimum wage in the UK is £7.20 per hour in 2017. Most of the foreign crews are from the Philippines with over 1400 believed to be employed in Scotland and Ireland. This can only happen outside of the territorial sea and so mainly the pelagic and demersal sectors will be involved. Seafish is a UK Non-Departmental Public Body (NDPB) set up by the Fisheries Act 1981 to improve efficiency and raise standards across the seafood industry. It is funded by a levy on the first sale of seafood products in the UK, including imported seafood in accordance with the Fisheries Act 1981. It has found it necessary to issue guidance to vessel owners on their responsibilities under the new UK Modern Slavery Act 2015 and human trafficking legislation [Seafish 2016].

Both literally and figuratively the working environment of the fishing industry is at the edge. It is at the edge in terms of physical location and of safety, employment practice, sustainability and in some respects, legality.

6.5 Innovation

Catch fishing is an industry under pressure. Widespread overfishing in excess of sustainable stocks has long been alleged and in many cases documented. The pressure of public opinion drives a search for sustainable solutions in management, methods and equipment. The demand for fish and fish product is strong and continuing to rise but the growth is entirely with farmed fish and the World Bank forecast zero growth in the catch industry up to 2030 (Figure 6.4). Fish prices are buoyant but management and environmental safeguards push up costs, leading to a search for efficiencies and cost saving technologies in order to stay competitive with surging farm fish supply and meat sources of animal protein. In addition to all these factors the fish catch industry is threatened with increasing spatial pressure as new 'Blue Growth' industries for energy, aquaculture and other uses take hold. Displacement from some traditional fishing areas seems certain. Everything points to an industry which is mature or even post-mature in its lifecycle but with a lot of potential life left in it, but only if it can meet the challenges through innovation and change.

These challenges fall to three main headings:

1. Innovation in sustainability
2. Innovation to meet technical and operational demands including cost reduction
3. Innovation in the market and marketing.

The implementation of an ecosystem approach to fisheries management is highlighted by the OECD [2017] as a key priority for fisheries innovation as well as improving the selectivity of gears; employing genetics and stock boundaries; introducing novel fishing techniques; reducing seabed impact; and mitigating the interaction with protected species and bycatch. Other necessary research is directed at the design of fishing vessels including fuel efficiency, emissions, maintenance, product conservation, safety and working conditions for the crew. The OECD have tracked and reported on patents in several countries as a means to measure activity in fisheries innovation [OECD 2017]. The United States leads in fish harvesting technology innovation with over 1000 patents closely followed by South Korea. Russia and South Korea lead in the field of 'New Products and Markets' with technologies helping the production of food from sea products such as fish meal.

Fisheries Innovation Scotland (FIS) is one of a number of organisations across the world designed to provide a formal structure of collaboration in fisheries research and innovation [FIS 2017]. The members of FIS are a diverse group of interests including government, scientists, industry, retailers and other key stakeholders. Marine Scotland, the Scottish government agency responsible for fishing, is working alongside the fishing industry and statutory agencies like Scottish Natural Heritage (SNH) and Seafish and also large food retailers and producers like Sainsbury's and Young's. An aim of FIS is to support the innovation objectives listed under the provisions of Article 26 of the European Maritime and Fisheries Fund (EMFF) which is described under 'investment' in the next section [EU 2017]. The presence of retailers in projects to enhance the sustainability of the fishing industry is a recent and important innovation. NGOs like the WWF and individuals, like the food broadcaster Hugh Fearnley-Whittingstall, have succeeded in mobilising public opinion to the extent that retailers want to show that their fish is sourced from sustainable fisheries. Other NGOs and charities have been formed to monitor and evaluate fisheries and issue certification of their fishing practices. The most prominent of these is the Marine Stewardship Council (MSC) who will inspect and certify those fisheries which employ sustainable methods in management. Increasing numbers of retailers have, in turn, committed to only buying fish from MSC certified fisheries or their equivalent. The effect is to integrate the industry from catch to plate and promote sustainable practice.

An example is the Orkney inshore fishery where the Orkney Sustainable Fisheries Project is aimed at MSC certification and beyond [Bell *et al.* 2015]. Orkney Sustainable Fisheries (OSF) is established as a cooperative consortium of local stakeholders. One of the first actions was to commission a pre-assessment for the creel fisheries (brown crab, European lobster and velvet crab) against the MSC standard for sustainable fishing [Hough, 2006], which identified three main issues: defining the extent of stocks, particularly the inshore and offshore components in brown crab; the lack of explicit objectives and effort controls; recording of catches and bycatch. The subsequent Orkney Shellfish Project has been established to respond to the licensing of areas of Pentland Firth and Orkney Waters for wave and tidal energy developments; and a second pre-assessment of the creel fisheries against the MSC standard. This identified the main issues as a lack of biological reference points, harvest control rules and monitoring of fishing effort [Bell and Gascoigne 2012]. The Crown Estate, a public body which manages UK assets including the seabed, funded the monitoring of spatial patterns of fishing effort in Orkney waters with vessel monitoring systems supplied by Marine Scotland, their interest being in developing a resource for wave and tidal energy developers in informing consenting activities. At the same time, the project has involved the development of a Fisheries Improvement Project to formally progress the brown crab fishery towards meeting the MSC standard for sustainable fishing, this being supported by WWF-UK and Marks & Spencer as a retailer working towards sourcing seafood products only from sustainable fisheries.

A key focus for innovation in all sectors of the fishing industry is selectivity. Catching the wrong species (by-catch including cetaceans and seabirds) or too many of the right species without quota, is wasteful and deeply unethical when it results in avoidable deaths and the discarding of unwanted or unauthorised catch. The reform of the EU Common Fisheries Policy (CFP) in 2012 introduced measures to prevent the practice of throwing unwanted catches overboard [Seafish 2017]. Introduced gradually, the general rule will be fully in place by 2019 by which time no commercial fishing vessel may return any quota species of fish, of any size, back to the sea once caught. Everything must be landed where it will be counted against quota with special rules for disposing of undersized or prohibited catch. This is the regulatory response but the industry is anxious to find technical solutions to the catching of the wrong species in the first place. The main instrument in the past has been mesh size in the nets but this is a crude and frequently ineffective method.

Fisheries regulation in Europe and the CFP has grown up over fifty years or so developing into a sophisticated but complex mechanism which underpins the whole approach of the industry to its work. The extent to which the CFP has helped conserve stocks is disputed although clearly it has its successes such as the conservation of North Sea cod and herring. Fishers in the UK, and in some other member states, have blamed the CFP for a downturn in their industry and its closure in some of the old fishing centres in England like Grimsby and Lowestoft. UK fishers were key drivers behind the campaign to leave the EU in the BREXIT referendum of 2016. They foresaw a chance to 'take back control' of UK waters with the exclusion of foreign vessels and a new management regime. At the time of writing it is impossible to say how this will work out. How far will the requirements of international law, transboundary relations with neighbours and the needs of free trade allow these sentiments of 'independence' to be realised? The negotiation could be long and hard and the fishers already fear their sacrifice in the interests of trade. However, looking beyond the chauvinism, there is a chance, just a chance, that root and branch innovation for the needs of modern fisheries management could introduce a more effective regime. In interviews with Shetland fishers there are ideas for high-tech solutions to selectivity which fail to get a hearing under the CFP because of its complexity as a mechanism, they say. There are both regressive and progressive ideas at play in the push for a UK fishery out of the EU.

A further area for research and innovation is the question of coexistence. New maritime industries, such as offshore wind power, can occupy very large areas of sea and threaten displacement of fisheries. Questions are raised of the opportunities for coexistence between wind farms and fishing or even enhancement of the fisheries. Most research and evidence to date has derived from the burgeoning offshore wind industry (Chapter 5) and the already well established operations around the coasts of Europe. Offshore wind turbines are usually sited in rows at distances apart of 500m or more so the waters between them might be used as nursery areas or possibly for trawls. The fishing obstacles are largely those of risk to the safety of power infrastructure and vessels and the apportionment of blame if things go wrong, a broken down vessel colliding with a turbine for example, or trawling through a power line. Other possibilities considered are the exploitation of turbine foundations as new habitats for crustaceans and the siting of fish farms within the confines of the wind farm.

6.6 Investment

Private sector investment in the fishing industry as a business follows a conventional model of equity and debt with little public support now available in the form of development grants for vessels and conventional equipment. There are no direct subsidies such as those available under the Common Agricultural Policy (CAP). Fishers with a good business plan and realistic projections of profit and cash flow will be able to access equity from shareholders and loans from banks to finance their operations and grow their businesses.

The target for public investment has changed. At its outset the CFP also supported investment to encourage growth in the output of European fisheries. Today, public investment, supported in some of its facets by private and volunteer investment, has evolved. It is focused on research, stock assessment, monitoring and evaluation, enforcement, infrastructure, sustainability and coexistence. Some of this public investment will be recovered in the form of levies and fees which will be reflected in the wholesale and retail prices of the end product.

Central governments will pay to be part of the international network of maritime law through conventions, treaties and institutions. The scientific work of ICES in undertaking the science of stock assessment will be met by governments paying to be members of the organisation with access to their results. Similarly work with international conventions like OSPAR aimed at the prevention of pollution and marine conservation in the NE Atlantic. On a national levels there is taxpayer funded investment into a host of promotional and regulatory organisations focused in the end at enforcement with fisheries protection vessels at sea and fisheries officers in landing ports.

The EU European Maritime and Fisheries Fund (EMFF) is established to provide grant aid in the promotion of sustainability in fisheries and to foster the implementation of the CFP [EU 2017]. It offers support under five headings:

1. Innovation in fisheries (Article 26) – improved equipment such as that needed for selectivity, techniques and management;
2. Conservation measures and regional cooperation (Article 37) – technical and administration measures and stakeholder participation across borders;
3. Reduced fishing impacts and protection of species (Article 38) – selectivity of species, elimination of discards, elimination of seabed damage;
4. Innovation for conservation (Article 39) – projects for sustainability and coexistence with protected predators;
5. Restoration of ecosystems and sustainability (Article 40) – wide ranging provisions from the collection of marine litter to compensation schemes, fisher education and Marine Protected Area (MPA) management.

The volunteer and charity sectors invest in fisheries monitoring and management raising funds through programmes of public awareness and providing services. The work of organisations like WWF and the MSC are effective in mobilising public opinion for investment in sustainable fishing. The MSC raises over £15m annually to finance its operations of inspecting and certi-fying fisheries. Of this 25% comes from volunteer and business donations and 75% comes from the selling of services, almost all of which is for the licen-sing of its logo on retail fish products [MSC 2017]. The act of certification also levers in funds from the major retailers and the fishing industry for more investment in sustainability. So, companies like Tesco, Sainsbury, Marks and Spencer, Waitrose and Lidl invest in research and projects which will increase in their sourcing capacity for the fish from sustainable sources which their customers demand. The 2016 results from the MSC identify 286 certified fisheries in 36 countries representing over 10% of global catch. About 40 fisheries are newly certified each year and nearly 100 are in assessment.

More investment comes to the fisheries affected by the new industries such as offshore wind power and aquaculture. The focus of this investment is coexistence aimed at reducing fishery objections to their use of marine space. It recognises the political power of fishing communities in the coastal regions where their industry is to be sited. Typical of this investment is the establishment of 'The West of Morecambe Fisheries Ltd' covering the waters between England and Ireland. It is a not-for-profit UK company established in 2013 and funded by the owners of several UK offshore wind farms including Dong Energy, Vattenfall, Scottish Power and Scottish and Southern Energy. It manages funding donated by offshore wind farm owners, provided for the purpose of supporting and developing commercial fishing activities [WMFL 2017]. The companies are cagey about the actual level of their support but the web-site is very professional and Dong energy recently donated £300,000 to a particular project. A 'going-rate' for community support donations from companies related to new onshore wind farms has been established at around £5000 per kw of capacity per year [Kerr *et al.* 2017]. At sea it varies but similar sums are sought by coastal communities for offshore farms in their vicinity. The donations are used to finance, set up and support Community Projects including those for the fishing industry that operates in the same areas as the wind farms. It works closely with relevant sectors of the fishing industry to invest in a number of Fishing Community Projects aimed at business, sustainability and safety.

6.7 Uncertainties and Concluding Remarks

The commercial capture of wild fish is based purely on harvest and harvest technologies. There is no nurture or production element. It depends entirely on the vagaries of the wild environment, such as natural climate change, and anthropological activities which directly affect stocks, such as fishing itself, or indirectly, such as pollution or human induced climate change. These anthropological factors are of prime consequence.

The demand for food fish is strong and growing with population growth and more sophisticated diets among developed and developing countries. However, price competition with rival sources from both animal and vegetable proteins will have an effect and the rapidly expanding availability of farmed fish by quantity and species is forecast to take up all the anticipated growth in fish consumption over the next decades. The World Bank anticipates zero growth in wild fish catch while farmed species output continues to grow at a rate of about 2.5% per annum.

Demand, therefore, is strong but catch is increasingly constrained by supply. Supply of the main finfish targets is constrained by significant and ongoing change in key factors:

- Stock availability – reduced availability in key species due to overfishing and climate change (natural and human induced);
- Reduction in permitted fishing targets – by specie and by quantity as stock assessment and quota allocation increases in sophistication and effect (output controls);
- Reduction in permitted fishing effort – (input controls)
- Increased sustainability measures in fishing methods – aimed at efficient selectivity of catch and reduced atmospheric emissions and seabed damage.
- Reduction in permitted fishing areas – developments in international and national marine governance and control reducing the open access nature of the High Seas and the marine commons. Also, increased spatial exclusion from the new 'Blue Growth' industries and MPAs.

These factors are mainly of concern to the bulk pelagic and demersal commercial fisheries which are very likely to see stasis in catch while facing increased costs in capture. Investment is aimed at preserving the industry at more or less current levels but reductions in activity are also highly possible. Growth in catch appears to be unlikely.

A different outlook applies to the inshore (small commercial) and artisanal fisheries. Almost every metre of coastline around the globe is fished to some degree, much of it at subsistence levels of activity. In these sectors are very high levels of formal and informal levels of employment and very significant social and economic issues affecting vulnerable and peripheral communities. In Europe, small commercial, owner operated fisheries in the inshore have flourished of late. Denied access to large finfish quotas they have specialised in non-quota species such as crustaceans and molluscs, which may have increased in stocks as finfish predators have been reduced. Firm evidence is awaited but the nephrops fishery has expanded enormously. These inshore fisheries are more likely to be affected by the spatial pressures of coastal activities and new industries but their methods are more adaptable to coexistence.

The catch fish industry is under pressure but it has been the cornerstone of the Blue Economy for centuries and remains so. No other sector is as widespread or as important in employment and social terms. As long as there are fish in the sea, there will be a business model to catch them, sustainably of course.

References

Bell, M.C. and Gascoigne, J. (2012). MSC pre-assessment. Orkney lobster, brown crab and velvet crab. Report by MacAlister Elliott and Partners Ltd to Orkney Sustainable Fisheries Ltd.

Bell, M., Johnson, K., Rydzkowski, K., Coleman, M., Crichton, S., Matheson, F. (2015). Science in support of industry-led initiatives in inshore fishery management in Orkney. Proceedings of ICES Annual Science Conference 2015. Copenhagen: International Council for the Exploration of the Sea.

Coull, J., (1996). Shetland: the land of sea and human environments. 66–67. In Waugh, J., Shetland's Northern Links Language and History. Edinburgh: The Scottish Society for Northern Studies.

Davis, M.E. (2001) Occupational safety and regulatory compliance in US commercial fishing. Washington DC: Archives of Environmental and Occupational Health.

Dietz, T., Dolsak, N., Ostrom, E., and Stern, P.C. (2002). The Drama of the Commons. Washington DC: National Academy Press.

EU. (2016). European Maritime and Fisheries Fund Guidance. Newcastle upon Tyne: Marine Management Organisation.

FAO. (2016). The State of the World Fisheries and Aquaculture. Rome: Food and Agricultural Organisation of the United Nations.

FIS. (2017). Fisheries Innovation Scotland. Projects in 2016 and 2017. www.fiscot.org [Accessed 14 July 2017].

HoL. (2016). Brexit: fisheries. House of Lords. European Union Committee. 8th Report of Session 2016–17. London: UK Parliament. www.parliament.uk/brexit-fisheries-inquiry [Accessed 14 July 2017].

Hough, A. (2006). *Pre-assessment report for Orkney Shellfish Fisheries (brown crab, lobster, velvet crab).* Report by Moody Marine Ltd to Orkney Sustainable Fisheries Ltd.

Kerr, S., Johnson, K., Weir, S. (2017). Understanding community benefit payments from renewable energy development. *Energy Policy,* 105 202–211.

Marine Scotland. (2016). Pilot Marine Spatial Plan for the Pentland Firth and Orkney Waters. Edinburgh: Scottish Government.

MMO. (2017). Marine Management Organisation. UK Sea Fisheries Statistics 2015. Newport: National Statistics Publication.

MSC. (2016). From sustainable fisheries to sea food lovers. Annual Report 2015–2016. London: Marine Stewardship Council. www.msc.org/annual report [Accessed 14 July 2017].

NEF. (2016). Who Gets to Fish? The allocation of fishing opportunities in EU Member States. London: New Economics Foundation. http://neweconomics.org/wp-content/uploads/2017/03/1513-NEF-Fisheries-Report-v9-FINAL.pdf [Accessed 14 July 2017].

Noble, T. (2003). Co-operating in fisheries management: trials and tribulations in Scotland. *Marine Policy* 27 433–439.

OECD. (2016). The Ocean Economy in 2030. Paris: OECD Publishing. http://dx.doi.org/10.1787/9789264251724-en [Accessed 14 July 2017].

OECD. (2017) Towards Innovation. www.oecd.org/fisheries-innovation/towards-innovation [Accessed 14 July 2017].

Seafish. (2016). Working on UK fishing vessels: the legal framework and support for fishers. Grimsby: http://www.seafish.org/media/publications/SeafishInsight_WorkingonUKFishingVessels_201610.pdf [Accessed 14 July 2017].

Seafish. (2017). The Landing Obligation explanatory booklet. Grimsby: Seafish http://www.seafish.org/responsible-sourcing/conserving-fish-stocks/discards/landing-obligation [Accessed 14 July 2017].

SFA. (2016). Shetland Fishermen Yearbook 2016. Lerwick: Shetland Fishermen's Association.

WMFL. 2017. West of Morecambe Fisheries Limited. Supporting the coexistence of commercial fisheries and wind power. http://www.westofmorecambe.com/co-existence/ [Accessed 14 July 2017].

World Bank. (2013). Fish to 2030 Prospects for Fisheries and Aquaculture. World Bank Report Number 83177-GLB. Washington DC: World Bank.

7

Offshore Oil and Gas

Irati Legorburu*, Kate R. Johnson and Sandy A. Kerr

Heriot-Watt University, Scotland
*Corresponding Author

Executive Summary

Driven by low selling prices, high production costs and the development of new onshore exploitation techniques, offshore oil and gas activities are experiencing a significant decline. The European sector is mainly composed of private companies that operate mostly at the global scale. However, the production from its territorial waters accounts for 9% and 13% respectively of the total oil and gas consumption in Europe, respectively. Thus, this decline can undermine the energy interests of the EU and especially, the economic activity of the North Sea countries (responsible for the production of virtually all of the oil and more than 80% of the gas).

Despite this negative outlook, the development of new and more efficient subsea exploitation systems can provide an important boost to the sector. However, in a Blue Growth context, the main importance of this industry relies on its important legacy of infrastructure, knowledge and experience (skills, business models, concepts of permanent occupation of the marine environment, etc.).

With this in mind, this chapter describes the main features of the offshore oil and gas industry along with the opportunities and barriers that it presents for the development of Blue Growth and MUS/MUP concepts.

7.1 Introduction

By value, technology and geopolitical status, the offshore oil and gas sector (O&G) is by far and away the most important sector in the contemporary Blue Economy. Offshore O&G came to prominence in the 1970s and currently

231

accounts for about 37% and 28% of the total O&G global production respectively (WOR, 2014). Companies continue to extend their areas of operations, with "Exploration and Production" (E&P) in ever more extreme and hostile areas. E&P is set to take off in the Arctic Ocean as the ice retreats; fields are already in production at the so called 'Atlantic Frontier' between Scotland and Faroe. The 1970s extreme of North Sea working at depths of up to 300m is replaced by a contemporary technology of working at depths in excess of 1500m. In contrast to the transient activities of fisheries and shipping, the offshore O&G sector introduced the concept of semi-permanent occupation of maritime space. It introduced the idea of fixed platforms at sea which could be supplied with materials and services for the production of O&G and a safe home for thousands of workers, hundreds of kilometres from land. The sector has led the way in maritime health and safety and in the development of risk assessed regulation to control operations and protect the environment.

However, the offshore O&G sector is also in decline. Recent, and possibly sustained, falls in the oil price render offshore production uneconomic compared to adequate low cost onshore resources and the rise of the 'fracking' process for onshore gas. Industry sources believe that the rapid advance of offshore technology peaked in the 1990s and has slowed very considerably. The whole (land and marine) oil sector has been driven by global dependence on fossil fuels as the main resource to supply a burgeoning energy demand. Many companies have employed successful business models and made their fortunes. The economic and political drivers have been with them. Others have been attracted to the sector by its successes but have not had the skills, or the luck, to flourish.

In 2015, the O&G sector has achieved maturity as the world approaches what is believed to be the 'peak oil' event. Pressure grows for emissions restraint and alternative sources of clean energy. Notwithstanding this, and in spite of efforts to move to new energy technologies such as renewable electricity, the use of fossil fuels continues to dominate energy supply and is forecast to continue to do so (well in excess of 50%) for the next fifty years or so. An as yet undetermined transformational technological event, perhaps in renewable and energy storage technologies, might possibly change this equation but current forecasts anticipate continuing dominance of fossils sourced primarily from terrestrial areas. The economic factors are, though, not the complete picture. Geopolitical factors have played a hugely significant role in O&G markets and will continue to do so. Oil has been used as a weapon by major producers to exploit their resources to the full and to punish those states they do not agree with. Wealthy states with smaller resources have therefore acted to exploit their own, even at uneconomic rates, for the purposes of energy security. Poor developing states have been anxious to develop any easily recoverable reserves to generate economic growth and foreign exchange. These will include the more accessible offshore resources.

7.1.1 The Offshore Oil and Gas Sector in the Development of Blue Growth

Although the offshore O&G industry may be at or past its peak, its products (not only fuels, but also e.g., synthetic materials) still will be necessary for the development of marine economic activities. In any case, its true value to Blue Growth is what it bequeaths at many levels. The successful offshore operators have established technologies, infrastructure and operational skills of enormous value to the Blue Growth sectors while, so far, demonstrating little appetite for diversification themselves. The O&G majors are among the largest multinationals in the world with significant capital to invest, many have started small preliminary investment in renewables but most have pulled back from serious participation. A few have gone further, like BP Solar or Statoil in the development of floating wind (Xing et al., 2014). However, with the depletion of traditionally exploited fields a new factor will enter into force in the short term: decommissioning. In the North Sea alone, 7% of the existing facilities are in the decommissioning process, and it is estimated that over the next 30 years this process will affect to 500–690 additional infrastructures (RAE, 2013).

With all this in mind, the great resource transferability from the O&G industry to the new Blue Growth sectors is clear, being:

- Infrastructures,
- vessels,
- technologies,
- operation procedures,
- human skills,
- supply industries and
- financial resources.

7.2 Market

7.2.1 Products

The need for energy has been the principal driver for the development of the O&G industry. While fuels needed by transport activities are the main oil products, gas is widely used in electricity generation and heating processes. However, O&G products and by-products have a wide applicability in day-to-day lives as they are used, among others, as raw materials for pharmaceuticals, chemicals, plastics, lubricants, waxes, tars, synthetic clothes, rubbers, paint or photographic films (WOR, 2014).

7.2.2 Market Trends

These are not good times for the O&G market. Imbalances between supply and demand, still tangible effects of the financial crisis, enforced environmental policies, changing consumer preferences or the development of more efficient transport systems have severely hit the industry, particularly in developed countries. However, and despite its marked and sustained slowdown, developing countries economies mean that global O&G consumption continues to increase, giving as a result two general global trends (Mitchell et al., 2012; BP, 2015):

- Non-OECD countries: Growth markets. Developing economies (mainly China) are responsible for the net growth in global consumption. However, these economies are facing an important deceleration, which is being reflected as a slowdown in the consumption growth rates of the sector.
- OECD countries: Non-growth markets. Opposite to developing economies, the O&G consumption rates in the OECD economies remain stagnant or even declining. Noteworthy in this regard are the cases of Japan and the EU, which have suffered, respectively, the largest O&G consumption declines over the last decades.

7.2.3 Prices

Hydrocarbon products are not common trading goods and complex factors influence their prices. Traditionally, their prices have been determined by the fundamentals of supply and demand, being directly influenced by factors like weather, changes in supply/demand patterns or the supply capacity of the producing countries. However, geopolitical and speculative factors have become of special relevance over the last decade. In geopolitical terms, the control over the production, distribution and prices provides economic and political power. Following the opening of the sector to financial markets, O&G products have become assets of great interest, strongly subjected to speculative interests. Although these factors are strongly interconnected, their individual influence on prices varies depending on specific political and economic situations or interests. As a result, prices in the sector are extremely volatile and unpredictable (NRCan, 2010).

7.2.4 Future Supply and Demand Gaps

As finite resources, existing O&G reserves can't meet the growing demand for energy in perpetuity. Disagreements exist between those who affirm enough reserves for the decades ahead, and the critical voices that warn about the near depletion of stocks (Owen et al., 2010). Considering the industry as a whole (onshore + offshore) both the discovery of new reserves (e.g. deeper offshore fields) and the development of non-conventional exploitation techniques (e.g., fracking, tar sands) will increase the availability of the resource, extending its potential supply capacity over time. However, these new reserves and non-conventional techniques are characterised by their higher exploitation costs. Therefore, the inability of the sector to commercially exploit its resources at prices assumable by the global economy may be a more crucial determinant, rather than the amount of reserves themselves (Owen et al., 2010). This might be of particular importance for offshore activities in which the trend towards exploiting even more hostile and remote areas implies a huge increase in operational costs. This may result in making them even economically unfeasible. In addition, the 2015 report by the UN Intergovernmental Panel on Climate Change (IPCC) clearly states that significant climate change will occur from carbon emissions before known reserves are exhausted. This has led many NGOs to campaign for a policy of *"keep the oil in the ground"*.

In addition, the O&G industry faces increased competition that can influence its future supply-demand trends (Mitchell, et al., 2012). In terms of intra-sectoral competition (i.e., Oil vs. Gas), the oil sector has largely relied on the transportation market. Lower prices of gas and improved air quality can be a driver for the development of gas-fuelled engines and encourage its replacement of oil as a principal fuel. On the other hand, the growing pressure from new fuels, new energy supply types and users requiring alternative non-fossil energy types, may further decrease the demand for O&G products (new biofuels and materials; electric vehicles; environmental protection policies; diversification of energy sources, e.g., renewables).

7.3 Sector Industry Structure and Lifecycle

7.3.1 Lifecycle

Although for the following decades O&G will remain as the main supplier for the global energy demand, the decline affecting the sector is particularly relevant for offshore activities. Following the depletion of traditionally exploited shallower fields, the production at deeper and more hostile areas presents important economic barriers. Even more significantly, the new non-conventional exploitation techniques (e.g., fracking) can redirect the focus of the industry towards onshore activities to the detriment of offshore production.

7.3.2 Industry Sectors and Segmentation

Depending on the processes involved, the O&G industry is divided into *upstream* (exploration, drill wells, production), *midstream* (transportation and storage) and *downstream* (refining and marketing) activities. Firstly, only upstream and midstream activities relate to offshore activities. And secondly, downstream processes are always onshore activities and therefore do not offer interesting alternatives for Blue Growth or potential combinations with other marine economic activities. Thus, considering the scope of this book, only upstream and midstream sectors activities will be considered (Table 7.1).

Table 7.1   Sectors and segments of the O&G industry

Sub-Sectors	Segments
Upstream	Major Companies
Search and exploration of resources, well drilling and extraction of raw materials.	Fully integrated: cover all the facets of O&G industry (upstream-midstream-downstream). Exploit large proven reserves, which require at the same time greater investment (as are also their returns).
Midstream	Small Companies
Transportation (pipelines, LNG/oil tankers) and storage of extracted raw materials.	More versatile, normally focused on exploration and production activities. Go after opportunities discarded by major companies, e.g.: (i) acquiring and exploiting depleted fields trying to squeeze some extra production at lower cost; (ii) exploring in areas where the probability for large discoveries is low; or (iii) operating in areas with uncertain fiscal and regulatory regimes. Invest just enough to reduce uncertainty.

7.3.3 Horizontal and Vertical Integration

Major oil companies usually have a fully integrated structure (vertically and horizontally). Given their huge resources they cover the whole O&G supply chain, from exploration and production of new reserves, to transportation, and, to the final refining and sale to the consumer (upstream-midstream-downstream). On the other hand, small companies do not have enough resources (or interest) to cover the entire supply chain. Usually they develop their activities in very specific segments of the industry (e.g., geophysical surveys, activities exclusively focused on production or transporting) and sell their products/services to third parties of the supply chain. Finally, mergers and acquisitions are common in the industry, so the release or subcontract of certain activities are a frequent practice of oil companies (horizontally within the different segments and vertically along the supply chain).

7.3.4 Centres of Activity

Currently, more than 600 active offshore extraction platforms exist in the EU-28, a value that significantly increases if those located in Norwegian waters are considered. The European offshore production constitutes 9% and 13.8% of the total O&G consumption, respectively. Therefore, the offshore production of hydrocarbons represents an important energy resource for Europe (JRC, 2015). Figures 7.1 and 7.2 show the distribution of the major O&G reserves and their associated infrastructure in the studied basins.

In the *Atlantic basin*, most of the exploration and production is developed in the *North Sea*. Practically all of the oil and more than 80% of the gas produced in Europe are produced by countries bordering the area (i.e., Norway, UK, Denmark, the Netherlands and Germany). Undeniably Norway, and the UK to a lesser extent, are the leading countries in terms of production. This is clearly reflected by the greater number of reserves and development of infrastructures within their territorial waters (Figure 7.1).

Compared to the North Sea, offshore production in the *Baltic* seems minimal. Production activities mainly develop along the Polish coast and represent only 0.1% of the total offshore production (Figure 7.1). However, this basin plays a very important role in strengthening the energy security of the EU. With a length of 1,224 km and a combined transport capacity of 55 bcm/yr (27.5 bcm per line), the Nord Stream twin pipeline crosses the Baltic Sea serving as a connection between the vast Russian gas reserves and the European markets (Nord Stream, 2014).

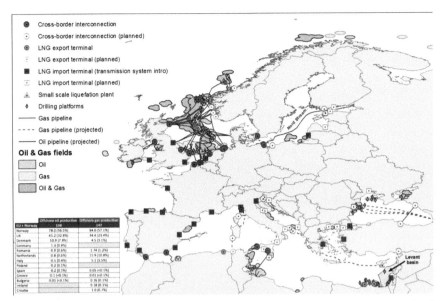

Figure 7.1 Distribution of main Oil and Gas fields and associated infrastructure in the Atlantic, Baltic and Mediterranean basins (Authors' compilation based in: ENTSOG, 2015; Lujala et al., 2007). Offshore oil and gas production values are given in million tonnes (JRC, 2015).

In regard to the European territories of the *Mediterranean*, traditional production areas have been located in Spanish, Greek, Maltese and Adriatic waters (mainly Italian). In this latter case, of special attention is the increase in the offshore production of Croatia. Although these activities can improve the energy self-sufficiency of the country, many critical voices warn about the danger to tourism from potential accidents as it is a tremendously important sector for the economy of the country. In any case, the main production areas in the Mediterranean are outside the territorial seas of the EU, being especially important the North African coast and the recent discoveries in the eastern Levant basin. These latter findings, partially located in Cypriot waters, have enabled the cooperation between the EU and some eastern Mediterranean countries (e.g., Israel, Lebanon). The agreements relate to issues such as, optimisation of exploitations, development of infrastructures, access to European markets, or pricing. Among the regarded options, the construction of the Cyprus-Greece pipeline or the building of a LNG terminal in Cyprus can be highlighted (EC, 2013). Romania and Bulgaria on the one hand

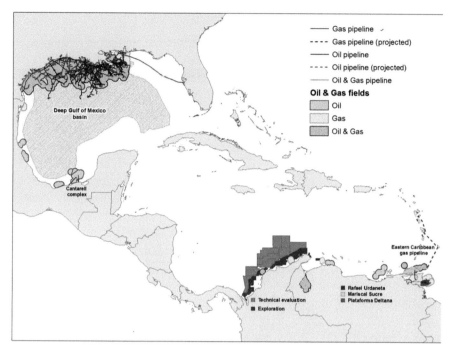

Figure 7.2 Distribution of main Oil and Gas fields and associated infrastructure in the Gulf of Mexico and Caribbean basin (Authors' compilation based in: ANH, 2016; BOEM, 2016; Lujala, et al., 2007; Petróleos de Venezuela SA; Theodora.com).

(intra-EU) and Turkey and Ukraine (extra-EU) on the other, have been the main hydrocarbon producers in the *Black Sea*. Historically, countries bordering the Black Sea have shown little interest in the exploitation of their massive energy resources. Importation (mainly from Russia) has been proven as an easy and cheap option for them. However, changes in the energy markets, the discovery of new reserves in the Bulgarian, Romanian and Turkish coasts or political tensions with Russia, are strengthening the development of offshore production in the region.

Finally, with countries like Mexico, the US, Colombia Venezuela or Trinidad and Tobago, the *Gulf of Mexico and the Caribbean Sea* have major actors in the global energy sector (Figure 7.2). However, most of these resources are outside the territorial waters of the EU or its associated overseas countries territories. Thus, the oil and gas activities carried out in the region may be of less interest for the development of EU's Blue Growth strategies.

7.3.5 Types of Ownership

In the same way that the demand for O&G presents two differentiated patterns (OECD and non-OECD economies), the ownership of O&G reserves also shows two main actors: private companies and National Oil Companies (NOCs). In any case, given the importance of the energy sector for the world economies, unregulated private companies do not exist. Even in the most developed economies, where O&G is supplied by private companies, the sector is strongly influenced by government policies (e.g., subsidies on exploitations and transport, taxes to consumption, price manipulation...).

Private companies have a primary objective to make profits for their shareholders. Typically, they exploit and produce their resources more quickly than NOCs, 10–12% depletion rates, compared to 3–5% for NOCs, (Mitchell et al., 2012). While their resources and infrastructure have a global coverage, their headquarters are normally located in developed economies and direct their production to competitive markets (OECD).

Although NOCs share about 86% of proven reserves, their production rate is comparable to that of private companies (55% of the total). Apart from their national economies, their main customers are located in emerging economies (non-OECD). NOCs normally belong to countries with a high economical reliance on their O&G exports. Hence, their production and reserve exploitation policies are highly conservative in comparison to those of private companies. The protectionism degree of governments towards their NOCs, closely relates to the diversification of their economies. As a result, there exist two types of NOCs (EIA, 2016). The *NOCs organised as corporations* have strategic and operational autonomy. Although mostly controlled by governments' interests, part of their shares are publicly traded and subject to private funding (e.g., Petrobras, Statoil, Gazprom). Thus, they are subjected to the rules of the Stock Exchange, and are characterised by their commercial objectives and income generation. The *NOCs that operate as an extension of government* are aimed to support national policies, both strategically and financially (e.g., Pemex, Saudi Aramco, Petróleos de Venezuela). Their objectives do not directly relate to the markets, as they seek to boost the national and foreign objectives of their countries (e.g., offering lower prices to domestic consumers or generating long term incomes for their economies). In any case, operation agreements between both types of companies (privates and NOCs) are a common practice in the sector that allows a joint venture arrangement where private companies operate NOC owned reserves.

7.3.6 Rules and Regulations

Since it is a source of important government revenue (e.g., by means of taxation, awarding of exploitation licenses, increased GVA, etc.), the O&G industry is crucial for the economies of producing countries. Regulations are applied to the economic activity itself and the industry is also subject to strong requirements on environmental safety. All this complexity is, at the same time, the main cause for investor's reluctance. They opt to invest in countries with favourable regulatory frameworks. Therefore, regulation can become a double-edged sword, as both strict and lax regulations may impair the economic interests of producing countries. As a result, the regulation in the sector is strongly influenced by constant challenges and opportunities in order to maintain a balance between national interests and concessions to the private sector (e.g., changing fiscal regimes, socio-political and environmental sensitivities, etc.). Annex 7.1 shows the main regulations affecting the sector in terms of economic activity, environmental protection and liability and compensation.

7.4 Working Environment[1]

7.4.1 Economic Climate

As already observed, economic and geopolitical factors have a major influence on the performance of this sector. The slow recovery of major economies (e.g., Europe, Japan, China) and current political conflicts in the Middle East and Russia-Ukraine (together with the sanctions imposed by the EU and US), fuel the mistrust of markets in the industry. As a result, the industry has to face an uncertain economic climate (Hays, 2015), in which producer countries adopt different response strategies.

NOCs are an important support for their economies. As an example, PEMEX revenues have accounted approximately for 35% of the Mexican federal government's budget, and PDVSA is the main company sustaining the Venezuelan economy. Therefore, the decline in demand and prices can cause a fatal impact in the socio-economic development of these countries. Thus, the attraction of foreign investments is part of the solution to get cash in both cases (e.g., potential denationalisation of certain fields, exploitation

---

[1]In general, the information provided in this section refers to the Oil and Gas sector as whole (inland + offshore activities). However, the main European Oil and Gas producers develop their activities at sea. Thus, at least for European countries these figures can be considered fairly representative of specifically offshore activities.

agreements, sale of international assets...). In the case of European private companies, the ageing of their reserves is an additional factor to be considered. Waiting for favourable regulatory and economic changes, these companies have opted to avoid or minimise new investments.

7.4.2 Employment, Skills and Migration

Figure 7.3 shows the direct employment created by the sector in some of the considered countries[2]. It provides a picture of the most important countries in the sector and its relative importance to their national economies: countries with higher production capacity, are those generating a greater number of direct jobs in the sector. The importance of the industry in terms of employment relies on its ability to create indirect employment. In the North Sea alone, it is estimated that each direct employment in the sector induces up to 7.5 other indirect jobs (ECORYS, 2013).

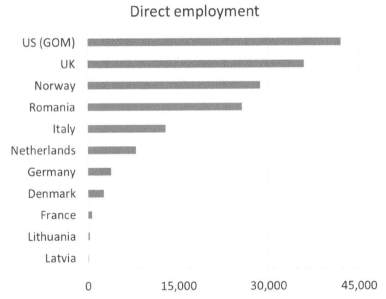

Figure 7.3   Direct employment derived from Oil and Gas exploration activities.

Source: EUROSTAT, 2016; Quest Offshore, 2011.

---

[2]The data in this section must be considered as indicative as:

- it has not been possible to find data for all the countries involved in offshore activities.
- depending on the sources, direct employment data can vary significantly.

In the coming years, skill shortages will be one of the main problems to be faced by the sector. The rejuvenation of the workforce (added to a poor transfer of knowledge), the retirement of experienced workers, the poor update on technological advances, or strict immigration laws that prevent the access to global talent are among the main causes for this shortage. Much of the expertise required in the sector relates to fields such as science, technology, engineering or mathematics (STEM). The industry is a highly male dominated industry and to balance the lack of skills, O&G recruiters are increasingly focused on the incorporation of women into the sector (Hays, 2015). Companies are increasingly recognising the high quality of women in STEM and they have an increasing presence in the workforce.

Regarding the migration and mobility of workers, European companies rely principally on their local workforce (Hays, 2013). Europe is characterised by its smaller reserves and by an industry dominated by private companies. These companies commonly operate at the global level, developing much of their production out of European territorial waters and favouring the displacement of workers outside their countries of origin. In addition, the high skills of its workforce can act as additional drivers for the mobility of European workers.

7.4.3 Economic Indicators

7.4.3.1 Contribution to GDP

Figure 7.4 shows the contribution to GDP of the rents derived from the extraction of hydrocarbons in the producing countries around the studied basins. Despite exceptions (e. g., Trinidad and Tobago, Ukraine, Netherlands, Israel), incomes derived from the exploitation of oil exceed those obtained through gas exploitation. Probably this is due to the fact that oil has been traditionally a more intensively exploited and marketed resource than gas, and consequently, more heavily taxed. However, it is likely that this pattern will change in the future: the depletion of oil reserves, along with changes in the preferences of the markets (lower prices of gas, replacement of oil as a primary fuel in transport) can help the expansion of the gas sector and increase the amount of rents collected by producing countries.

Driven by their higher amount of reserves and the lower diversification of their economies (a probable consequence of the former), Caribbean and North African countries are those with a higher reliance on the Oil and Gas sector.

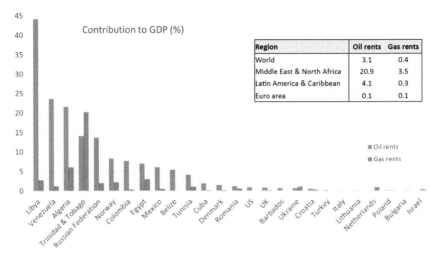

Figure 7.4 Contribution of O&G rents to individual and regional economies.

Source: World Bank, 2015.

7.4.3.2 Wages

Exceeding a global average of $81,000 annually, salary is one of the main attractions for workers in this sector. The countries bordering the North Sea, the US, Colombia and France are at the top of the list, exceeding that average for either their local or imported workforce (Hays, 2013).

In contrast to Norway where salaries of local workers may be up to 60% higher, the remainder of the North Sea countries, US and France present a balance in the wages for both types of workers. These cases should be considered exceptions and indicative of their highly skilled local workforce. In the rest of the countries the salaries of foreign workers are significantly higher, which may be due to two main reasons:

- The allocation to foreign subsidiaries or exploitations of workers from private US and European companies.
- Attempts to attract talent by countries with much production capacity but with a lack of skilled labour.

The bonuses received by the workers are another important aspect to be considered in relation to wages. Companies commonly offer incentives in order to ensure and maintain their skilled workers. Almost 80% of the staff in North Africa and South and North America receives some kind of bonus, while in Europe this value drops to 60%. Bonuses, health plans, home allowances or retirement plans are among the most common incentives (Hays, 2013).

7.4.3.3 Export potentials

Table 7.2 shows major O&G exporting and importing countries. At the EU (+ Norway) level, the only countries with a certain gas export potential are Norway and the Netherlands. In fact, despite the production activity developed by some Member States, the EU as a whole, is a net energy importer. Outside its territorial waters, the Caribbean and Mediterranean basins are those with a higher export potential. In the Caribbean, the development of new offshore exploitations can strengthen the role of the existing exporting countries of Mexico, Venezuela and Trinidad and Tobago. In this latter case, the construction of the Eastern Caribbean Gas Pipeline which will ensure the supply of gas from Trinidad and Tobago to the Eastern Caribbean Islands, will reinforce its role as gas supplier in the region. The recent discovery of huge gas reserves in the Eastern Mediterranean, not only increases the production capacity of the basin (mainly developed in North Africa) but also the export potential of the Eastern Mediterranean countries. In this sense, the agreements signed by the EU and these countries (e.g., Israel), involve a series of advantages for the EU in terms of imports-exports, which might be helpful to ensure its energy security.

7.4.4 Infrastructure and Support Services

Given its complexity and the risks involved, the oil and gas industry requires a large amount of supporting services. Although some large companies integrate these services within their structures, contracting third-parties for support services is a common practice in the sector. Following the NACE classification of economic activities, these services include a variety of

Table 7.2 Top ten of exporter and importer countries

Crude Oil		Natural Gas	
Net Exporters (Mt)	Net Importers (Mt)	Net Exporters (bcm)	Net Importers (bcm)
Saudi Arabia (271)	US (442)	Russia (203)	Japan (123)
Russia (239)	China (269)	Qatar (121)	Germany (76)
Nigeria (124)	India (185)	Norway (103)	Italy (62)
Iraq (119)	Japan (179)	Canada (54)	Korea (53)
UAE (118)	Korea (128)	Algeria (45)	China (49)
Kuwait (103)	Germany (93)	Turkmenistan (45)	Turkey (45)
Venezuela (93)	Italy (74)	Netherlands (40)	France (43)
Canada (90)	Spain (60)	Indonesia (35)	UK (39)
Angola (84)	Netherlands (57)	Australia (26)	US (37)
Mexico (66)	France (57)	Nigeria (22)	Spain (30)

Source: IEA, 2014.

additional industries, which among others, relate to shipping, transport, port services, R&I, construction and engineering, wholesale or health and safety (EUROSTAT, 2008).

7.5 Innovation

7.5.1 Innovative Aspects and New Technology

The depletion of the more accessible offshore reserves (<400m depth) has pushed the search for hydrocarbons towards deepwater ($\sim$1500m) and ultra-deepwater (>1500m) areas. The use of the most advanced geophysical exploration techniques has enabled the detection of vast deposits at depths of up to 12 km. According to recent estimates, these deepwater/ultra-deepwater deposits account for more than 50% of the newly discovered larger offshore fields (i.e. fields with an estimated minimum recoverable reserve of 170 billion barrels). However, the high costs of production at such deep locations, puts in risk the economic viability of these deepwater/ultradeep water reserves (WOR, 2014).

In this sense, the development of subsea completion systems offers a series of advantages and alternatives to the traditional use of large platforms. Integrating several components for the processing of oil and gas (compressors, pumps, and separators), these systems are directly deployed onto the seabed, and underwater robots connect the different components to form large production ensembles (Devold, 2013). Among the advantages provided by these subsea systems, the following are innovation areas currently being explored:

- Simplification and efficiency improvement of the extraction, cleaning and processing processes: improves the performance of pumps and compressors and avoids the need for pumping to drilling platforms.
- Reduction of the amount of offshore production infrastructures.
- Increase of the exploitation radius: it is now possible to deploy within a wider radius several wells which pump to a common production station.
- Reduction of operating costs.

Although several fields operated by these subsea systems already exist (e.g., Gulf of Mexico, South America, Norway), its full commercial development still requires a number of technological innovations. In traditional platforms the maintenance of production infrastructures (pumps, compressors, etc.) may be relatively simple. However, these tasks turn highly complex when working subsea and at such great depths. To solve these issues, much of the innovation work in the offshore industry is focused on the development of robust, highly reliable and commercially operative submarine production systems (WOR, 2014).

7.5.2 Decommissioning and Cross-Sectoral Opportunities

Decommissioning is the dismantling process of O&G infrastructures once the exploitations reach the end of their lifecycle. Given the rapid decline of the reserves in the North Sea, most of the information on decommissioning relates to that area. It is estimated that all of the existing facilities will require decommissioning over the next 30 years (RAE, 2013). These operations will not only require strong economic investment (estimations in the North Sea exceed £30billion over the next 30 years) but also great human and technical capital. In any case, it can be expected that with the future depletion of existing exploitations, decommissioning will also acquire an increasing importance all over the world. However, it presents some interesting characteristics and possibilities for Blue Growth:

a) From a strict point of view, decommissioning is not considered a sector within the O&G industry. However, as a result of the decline of the sector, it may emerge forcefully as a new offshore and highly technical activity that may absorb and replace the loss of highly skilled employment from E&P activities.

b) Development of MUS/MUP activities: existing offshore O&G platforms can turn into valuable assets, as they can provide the infrastructure needed for the combination of maritime activities. However, based on previous experiences from the Gulf of Mexico (BOEM, 2007) the success of these combinations may vary greatly.

b.1) Active platforms: apart from being the owners of the platforms, oil companies assume elevated risks and costs in their production activities. For this reason, it cannot be forgotten that in any combination including the use of any active facility, the interests of the O&G industry will always predominate against additional industries. Thus, oil companies may be reluctant to combine and share their infrastructures with sectors that add risks to their operations without obtaining any direct benefit (e.g., aquaculture, immature renewable technologies). As an exception, the combination with wind energy can arise more interest, since the combination of these fully developed technologies can provide short term benefits to all parties.

b.2) <u>Obsolete platforms</u>: this seems to be the most suitable option for the combination of activities, since it reduces either the power positions between industries or the risks associated with the oil industry (e.g., spills, contamination of farmed species, etc.). It can also be an incentive for oil companies, which can consider it as an option to delay and reduce the expenses of the future decommissioning of their infrastructures (rental agreements, leases, etc.). However, this option also poses a series of challenges, related mainly to the regulatory framework. Despite some exceptions that enable derogation (e.g., sub-structures weighing more than 10,000 tonnes), most of regulations on decommissioning dictate the complete removal of all the infrastructures once they become obsolete (e.g., UNCLOS Article 60 (3); OSPAR Decision 98/3). Therefore, the possible re-use or reconversion of obsolete platforms must be regarded as a case-by-case study of the available options and applicable regulations.

7.6 Investment

Government incentives, public donors (e.g., EU) and private investors are the main funding source for oil companies (ECORYS, 2013).

The O&G industry is very lucrative not only for companies, but also for Governments, who receive substantial revenue through the taxation derived from the whole sector chain (from producing companies to final consumers). To ensure these revenues and attract and retain the investment in the sector, Governments often provide support to oil companies (Table 7.3).

Funding through their own reserves, private equity funds, bank loans or bonds are the main forms of private investment. While government investments seek to secure revenues for the development of their national economies, private investments try to maximise benefits. Thus, some private investors may opt for higher risk investments (in more hostile areas or new explorations), which provide the opportunity for greater benefits.

7.7 Uncertainties and Concluding Remarks

Although the dominance of the O&G industry as the principal energy supplier is expected to continue in the future, its offshore activities are in decline. The depletion of the more accessible reserves has driven the search and

Table 7.3 Common types of Government Interventions in Energy Markets

Intervention Type	Description
Natural resource access	Policies governing the terms of access to domestic onshore and offshore resources (e.g., leasing)
Cross-subsidy	Policies that reduce costs to particular types of customers or regions by increasing charges to other customers or regions
Direct spending	Direct budgetary outlays for an energy-related purpose
Government ownership	Government ownership of all or a significant part of an energy enterprise or a supporting service organization
Import/export restriction	Restrictions on the free market flow of energy products and services between countries
Information	Provision of market-related information that would otherwise have to be purchased by private market participants
Lending	Below-market provision of loans or loan guarantees for energy-related activities
Price control	Direct regulation of wholesale or retail energy prices
Purchase requirements	Required purchase of particular energy commodities, such as domestic coal, regardless of whether other choices are more economically attractive
Research and development	Partial or full government funding for energy-related research and development
Regulation	Government regulatory efforts that substantially alter the rights and responsibilities of various parties in energy markets or that exempt certain parties from those changes
Risk	Government-provided insurance or indemnification at below-market prices
Taxes	Special tax levies or exemptions for energy-related activities

Source: World Bank, 2010a.

exploitation of hydrocarbon resources towards more remote and therefore, more expensive areas. This, together with the development of new onshore techniques and the general fall of prices, can turn offshore activities economically unfeasible. While the big European companies operate at the global scale, the production in European territorial waters accounts for 9% and 13.8% of the total oil and gas consumption of the EU, respectively. Within territorial waters, most of the activity is developed in the North Sea, being Norway and UK by far the principal producers. The Caribbean is one of the main producers worldwide and the Mediterranean holds recently discovered enormous deposits. However most of these deposits are located outside the EU's territorial waters. Therefore, the decline of the North Sea reserves may limit even more the supply capacity of the EU and increase the need for importation of hydrocarbons.

In any case, the decline of the O&G sector also presents a series of opportunities and challenges for the development of BG industries, which principally rely on two fundamental aspects of the industry: skills and infrastructure.

- Skills. The extensive working experience in the marine environment, has resulted in a competitive industry which holds a highly skilled workforce. In this sense, the high human skill transferability and the experience dealing with adverse situations (both environmental and financial), are of great interest and a good example for the development of new offshore economic activities.
- Assets. The oil industry has developed and integrated technologies, operational models and equipment adapted to harsh marine environments. These include: vessels (e.g., platform supply vessels, tankers), underwater scanning and surveying methods (e.g., ROVs and AUVs), complex engineering techniques (e.g., floating anchoring systems, deep sea drilling, subsea systems) or personnel trained to work at sea. All these assets are of value for the future development of new offshore industries, especially for those that require large and challenging technical works (e.g., deployment of renewable energy devices, deep sea mining, offshore aquaculture, etc.).
- Infrastructure. The sector has many offshore installations, which could be an important support for BG sectors, and more specifically, for the development of MUS/MPP concepts. However, most of the current marine legislation dictates the dismantling of all the existing infrastructures once they reach the end of their lifecycle. Despite certain exceptions that permit for derogation, decommissioning is an extremely complex process. These difficulties not only rely on the huge financial and technical requirements, but also in the possible environmental and socioeconomic impacts (e.g., pollution, conflicts with fisheries/aquaculture, restrictions on the use of space, ecological impacts). At national levels, the development degree of policies and guidelines on decommissioning, varies depending on the maturity of the O&G industry and the previous experiences of countries. In this way, countries like Norway and UK have regulatory provisions on decommissioning in their legal frameworks. These requirements range from constitutional provisions to specific requirements (World Bank, 2010b). The creation of a common and clear regulatory framework not only will allow operators to know compliance requirements, but it can also set the conditions that will allow the conversion of existing infrastructures. Thus, for the moment,

the reuse for new purposes of an existing O&G infrastructure, will be subjected to a case by case study, in which either the type of infrastructure or the regulatory framework to which it is subject must be considered.

Annex 7.1 – Regulation in the Oil & Gas industry

Economic activities (reserves, licenses, exploration and production...)	EU	• Directive 94/22/EC on the conditions for granting and using authorisations for the prospection, exploration and production of hydrocarbons • Decision 1999/280/EC regarding a Community procedure for information and consultation on crude oil supply costs and the consumer prices of petroleum products • Decision 2003/796/EC on establishing the European Regulators Group for Electricity and Gas • Regulation (EC) 715/2009 on conditions for access to the natural gas transmission networks • Directive 2009/73/EC concerning common rules for the internal market in natural gas and repealing Directive 2003/55/EC • Directive 2009/119/EC imposing an obligation on Member States to maintain minimum stocks of crude oil and/or petroleum products • Regulation (EU) 994/2010 concerning measures to safeguard security of gas supply • Regulation (EU, Euratom) 617/2011 concerning the notification to the Commission of investment projects in energy infrastructure within the European Union and repealing Regulation (EC) No. 736/96
	US	• Outer Continental Shelf Lands Act (OCSLA) • Oil and Gas Royalty Management Act • Petroleum Marketing Practices Act
	Mexico	• Ley de Hidrocarburos
	Venezuela	• Ley Orgánica de Hidrocarburos • Ley Orgánica de Hidrocarburos Gaseosos

(*Continued*)

Annex 7.1 Contniued

	Colombia		• Ley 1274 de 2009 por la cual se establece el procedimiento de avalúo para las servidumbres petroleras
	Trinidad and Tobago		• The Petroleum Act • The Petroleum Regulations • The Petroleum Taxes Act
Environmental protection	Regional conven-tions	OSPAR (North East Atlantic)	• Annex III on elimination of offshore pollution sources • Recommendation 2010/18 on the prevention of significant acute oil pollution from offshore drilling activities
		HELCOM (Baltic)	• Annex VI on prevention of pollution from offshore activities
		Barcelona (Mediter-ranean)	• Protocol for the protection of the Mediterranean sea against pollution resulting from exploration and exploitation of the continental shelf and the seabed and its subsoil
		Cartagena (Caribbean)	• Oil spills protocol
	EU		• Directive 2008/56/EC. Marine Strategy Framework Directive • Directive 2013/30/EU on safety of offshore oil and gas operations.
Liability and compensation for damages	International		• International law principles
	Barcelona convention (Mediterranean)		• Protocol for the protection of the Mediterranean sea against pollution resulting from exploration and exploitation of the continental shelf and the seabed and its subsoil • Guidelines for the determination of liability and compensation for damage resulting from pollution of the marine environment in the Mediterranean sea area (not binding)
	EU		• Directive 2004/35/EC on environmental liability with regard to the prevention and remedying of environmental damage • Directive 2013/30/EU on safety of offshore oil and gas operations.

References

ANH (2016). Mapa de Tierras. Agencia Nacional de Hidrocarburos. (http://www.anh.gov.co/Asignacion-de-areas/Paginas/Mapa-de-tierras. aspx).

BOEM (2007). Alternate Uses of Existing Oil and Natural Gas Platforms on The OCS. In: Final Programmatic Environmental Impact Statement for Alternative Energy Development and Production and Alternate Use of Facilities on the Outer Continental Shelf. Bureau of Ocean Energy Management (http://www.boem.gov/Renewable-Energy-Program/Regulatory-Information/Guide-To-EIS.aspx).

BOEM (2016). Geographic Mapping Data in Digital Format-Pipelines. Bureau of Ocean Energy Management. (https://www.data.boem.gov/ homepg/data_center/mapping/geographic_mapping.asp).

BP (2015). BP Statistical Review of World Energy. 48 pp. (http://www.bp. com/en/global/corporate/energy-economics/statistical-review-of-world-energy.html).

Devold, H. (2013). Oil and gas production handbook. An introduction to oil and gas production, transport, refining and petrochemical industry. ABB AS. 108pp. (http://dspace.bhos.edu.az/xmlui/bitstream/handle/1234567 89/944/Oil%2Band%2Bgas%2Bproduction%2Bhandbook.pdf?sequ ence=1).

EC (2013). Implementation of the Communication on Security of Energy Supply and International Cooperation and of the Energy Council Conclusions of November 2011. COM(2013) 638 final, Brussels. 15 pp. (http://eur-lex.europa.eu/legal-content/EN/TXT/?uri=SWD:2013: 0334:FIN).

ECORYS (2013). Study on Blue Growth and Maritime Policy within the EU North Sea Region and the English Channel. Annex III B – Sector Analysis – Offshore oil and gas. Rotterdam/Brussels. 23 pp. (https://web gate.ec.europa.eu/maritimeforum/sites/maritimeforum/files/Annex%20 III%20B%20-%20Final%20sector%20Analysis%20oil%20an %20gas.pdf).

EIA (2016). Oil: Crude and petroleum products explained. Where our oil comes from. https://www.eia.gov/energyexplained/index.cfm?page=oil_ where (last accessed May 2017).

ENTSOG (2015). Transmission capacity map 2015. European Network of Transmission System Operators for Gas (http://www.entsog.eu/public/

uploads/files/maps/transmissioncapacity/2015/ENTSOG_CAP_MAY20 15_A0FORMAT.pdf).

EUROSTAT (2008). NACE Rev. 2. Statistical classification of economic activities in the European Community. Eurostat Methodologies and Working Papers, Luxembourg. 369 pp. (http://ec.europa.eu/eurostat/doc uments/3859598/5902521/KS-RA-07-015-EN.PDF).

EUROSTAT (2016). Structural Business Statistics-Main Indicators. Persons employed by NACE Rev.2: Extraction of crude petroleum and gas. (http://ec.europa.eu/eurostat/web/structural-business-statistics/data/main-tables).

Hays (2013). Oil and Gas global salary guide – Review of 3013 outlook for 2014. 36 pp. (http://www.hays.com/cs/groups/hays_common/@og/@co ntent/documents/promotionalcontent/hays_920901.pdf).

Hays (2015). Oil and Gas Global Salary Guide 2015. 44 pp. (http://hays.com/ oil-and-gas/SalaryGuide/2015SalaryGuide/index.htm).

IEA (2014). Key World Energy Statistics, 2014. International Energy Agency. 82 pp. (https://www.iea.org/publications/freepublications/publi cation/key-world-energy-statistics-2014.html).

JRC (2015). Offshore Oil and Gas production in Europe. Joint Research Centre-European Commission. http://euoag.jrc.ec.europa.eu/node/63 (last accessed August 2015).

Lujala, P., Rød, J. K., Thieme, N. (2007). Fighting over oil: A new dataset. *Conflict Management and Peace Science* 24(3): 239–256.

Mitchell, J., Marcel, V., Mitchell, B. (2012). What next for the Oil and Gas industry? London, Chatham House (The Royal Institute of International Affairs), London. 128 pp. (https://www.chathamhouse.org/publications/ papers/view/186327).

Nord Stream (2014). Secure Energy for Europe. The Nord Stream Pipeline Project (2005–2012). 139 pp. (http://www.nord-stream.com/press-info/ library/).

NRCan (2010). Review of Issues Affecting the Price Crude of Oil. Ottawa Petroleum Resorces Branch-Energy Sector. Natural Resources Canada-Ressources Naturelles Canada, Ottawa. 37 pp. (http://www.nrcan.gc.ca/ energy/crude-petroleum/4557).

Owen, N. A., Inderwildi, O. R., King, D. A. (2010). The status of conventional world oil reserves – Hype or cause for concern? Energy Policy 38(8): 4743–4749.

Quest Offshore (2011). United States Gulf of Mexico Oil and Gas Natural Gas Industry Economic Impact Analysis – The Economic

Impacts of GOM Oil and Natural Gas Development on the U.S. Economy. Prepared for the American Petroleum Institute (API) and the National Ocean Industries Association (NOIA). 152 pp. (http://gulfeconomicsurvival.org/pdfs/QuestGOMEconomicStudy.pdf).

RAE (2013). Decommissioning in the North Sea. A report of a workshop held to discuss the decommissioning of oil and gas platforms in the North Sea. Royal Academy of Engineering, London. 15 pp. (http://www.raeng.org.uk/publications/reports?q=decommissioning).

WOR (2014). Oil and Gas from the sea. World Ocean Review. Marine Resources-Opportunities and Risks. 8–51 pp. (http://worldoceanreview.com/en/wor-3-overview/oil-and-gas/).

World Bank (2010a). Subsidies in the Energy Sector: An overview. Background Paper for the World Bank Group Energy Sector Strategy. July 2010. 115 pp. (http://siteresources.worldbank.org/EXTESC/Resources/Subsidy_background_paper.pdf).

World Bank (2010b). Towards Sustainable Decommissioning and Closure of Oil Fields and Mines: A toolkit to Assist Government Agencies. World Bank Multistakeholder Initiative, 186 pp. (http://siteresources.worldbank.org/EXTOGMC/Resources/336929-1258667423902/decommission_toolkit3_full.pdf).

World Bank (2015). World development indicators. http://databank.worldbank.org/data/reports.aspx?source=world-development-indicators&Type=TABLE&preview=on (last accessed September 2015).

Xing, Y., Karimirad, M., Moan, T. (2014). Modelling and analysis of floating spar-type wind turbine drivetrain. *Wind Energy* 17: 565–587.

8

Shipping: Shipbuilding and Maritime Transportation

Irati Legorburu*, Kate R. Johnson
and Sandy A. Kerr

Heriot-Watt University, Scotland
*Corresponding Author

Executive Summary

Shipbuilding and maritime transportation are the main sectors around which the shipping industry is built. Despite their great differences, both sectors are closely related, showing a strong and direct dependency on the performance of international markets.

Clearly dominated by Asian countries, the industry is highly competitive and globalised. To face this competition, the European shipbuilding industry has adopted a specialisation strategy and focused its activities to the construction of high value-added vessels. Largely thanks to its location along major trade routes the European maritime transportation companies have a leading position in the global industry.

The European shipping industry, and more specifically the shipbuilding sector, offers a number of important opportunities for the development of Blue Growth sectors. The need for highly-specialised new vessels is in line with the technological requirements of many of the BG sectors (e.g., development of renewables, seabed mining or biotechnologies). Also, its long working experience may turn into an important source of knowledge for emerging maritime industries. Underlining the characteristics of the industry in the different studied basins, this chapter describes the main features and socio-economic impacts of the European shipping industry (markets, industrial structure, employment, skills, etc.).

8.1 Introduction

Accounting for about 80% of global trade by volume, maritime transportation is the most important conduit for international trade. Population growth, increasing standard of living, industrialisation, exhaustion of local resources, road congestion, and elimination of trade barriers, all contribute to the continuing growth in maritime transportation (Christiansen et al., 2007). Thus, the shipping industry has a crucial role in the increasingly globalised economy.

Shipbuilding and maritime transportation are the two main activities around which the shipping industry is built – Asian countries are the current leaders (ECORYS, 2009). This is attributable to: (i) the historic shipbuilding tradition of Korea and Japan; and (ii), the rapid economic development of China. However, over the last years, new countries have emerged as potential shipbuilding nations (e.g., Brazil, India, Philippines, and Vietnam).

Together, China, Japan and S. Korea account for more than the 80% of the market for new orders. Ship construction is a long process, in which vessels are delivered several years after order. Thus, taking into account the order book increase of Asian shipyards (mainly China), their dominance is expected to continue in the following years. However, despite of the leading position of Asian countries in the industry, Europe remains as an important contributor to the development of this sector (SeaEurope, 2013a).

8.1.1 Sector Description: Shipping Cycles

The performance of the shipping industry is highly influenced by markets, as it is subjected to the constant changes of world trade volume. As pointed out by Stopford (1997), this is clearly explained with a simple example: "If the active merchant fleet is 1000 m. deadweight tonnage (dwt), and seaborne trade grows by 5 per cent, this will generate demand for an additional 50 m. dwt of new ships. If, in addition, 20 m. dwt of ships are scrapped, the total requirement for new vessels will be 70 m. dwt. If, however, instead of growing by 5 per cent seaborne trade remains at the same level, then there will be no need to expand the fleet and demand will be only 20 m. dwt. Taking the argument a step further, if seaborne trade falls by 5 per cent there will not be any demand for new ships". Thus, shipping is a highly cyclical industry, turning it into an irregular industry.

To understand how the shipping industry works, a good knowledge of these cycles is needed. As a result of shipping cycles the supply and demand for ships is balanced (Stopford, 1997). If the supply is low, the market rewards investors willing-to-pay high freight rates. In the contrary, if the supply is high, the market squeezes the cash flow until the owners waive the offer and ships are scrapped. Therefore, cycles not only affect shipbuilding activities, but also maritime transportation businesses. As shown in Table 8.1, shipping cycles consist of 4 stages.

Table 8.1 Stages and characteristics of shipping cycles

Stage	Characteristics	Consequences
1: Trough	• Evidence of shipping overcapacity. • Freight rates fall to the operating cost of the least efficient ships in the fleet • Sustained low freight rates and tight credit create a negative net cashflow which becomes progressively greater	• Ships queue up at loading points and vessels at sea slow steam to save fuel and delay arrival • Shipping companies short of cash are forced to sell ships at distress prices, since there are few buyers. • The price of old ships falls to the scrap price, leading to active demolition market.
2: Recovery	• Supply and demand move towards balance • Markets remain uncertain and unpredictable. • Liquidity improves	• The first positive sign of a recovery is positive increase in freight rates above operating costs, followed by a fall in laid up tonnage. • Spells of optimism alternate with profound doubts about whether a recovery is really happening (sometimes false recovery periods!). • Liquidity improves second-hand prices rise and sentiment firms.
3: Peak/Plateau	• All the surplus has been absorbed • Freight rates are high, often 2–3 times operating costs. • Only untradeable ships are laid up • Owners become very liquid • Second-hand prices move above 'book value' and prompt modern ships may sell for more than the newbuilding price	• Markets enter a phase where supply and demand are in tight balance: the peak may last a few weeks or several years, depending on the balance of supply/demand pressures. • The fleet operates at full speed • Banks are keen to lend • There are public flotations of shipping companies. • The shipbuilding order book expands, slowly at first, then more rapidly

(*Continued*)

Table 8.1   Continued

Stage	Characteristics	Consequences
4: Collapse	• Supply overtakes demand • Factors such as the business cycle, the clearing of port congestion and the delivery of vessels ordered at the top of the market cause the downturn • Spot ships develop in key ports • Freight rates fall • Liquidity remains high	• Markets move into the collapse phase • Sentiment about these factors can accelerate the collapse into a few weeks • Ships reduce operating speed and the least attractive vessels have to wait for cargo • Sentiment is confused, changing with each rally in rates

Source: Stopford, 1997.

8.1.2 Importance of the Shipping Industry for the BE and BG Sector

The shipping industry directly contributes €56 billion to EU GDP (Oxford Economics, 2015). Ships and maritime transportation are necessary for most of the activities related to BE/BG, e.g: Sea mining and fishing activities; construction and maintenance of offshore infrastructures (oil & gas and MUP platforms, offshore renewable and aquaculture facilities); tourist transportation; on board biotechnological researches; etc. Therefore, a healthy and productive shipping industry will facilitate the successful development of BE/BG objectives.

8.2 Market

Due to their high reliance on market's performance, product demand and prices in shipping are highly volatile (SeaEurope, 2013a; Stopford, 1997). As a result, the current poor economic situation has strongly impacted the shipping industry, which is reflected in the dramatic demand decrease for newbuilding and the low levels of freight rates (SeaEurope, 2013a; UNCTAD, 2014).

8.2.1 Product Demand and Price

Despite market fluctuations the cost of constructing a ship usually breaks down as follows (ECORYS, 2009): materials account for around 53% of shipbuilding costs, while overhead and direct labour costs represent around 47%. The openness and competitiveness of the shipbuilding industry is an additional factor influencing the prices of ships, which depends on the amount of shipyards competing for a given order (Stopford, 1997).

Regarding transportation, despite the moderate growth in world trade volume, freight rates remain low. This is due to: (i) the poor world economic development; (ii) the volatile demand; and (iii) the persistent supply overcapacity of the sector (UNCTAD, 2014).

8.2.2 Market Trends

Following the performance of global markets, the increase in demand for shipping activities is being driven by developing economies (principally Asian). This is occurring at the expenses of western economies, where despite signs of recovery, the future of many developed economies is still uncertain (DNV, 2012).

8.2.2.1 Shipbuilding

There are two main factors responsible for the dominance of Asian shipyards over the Europeans yards (ECORYS, 2009):

- Labour costs: Europe, Japan and S. Korea have similar labour costs, which are significantly higher than those from China.
- Steel price: Steel is the main raw material for shipbuilding and the one that determines to a greater extent the final price of ships. The global steel production and consumption is dominated by Asian countries (principally China). This creates a disadvantage for European shipyards that have to pay higher prices for raw materials.

In order to face this adverse environment the European shipbuilding industry has adopted a clear specialisation strategy, by focusing in the construction of high value-added technical and complex ships (SeaEurope, 2013b). Such a strategy permits the European shipbuilding industry to reduce the effect of having higher labour costs (ECORYS, 2009). Finland, France, Italy and the UK specialise in passenger vessels and ferries, Denmark in container ships and the remainder of the countries show a more diversified portfolio (IKEI, 2009).

8.2.2.2 Transport

The global maritime trade of goods continues increasing. Geographically, the growth rate of trade routes that connect developing economies (i.e., Middle East-Asia, South America-Africa-Asia, Europe-Middle East) has more than doubled that of mainline trades that connect with developed economies (i.e., Asia-North America, Asia-Europe).

As a result of the huge orderbook for newbuildings made during the economic boom in the early 2000s, the global fleet is characterised by its important supply overcapacity. Growth rates of developed countries remain low and the increasing demand for maritime trade of developing economies is not able to balance the supply and demand sides of the maritime transport market.

However, overcapacity does not affect equally all maritime transportation markets (DNV, 2012). Driven by the Oil and Gas sector, there is an increasing demand for Liquefied Natural Gas (LNG) tankers and specialised offshore vessels. Also, the demand for cruise tourism shows an increasing pattern. With regard to the latter, the sea basins considered may be well positioned as the cruise tourism market is mainly developed in the Caribbean, followed secondly by the Mediterranean and the remainder of European sea basins in the third position (CLIA, 2015).

8.2.3 Future Supply and Demand Gaps

Closely related with the growth of world economy, the future of the shipping industry remains uncertain. However, the need for new builds in the future will not only rely on the economic environment, but also in the regulatory framework. From the economic point of view, overcapacity times are characterised by the lack of investments in new builds and active scrapping markets, which last until the supply and demand are balanced. Therefore, a low demand for shipbuilding products must be expected at this point. In contrast, the upcoming regulatory framework can imply a source for new build demand. These regulations, mainly related with environmental and energy efficiency issues (e.g., polluting/greenhouse gases, ballast waters, efficient fuels), require the renewal or adaptation of the fleet (see Section 8.5). Given the poor economic situation, the investment capacity on technological development of the shipping industry is limited. In this context, the future for shipping companies appears extremely challenging as they will have to face a period where clear strategic decisions will be needed (DNV, 2012).

8.3 Sector Industry Structure and Lifecycle

8.3.1 Lifecycle

Shipping is an old and strongly fluctuating industry, which for many years has suffered from the image of being a declining industry. However, current industry requires ever larger ships and more sophisticated, safe and environmentally-friendly ships (IMO, 2012). In line with the growing demand for highly advanced vessels, new technologies and practices have emerged over the last years. These mainly relate to the improvement of naval architecture and engineering or the implementation of green shipping practices (that reduce among others fuel consumption and pollution from shipbuilding and transportation activities). Therefore it can be considered that the shipping industry is facing a new growth stage (OECD, 2016).

8.3.2 Industry Sectors and Segmentation

The sectors, sub-sectors and segmentation of the shipping industry are summarised in Table 8.2.

The shipbuilding industry consists of four main sectors, i.e., ship construction, marine equipment, scrapping and naval ships. Consequently six segments are defined within the sector: tankers, dry bulk, container, passenger, specialised and mega yachts.

Maritime transportation is defined in four main subsectors: deep sea, short sea, domestic ferries and cruises.

8.3.3 Horizontal and Vertical Integration

Shipbuilding and maritime transport are separate industries that are inextricably linked and mutually dependent. In both, the degrees of vertical and horizontal integration are high (Figure 8.1).

Regarding the shipbuilding industry, the horizontal cooperation between shipyards is a very common practice. Shipbuilding requires a huge production capacity, which often is beyond the production capacity of the main contractor. As such, the important flow of subcontracts among main contractors permits them to maintain their production balances. In addition, it must be noted that the ship construction process is characterised by long development phases followed by long manufacturing phases. As a result, shipyards have to face periods where they operate at full capacity while in others they have to manage their capacities. Therefore, the horizontal co-operation between shipyards, permits them to manage orders and manufacturing personnel in times of temporary over or under-capacity (Balance, 2014). Vertical co-operation between the ship construction and the marine equipment sector exist due to the high complexity and fragmentation of the products needed in shipbuilding. This principally occurs into two main forms: supplies and subcontracts (Balance, 2014).

Table 8.2 Sectors and segments of the shipping industry (Author's compilation based on Stopford, 1997; Christiansen et al., 2007; ECORYS, 2009)

Shipbuilding	Sectors	Ship construction (incl. Newbuilding and Repair & Conversion)	Builds the hull and basic structures of a ship. In order to increase efficiencies in shipbuilding, there is an increasing trend of splitting new building and repair activities in different shipyards. As a result, a geographical displacement of ship repair activities is occurring towards areas close to the major transportation routes.
		Marine equipment	Defined as *"the supply industry to the shipyards"*. Increasing outsourcing and subcontracting of shipbuilding activities gave rise to the marine equipment industry. Accounts for a large part of the value-added of a ship (could be as high as 70–80%). Given the wide variety of products and services provided by the marine equipment industry it is considered a very heterogeneous sector.
		Ship scrapping	In charge of dismantling ships. It is a very basic industry in which either companies or their markets are normally located in developing countries (lower labour costs). Obtained steel panels use to be rerolled and reused in the local markets as raw materials (e.g., construction industry). Due to these characteristics this sub-sector is not of special relevance for European shipyards.
		Naval vessels	As it is dominated by political and strategic factors, it differs from the competitiveness point of view. In contrast with the previous sectors it is characterised by its relatively stable market.
	Segments	Tankers	Designed to transport liquids and gases in bulk (e.g., oil, gas, juice, wine, etc.). There exist different types of tankers depending on their size: Panamax (up to 70,000 DWT); Aframax (70,000–120,000 DWT); Suezmax (120,000–200,000 DWT); Very Large Crude Carriers (200,000–325,000 DWT); and, Ultra Large Crude Carriers (325,000–550,000 DWT).
		Dry bulk carriers	Less sophisticated, but highly efficient. Intended for the transport of dry unpacked cargo (e.g., coal, cement, mineral ores or grain). Classified depending on their size characteristics: Handies (10–49,999 DWT); Panamax (50,000–79,900 DWT); and, Capesize (>80,000 DWT).
		Container ships	A revolution for the shipping industry. Containerisation of goods facilitated the mechanised handling of the cargo and reduced burglary. Despite of some exceptions (e.g., cars), containers are the cargo standard unit for almost every manufactured item. Standard containers are 20 feet and 40 feet in length.

Maritime Transportation		
Sectors	Passenger ships	Two main categories: cruise ships and ferries (i.e., "fun" or "function"). Cruise ships are designed for leisure purposes and ferries to move people (and, vehicles) on regular itineraries quickly and cheaply. A wide variety of ferry types exist: ranging from small passenger ships to big Ro-Ro ferries (Roll-on Roll-off), that have the capacity to carry thousands of passengers and hundreds of vehicles.
	Specialised vessels	Vessels with some onboard machinery/equipment to perform specific tasks related to different marine industries (e.g., offshore vessels, dredgers, chemical tankers or LPG-LNG carriers).
	Mega-Yachts	Luxury yachts of 24 meter or more in length. Professionally crewed, very expensive and privately owned sailing or motor ships.
	Deep sea	Inter-continental transportation that employs the larger size vessels. Deep-sea vessels spend long periods of time at sea.
	Short sea	Intra-continental transportation that employs the smaller size vessels. It redistributes cargo delivered to continental centres (e.g., Hong Kong or Rotterdam) by deep sea vessels, competing with land based transport. Subject to many political restrictions (e.g., cabotage).
	Domestic ferries	Transport of people, vehicles and cargo. Often used as shuttle service between ports. Short routes are served by small vessels, while large liners are used in longer routes.
	Cruises	Passenger transport with leisure purposes. Transportation on itself may not be the principal purpose, as ship's amenities are part of the experience.
Segments	Liner	Operate according to a published itinerary and schedule. Usually control container and general cargo vessels.
	Tramp	Operate with no fixed itinerary and schedule. Transport the available cargo under contracts of affreightment. Usually control tankers and bulk carriers.
	Industrial	Operators that own the cargoes shipped and control the vessels used in transportation. Strive to minimise the cost of shipping the cargo of vertically integrated companies (e.g., oil, chemicals, ores...).

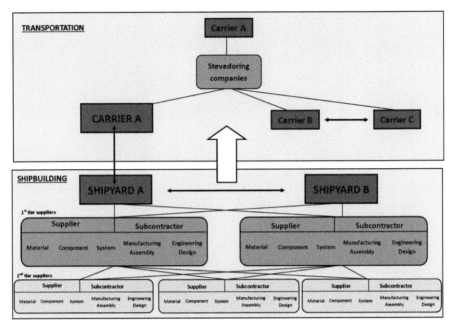

Figure 8.1 Vertical and Horizontal integration among industries in the shipping sector. Author's interpretation after Balance 2014 and Cariou 2008.

Considering maritime transportation, the size of carrying companies is an important factor determining horizontal and vertical cooperation issues. Although depending on market conditions the strategies vary, two main approaches are predominant. On the one hand, the biggest carrying companies opt for direct investment in new vessels in order to expand their market share. On the other hand, strategic alliances, slot exchange agreements or mergers and acquisitions permit small companies to increase the quality of their services (in terms of e.g., frequency, spatial extension) without investments. Finally, the transportation industry relies strongly on the "Hubs and Spokes" system, which uses strategically located points as centres for the further redistribution of goods in a given region. Thus, it implies a huge need for efficient transhipments, for which stevedoring companies become crucial to ensure and support the redistribution of goods at smaller spatial scales (Cariou, 2008).

8.3.4 Centres of Activity in Europe and Caribbean

Figure 8.2 shows some relevant statistics for the European shipping sector, considering the whole sector, the Mediterranean and Atlantic countries are of special relevance.

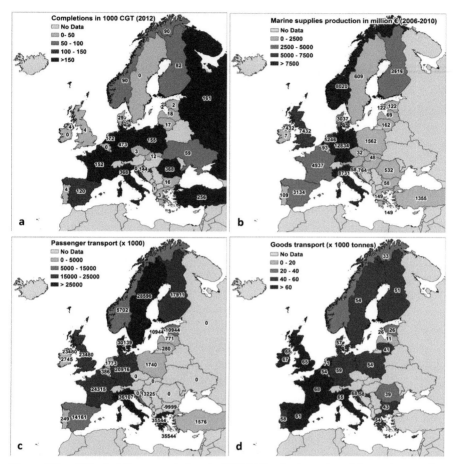

Figure 8.2 a) Newbuildings completions in 2012; b) Production value of the marine supplies industry in 2012; c) Passenger transport; d) Goods transport.

Sources: Balance, 2014; EUROSTAT; SeaEurope, 2013a.

Considering shipbuilding, Germany and Italy are leading countries either in the ship construction and marine equipment subsectors, followed by countries such as Turkey, France, Spain, Romania or Poland. In general terms, leading countries in ship construction also have an important marine supply industry. As an exception to this generality, countries like the UK, Sweden or Greece can be mentioned, which in contrast to their low activity in ship construction, are highly active in the supply industry. This might be due to: (i) the high technological development level of UK and Sweden; and (ii) the geographical location of Greece (close to important freight transportation and cruise routes). Although there is no data available about North-African countries bordering the Mediterranean, their activity in shipbuilding can be considered as minor (SeaEurope, 2013a). However, given their location, their activities are presumably oriented to the repair of ships crossing the principal Mediterranean transportation routes.

In relation to passenger transportation, some specific aspects can be highlighted. Apart from being located in main cruise routes, the high numbers of passenger transport in countries like Italy, Croatia, Greece, Sweden and Denmark could be due to the interisland transportation of national passengers in the case of the former and to the use of maritime transport as a way to connect Nordic countries and central Europe for the latter.

Estimates of activity in the Caribbean basin are more difficult tasks. On the one hand, the high number of non-unified countries, makes it difficult to obtain reliable information on the industry. On the other hand, Caribbean shipping companies are commonly subsidiaries or partners of larger foreign companies, as is probably their economic performance. However, as a general picture, several aspects can be highlighted: (i) most of the shipping activity will probably be built around tourist cruises; (ii) the only country with a relatively significant weight in the global shipbuilding industry is the USA (SeaEurope, 2013a); (iii) after being in a dormant state for the last 20 years, the Mexican institutions and business corporates are making efforts to reactivate the national shipbuilding industry (principally with a view to support the national oil & gas industry); and, (iv) given the geographical location along major transportation and cruise routes, most of the shipbuilding activity in non-US countries probably relies on repair facilities, principally for cruises and yachts (ECORYS, 2009).

8.3.5 Nature of Ownership

Rather than from a country basis, ownership of shipyards must be considered from a globalised perspective. In order to rationalise their production and make use of global competitive advantages (e.g. lower labour costs, technological advancements, expansion of transport routes), mergers and acquisitions between companies are common. leading to the emergence of major conglomerates dominating the world industry. As an example some of the most important European shipyards could be mentioned, which despite of being located in Europe their ownership has partially changed to Asian hands.

Another characteristic aspect of ownership in maritime transportation is the difference between the "beneficial ownership location" and the "ultimate owner's nationality". While the former relates to the country in which the company that has the main commercial responsibility for the vessel is located (i.e., the registration flag), the latter refers to the nationality of the ship's owner, independent of the location of the vessel. Nowadays, most of the ships of the world fleet have a flag of registration different to the economy/country of their owner. The registration under a flag of a different country (also known as flag of convenience), permits reduced operating costs or can avoid the more restrictive regulations of the shipowner's country. In terms of owner's nationality, as observed for the shipbuilding industry, Asian and European countries appear as market leaders (Table 8.3).

8.3.6 Rules and Regulations

As a global industry, shipping is regulated by a series of international regulations. These conventions have been agreed in international forums such as the International Maritime Organisation (IMO), the World Trade Organisation (WTO) or the Organisation for Economic Co-operation and Development (OECD). They are ratified by the signing parties, to establish, among others, the rules and regulations in: technical matters; maritime safety and security; marine pollution; and liability and compensation (Annex 8.2).

Table 8.3   Top 20 fleets by beneficial ownership location and ultimate owner's nationality

Top 20 Fleets by Beneficial Ownership	Top 20 Fleets by Owner's Nationality
1. Panama	1. Japan
2. Liberia	2. Greece
3. Marshal Islands	3. Germany
4. Hong Kong, China	4. China
5. Bahamas	5. United States
6. Singapore	6. United Kingdom
7. Greece	7. Norway
8. Malta	8. Republic of Korea
9. China	9. Denmark
10. Cyprus	10. Hong Kong, China
11. Italy	11. Taiwan Province of China
12. Japan	12. Singapore
13. United Kingdom	13. Italy
14. Germany	14. Russian Federation
15. Norway	15. Canada
16. Republic of Korea	16. Turkey
17. United States	17. Malaysia
18. Isle of Man	18. India
19. Denmark	19. France
20. Antigua and Barbuda	20. Belgium

Source: IMO, 2012.

8.4 Working Environment

8.4.1 Economic Climate

Considering the current global economic situation, shipping cycles provide a good explanation of the close relationship between the economic climate and the performance of the shipping industry. Considering activities separately, the economic downturn has slowed the increase in global trade, but in the case of shipbuilding the impact has been very strong. Besides, instead of being an immediate impact, the economic downturn has taken some time to hit shipbuilding. This has been due to two main reasons:

(i) The long construction and delivery times of ships (2–4 years). Demands for new orders usually take place during economic prosperity, while their delivery can coincide with depression periods (with the subsequent high risk of overcapacity).

(ii) The new ordering boom occurred over the past decade. Although shipbuilders have had to face an increasing number of order cancellations, already committed contracts have soften to some extent the immediate impact of the economic crisis.

8.4.2 Employment, Skills and Migration

The shipbuilding and maritime transport industries employ more than 200,000 people in Europe, a value that increases significantly if indirect employments derived from outsources and subcontracts are entered into accounts (ECORYS, 2009; EC, 2011). The women workforce accounts only for 2% of the total (ITWF, 2015) and is mainly active in the cruise and ferry sectors.

In relation to its workforce, the European shipping sector faces two main challenges (Hart and Schotte, 2007):

(i) Ageing of the workforce. Although ageing affects the whole European workforce, this is slightly higher in the case of the shipping industry. With a large part of the workforce over 50 years old, a high loss of employees due to retirement might be expected in the coming years.

(ii) Specialisation of the workforce. Given its high level of specialisation, the European shipbuilding industry requires a highly specialised workers. However, as a result of either the ageing/retirement of the workforce or the further technological specialisation degree of the industry, a need for skilled personnel must be expected in the future. In relation to seafarers, a demand of 45,000 officers and 145,000 ratings has been reported recently (EC, 2011).

The downturn in the shipping sector does not help to overcome these labour challenges, and as such the European industry might face a lack of national workforce in the future. Moreover, the contribution of workers from outside the EU is increasing (e.g., Philippines, Ukraine, Russia...), due to cheaper employment costs. At the EU level, eastern countries have a surplus of cheaper and younger workforce (e.g., Poland, Bulgaria and Romania).

8.4.3 Economic Indicators

8.4.3.1 Gross Domestic Product

In comparison with other economic sectors, EU's shipping industry is a highly productive sector (Table 8.4): Accounting for 1% of EU's GDP, its contribution was estimated in €147 billion in 2013. Each worker contributes with €85,000 to EU GDP (EU average €53,000) and the GDP multiplier of the industry is 2.6. This means that for every €1 million of GDP the industry creates another €1.6 million is created elsewhere in the EU economy (Oxford Economics, 2015).

8.4.3.2 Wages

Figure 8.3 shows the wages for different categories of workers in the shipping sector. The first three refer to the transportation subsector, and the latter, to the shipbuilding sector. The range of wages is very large in Europe, defined by high wages in Western Europe, falling the further east in Europe. In the specific case of shipbuilding, it is estimated that in Europe wages account for 21–23% of the total cost of a ship (ECORYS, 2009).

Given the following reasons these figures should be treated with caution (EC, 2011): influence of additional elements of working conditions (leave ashore, voyage length, specific national fiscal facilities); lack of published information on real seafarers' salaries at national level; difficulties to understand the applicability of collective agreements and bonuses; or aspects

Table 8.4 Economic impact of the EU shipping industry (from Oxford Economics, 2015)

Impact	People Employed	Contribution to EU GDP
Direct	615,000	€56 billion
Indirect	1,100,000	€61 billion
Induced	516,000	€30 billion
TOTAL	2,200,000	€147 billion

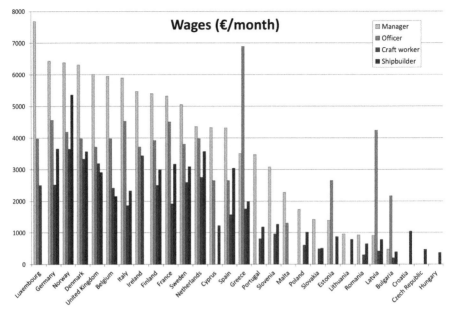

Figure 8.3   Wage distribution among EU countries for different working categories in the shipping sector.

Sources: EC, 2011; EUROSTAT.

related to the flag of convenience. With these considerations in mind and despite differences among categories, the results reflect clear differences, with western countries having higher salaries.

8.4.3.3 Export potentials

The export potentials of the European shipbuilding industry can be discussed in two main aspects:

(i) <u>Skills:</u> due to its high specialisation level, the workforce of the European shipbuilding sector is highly skilled. In fact, although mainly driven by their wage differentials and labour youth, Eastern countries act as net exporters of skilled workers (mainly to western European countries and US) (t'Hart and Schotte, 2007).

(ii) <u>Shipping products:</u> Considering CESA (Community of European Shipyards Association) countries, two main patterns can be observed, which can be related to labour costs and the specialisation on high value-added segments, respectively (CESA, 2013b). On the one hand, eastern countries act generally as net exporters. Favoured by their low labour costs, the market share of these countries is mainly focused in non-complex vessels. As such, they can be seen as valuable "low-cost" countries for buyers which do not need highly technical vessels. On the contrary, countries with high labour costs like Finland, Germany or France also show high export rates, which are due to their important specialisation in high value-added segments. The same occurs to the marine equipment sub-sector, where despite the current growth of Asia in this field, European companies are leaders and act as net exporters (Balance, 2014).

8.4.4 Infrastructure and Support Services

Considering that maritime transport has been the main form of trade between developed economies, these countries have large infrastructures of support and distribution. However, emerging economies are expanding quickly, at the expense of the slowdown in developed countries. The globalisation of the economy requires new epicentres of distribution and the use of even bigger ships for the optimisation of goods transport. Therefore, the existing shipping infrastructure might be in trouble arising from the difficulty to host and manage the growing demand outside the historical trade routes. In this sense, the development of offshore shipping infrastructures appears to be an effective solution. They would also potentially boost blue growth sectors, since they would have the ability to provide the necessary infrastructure for the development of different offshore economic activities.

8.4.5 Cluster and "Triple Helix" Features

By adopting the "triple helix" approach (cooperation between government-industry-university), maritime clusters can serve as a basis for guiding the economic and sector policy development, which would facilitate the growth of maritime activities (DSA, 2010). Ideally, these clusters require some essential "core sectors" around which surrounding industries depend for demand and activity. Due to the strong linkage with other maritime industries (e.g., offshore oil and gas, offshore renewables, cruise tourism, capture fisheries or marine aquaculture), the shipbuilding industry has a crucial role for the future development of Blue Growth (OECD, 2016).

8.5 Innovation

Table 8.5 shows the main drivers, market opportunities, barriers and technological responses for innovation in the shipping sector. Driven principally

Table 8.5 Drivers, market potential, barriers and technological responses for innovation in the shipping sector

	Trend	Drivers	Market Potential	Technological Barriers			Technological Response
				Development	Scaling Up	Expression of Demand	
Market Trends	1. Fuel efficiency & cost reduction	Increased competition: pressure to reduce operating costs; Increasing oil prices	Fuel efficient systems; Alternative fuel types[1]	Financial: High perceived risk; Difficulties to fulfil technical standards; Lack of skilled labour	Reluctance of shipping companies: hamper innovation (lack of learning by doing)	Conservatism of shipping companies; Lack of newbuilding: no incorporation of new technologies	Improved designs (hull, reduced resistance); Propellers; Propulsion Improving Devices; Wind assistance devices; Optimised operation; Improved engine efficiency; Waste heat recovery systems
	2. Environmental awareness and CSR	Increased environmental awareness of consumers	Limited. Almost always linked to a direct business case: more likely to take place in segments operating close to consumers	Technologically do not differ significantly from those described for Trends 1, 3, 4, 5, 6. Due to the lack of direct regulatory pressure, the main barrier relies on the willingness of shipping companies (which can see the increasing environmental awareness of consumers as a potential investment/business opportunity.			Similar to those described for Trends 1, 3, 4, 5, 6. Positive factor: Environmental awareness and CSR development more progressed in Europe (potential "home market")

Regulatory Trends*						
3. NO_x abatement	Global regulation: e.g., IMO, MARPOL EU regulation: e.g. Air Quality Directive National/local regulation: e.g., tax levies in Norway	Selective Catalytic Reduction[2] Dual fuel engines LNG engines[3] Exhaust Gas Recirculation[2]	Relatively few. Active European industries in the development of green innovations	Reluctance of shipping companies: hamper innovation (lack of learning by doing)	Regulatory/ market uncertainty Lack of supporting infrastructure	Development of new LNG engines Installation of SCR systems Retrofit of existing engines
4. SO_x abatement	Global regulation: e.g., IMO, MARPOL EU regulation: e.g. Directive (2005/33/EC)	Low sulphur content fuels (MDO, LNG) Scrubbers[4]	Relatively few. Active European industries in the development of green innovations		Regulatory uncertainty	Development of new LNG engines Use of on board scrubbers
5. CO_2	Global regulation[5]: e.g., Energy Efficiency Design Index (EEDI), Ship Energy Efficiency Management Plans (SEEMPs) EU regulation: Transport White Paper	Electrical energy efficient technologies	Similar to those described for Trend 1.		Regulatory uncertainty	Improved fuel efficiency of engines Low carbon content fuels Reduced engine power

(Continued)

Table 8.5 Continued

	Trend	Drivers	Market Potential	Technological Barriers			Technological Response
				Development	Scaling Up	Expression of Demand	
Regulatory Trends*	6. Ballast water and sediment treatment	Global regulation: Convention for the Control and Management of Ships' Ballast Water and Sediments	Ballast water management systems[6]	Relatively few. Active European industries in the development of green innovations		Regulatory uncertainty	Integration of ballast treatment systems on board which kill organisms and bacteria
Others	7. Offshore renewable energy	European Renewable Energy Directive (2009/28/EC)	Specialised support vessels Platforms and foundations for turbines	Lack of yard infrastructure and skills	Limited. Barriers presumably related to the state-of-the-art of the renewable sector rather than to the shipping sectors	Regulatory and budgetary uncertainty	Adaptation of technologies and vessels to make them suitable for even deeper and further offshore areas

	Trend	Drivers	Climate Change		Barriers		Technological Response
			Market Potential	Access to the Arctic Route	Financing	Exploration of Oil & Gas	
Others	8. The Arctic dimension	Opening up of cross-Artic routes Increased access to Oil & Gas and other mineral resources	Icebreakers Ice strengthened vessels Zero-emission ships	Uncertainty in route accessibility High icebreaking fees: current routes remain commercially preferred	High cost of icebreakers: lack of financing	Environmental concerns Lack of financing	Improved design of icebreakers and ice strengthened vessels

Source: DNV, 2012.

*See Annex 8.2 for main regulations. DNV, 2012.

[1] The demand for marine distillates could be as high as 200–250 million tonnes annually.

[2] At least 30–40% of newbuilds will be fitted with EGR or SCR by 2016.

[3] In the next 8 years, more than 1 in 10 newbuildings will be delivered with gas fuelled engines.

[4] Scrubbers will be a significant option after 2020.

[5] Newbuildings in 2020 will emit up to 10 to 35% less CO_2 and EEDI will be the driver for more than half of the reduction.

[6] Ballast water treatment systems will be installed on at least half of the world fleet.

by market and regulatory trends, these relate principally with green shipping opportunities. Also the development of the offshore renewable sector as a driver for innovation in shipping is shown and can be considered indicative of the high potential for knowledge sharing between both industries. In this sense, it can be assumed that given its high cross-sectorality, the development of BG as a whole will also imply an important boost for innovation in shipping and skill transfer among sectors. Finally, several potential aspects for innovation that arise as a consequence of climate change are also shown.

As previously mentioned, the lack of skilled workforce may be a barrier to innovation in the European shipping sector. It is at this point where the adoption of the triple helix approach can be of great importance. It can help to improve the sector's image, promoting the specialisation of labour and innovation.

8.6 Investors

Table 8.6 summarises the main investment and financing facilities in the shipping industry.

Table 8.6   Investment and financing facilities

Category	Typical Features	Types
Private funds	• Main source of start-up capital	• Owners private funds • Private equity firms
Commercial bank finance	• Most important source of ship finance • Provide access to capital while leaving borrowers with full ownership of the business • Specialised financial activity	• Mortgage-backed loan • Corporate bank loan • Loan syndications and asset sale • Shipyard credit schemes • Mezzanine finance • Private placement of debt and equity
Capital markets	• Large companies raise finance by issuing securities. • Offers wholesale finance and a quick way of raising money • Difficult funding source for small companies	• Public offering of equity • Bond issue

Category	Typical Features	Types
Special purpose companies	• Standalone structures, set up for particular transactions • Reduction of finance costs by transferring ownership of vessels to • a company which can use its depreciation to obtain a tax break	• Limited partnerships

Source: Stopford, 2009.

8.7 Uncertainties and Concluding Remarks

Table 8.7 summarises the strengths, weaknesses, opportunities and threats (SWOT) of the European shipping industry.

Table 8.7 SWOT analysis of the European shipping industry (Modified from ECORYS 2009)

Strenghts	Weaknesses
• Level of innovation • Innovative SMEs and strong position of marine equipment industry • Strong linkages yards & marine equipment: Efficiency • Spillovers between shipping and BG sectors • Specialisation in niche markets	• Cost levels (wage and steel) • Potential difficulties in knowledge protection (especially among SMEs) • Access to finance • Fragmented government responses • Access to skilled labour
Opportunities	**Threats**
• New segments, continuous innovation • Greening of the industry • Existing transport policies • Enhanced requirements regarding shipping standards	• Demand shift from European to Asian buyers • Competitors moving up to the ladder • SMEs not surviving the crisis • Support from competitor's governments to their industry • Critical mass required to maintain/refresh high skilled workforce. Europe may be too small compared to competitors. Ageing workforce • Price competition in light of economic crisis

8.8 Conclusions

Although the development of new maritime industries needs a strong shipping industry, the structure of this sector at the global scale presents a number of features that can influence its productivity at the European level. The shipbuilding industry is clearly dominated by Asian countries (Japan, Korea, China), while the demand for new construction is driven by developing economies, which are growing at the expense of the slowdown in developed economies. The sector is characterized by its volatility and reliance on global markets, and thus, good economic times favour its fast growth, while crisis times hit it strongly. In addition, the competitiveness of shipping companies is strongly linked to labour costs, which are higher in the European sector. In order to face its Asian competitors, the European industry has adopted a clear specialisation strategy in higher value-added and technologically complex segments. However, given the wage differences between Eastern and Western countries the global specialisation strategies are replicated at the intra-European level. Eastern countries (intra-and extra-EU) act as "low cost" countries, building ships of lower value-added and technological level. On the other hand, the western countries, overcome the fact of having higher wage costs by specialising in high value-added and technological segments. Given that the entry into the EU of some of these "low cost" countries is relatively new (or it might be expected by the mid-term future), the impact of these inequalities in the sector is unknown.

Related to maritime transport activities, the growing demand of developing countries requires new epicentres for the distribution of goods. The need to transport more goods, can lead to problems of lack of infrastructure and logistics, either inside or outside the transportation mainlines. From a socio-economic point of view, the ageing of the labour force together with the poor image of the sector can put at risk the renewal of the qualified labour force. From the whole labour force, women only represent 2%, their presence being limited in practice to the cruise and ferry segments. However and despite the poor global economic situation, new market trends and the international regulatory framework can act as a boost to the sector. Moreover, the promotion of maritime clusters can improve the image of the sector, and: (i) make it more attractive for future workers; (ii) increase the transfer of knowledge between maritime economic sectors; and (iii) encourage the involvement of women.

Annex 8.1 – List of CESA Members

- Belgium
- Bulgaria
- Croatia
- Denmark
- Finland
- France
- Germany
- Greece
- Italy
- Lithuania
- Netherlands
- Norway
- Poland
- Portugal
- Romania
- Spain
- United Kingdom

Annex 8.2 – International Regulation in the Shipping Sector

Maritime safety and security and ship/port interface	International Convention for the Safety of Life at Sea (SOLAS)
	International Convention on Standards of Training, Certification and Watchkeeping for Seafarers (STCW)
	Convention on the International Regulations for Preventing Collisions at Sea (COLREG)
	Convention on Facilitation of International Maritime Traffic
	International Convention on Load Lines (LL)
	International Convention on Maritime Search and Rescue (SAR)
	Convention for the Suppression of Unlawful Acts Against the Safety of Maritime Navigation (SUA)
	Protocol for the Suppression of Unlawful Acts Against the Safety of Fixed Platforms located on the Continental Shelf
	International Convention for Safe Containers (CSC)
	Convention on the International Maritime Satellite Organization (IMSO C)
	The Torremolinos International Convention for the Safety of Fishing Vessels (SFV)
	International Convention on Standards of Training, Certification and Watchkeeping for Fishing Vessel Personnel (STCW-F)
	Special Trade Passenger Ships Agreement (STP)
	Protocol on Space Requirements for Special Trade Passenger Ships

(Continued)

Annex 8.2: Continued

Marine pollution	International Convention for the Prevention of Pollution from Ships (MARPOL)
	International Convention Relating to Intervention on the High Seas in Cases of Oil Pollution Casualties (INTERVENTION)
	Convention on the Prevention of Marine Pollution by Dumping of Wastes and Other Matter (LC)
	International Convention on Oil Pollution Preparedness, Response and Co-operation (OPRC)
	Protocol on Preparedness, Response and Co-operation to pollution Incidents by Hazardous and Noxious Substances, 2000 (OPRC-HNS Protocol)
	International Convention on the Control of Harmful Anti-fouling Systems on Ships (AFS)
	International Convention for the Control and Management of Ships' Ballast Water and Sediments
	The Hong Kong International Convention for the Safe and Environmentally Sound Recycling of Ships
Liability and compensation	International Convention on Civil Liability for Oil Pollution Damage (CLC)
	Protocol to the International Convention on the Establishment of an International Fund for Compensation for Oil Pollution Damage
	Convention relating to Civil Liability in the Field of Maritime Carriage of Nuclear Material (NUCLEAR)
	Athens Convention relating to the Carriage of Passengers and their Luggage by Sea (PAL)
	Convention on Limitation of Liability for Maritime Claims (LLMC)
	International Convention on Liability and Compensation for Damage in Connection with the Carriage of Hazardous and Noxious Substances by Sea (HNS)
	International Convention on Civil Liability for Bunker Oil Pollution Damage
	Nairobi International Convention on the Removal of Wrecks

References

Balance, 2014. Competitive position and future opportunities of the European marine supplies industry. 128 pp.

Cariou, P., 2008. Liner shipping strategies: an overview. International Journal of Ocean Systems Management 1(1): 2–13.

Christiansen, M., Fagerholt, K., Nygreen, B., Ronen, D., 2007. Maritime Transportation. In: Barnhart, C. and Laporte, G. (Eds.), Handbooks in Operations Research and Management Science, Vol. 14: 189–284. Elsevier.

CLIA, 2015. Cruise industry outlook. Cruising to New Horizons and Offering Travelers More. 40 pp.

DNV, 2012. Shipping 2020 report. 68 pp.

DSA, 2010. The economic significance of maritime clusters. Lessons learned from European empirical research. Working paper published by the Danish shipowners' association. 86 pp.

EC, 2011. Study on EU seafarers employment. Final Report. European Commission; Directorate General for mobility and transport; Directorate C-Maritime transport. 109 pp.

ECORYS, 2009. Study on competitiveness of the European shipbuilding industry. Final report. 239 pp.

IKEI, 2009. Comprehensive sectoral analysis of emerging competences and economic activities in the European Union: Building and Repairing of Ships and Boats. 143 pp.

IMO, 2012. International Shipping Facts and Figures – Information Resources on Trade, Safety, Security, Environment. 47 pp.

ITWF, 2015. International Transport Workers' Federation (http://www.itfsea farers.org/ITI-women-seafarers.cfm).

OECD, 2016. The Ocean Economy in 2030. OECD Publishing, Paris. 256 pp. http://dx.doi.org/10.1787/9789264251724-en

Oxford Economics, 2015. The economic value of the EU shipping industry – update. A report for the European Community Shipowners' Association (ECSA). 24 pp.

SeaEurope, 2013a. Shipbuilding Market Monitoring. Report N° 30. Ships and Marine Equipment Association. 43 pp.

SeaEurope, 2013b. Annual Report 2011–2012. Ships and Marine Equipment Association. 68 pp.

Stopford, M., 1997. Maritime Economics (2nd Ed.). Routledge, New York. 593 pp.

Stopford, M., 2009. Maritime Economics (3rd Ed.). Routledge, New York. 840 pp.

t'Hart, P., Schotte, D., 2007. Demographic change and skills requirements in the European shipbuilding and ship repair industry. European Shipbuilding Social Dialogue Committee. 42 pp.

UNCTAD, 2014. Review of maritime transport. United Nations, 136 pp.

Volk, B., 1994. The shipbuilding cycle-a phenomenon explained? Institute of Shipping Economics and Logistics. 208 pp.

9

Tourism

Dimitrios Pletsas, Sara Barrento, Ian Masters*
and Jack Atkinson-Willes

Swansea University, Wales
*Corresponding Author

9.1 Introduction

Coastal and Marine Tourism & Leisure (T&L) is one of the Blue Economy (BE) sectors that can help unlock the potential of multi-use of space at sea by engaging with Blue Growth (BG) sectors such as Aquaculture and Marine Renewable Energy among others. BG aims to exploit opportunities in the offshore areas, coastal T&L is mainly based onshore with very few exceptions (i.e. cruise ships) that also depend on land infrastructure (ports) to serve their customer segments.

The current chapter provides insights into the trends that exist in the T&L sector. Supply and demand gaps will be identified by studying the supply chain of T&L. Regional variations in four different basins (**Baltic, Atlantic, Mediterranean** and **Caribbean**) will be discussed. Life-cycle learning from each sub-sector will be gained keeping note of any different life stages observed for the same subsector at different regions (i.e. local, national, EU, Caribbean and international) and any geographic focal points for development. Also social trends will be identified including: employment, migrations associated with BG opportunities and associated pressures (e.g. social; public services and infrastructure).

Definitions

The definition of tourism has evolved over time, and there is still no consensus about the definition of tourism. For the purpose of this chapter we will use the definition for tourism and leisure set by the United Nations World Tourism Organization (UNWTO, 2016b).

Tourism is *"a social, cultural and economic phenomenon which entails the movement of people to countries or places outside their usual environment for personal or business/professional purposes. These people are called visitors (which may be either tourists or excursionists; residents or non-residents) and tourism has to do with their activities, some of which involve tourism expenditure."*

Leisure industry is *"the segment of business focused on entertainment, recreation, sports and tourism related products and services."*

The EU further breaks down tourism within the BE context as maritime and coastal tourism (ECORYS, 2013).

- **Maritime tourism** covers tourism that is largely water-based rather than land-based (e.g. boating, yachting, cruising, nautical sports), but includes the operation of landside facilities, manufacturing of equipment, and services necessary for this segment of tourism.
- **Coastal tourism** covers beach-based recreation and tourism (e.g. swimming, surfing, sun bathing), and non-beach related land-based tourism in the coastal area (all other tourism and recreation activities that take place in the coastal area for which the proximity of the sea is a condition), as well as the supplies and manufacturing industries associated to these activities.

Some examples of maritime and coastal leisure activities include those summarized in Figure 9.1. All basins – Atlantic, Baltic, Caribbean and Mediterranean – provide the above marine activities, or aim to provide it in a very near future. This chapter will focus around Tourism and Leisure sub-sectors under three themes as shown in Table 9.1.

9.2 Market

9.2.1 Market Key Facts

Tourism is one of the most important and fastest growing economic activities worldwide – it grows faster than the wider global economy, it is resilient to change, and benefits both developed and developing countries (UNWTO, 2016b, 2016c).

International tourism now represents 7% of the world's exports in goods and services, ahead of food and car industries, and ranking third just after fuel and the chemical industry (UNWTO, 2016a). Southern Europe and the

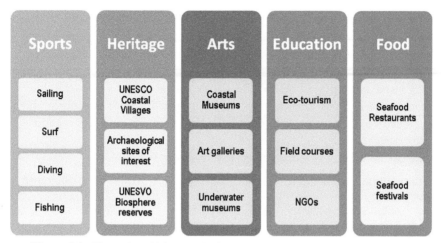

Figure 9.1 Examples of leisure activities related to marine and coastal tourism.

Table 9.1 Definition of tourism and leisure sub-sectors

Tourism & Leisure Sub-Sectors	Definition
Theme 1: Selling directly to the customer	The **sale of products and services** from a supply chain company (e.g. hotel owner, cruise operator, online or high-street travel agent) **directly to consumers**: • Business tourism – business trip package, flight, local transportation, hotel, conference/exhibition. • Pleasure tourism-holiday package, flight, local transportation, resort/hotel. • Leisure – beach/water sports, recreation, heritage, arts, entertainment, education, shopping, gastronomy.
Theme 2: Selling directly to corporate	The **sale of products and services from a supply chain company to another** also known as "Tier 1 suppliers" prior to reaching the consumer: e.g. hotel owner sells to tour operator; water sports centre sells to hotel; flight operator sells to tour operator; software/app designer sells to online travel agent.
Theme 3: Vertical integration	A combination of the above where **companies have ownership** of large parts **of the Tourism and Leisure supply chain**: e.g. Online travel agent has shares in airline and/or holiday destination hotel/resort and/or other product and services companies (Monarch group, TUI group etc.); Cruise liner owner sells cruises and has stakes in destinations.

Mediterranean still retain the biggest market share (51%) followed by the Americas including the Caribbean (16%). But the market is getting ever more competitive and these shares have been decreasing in the last 20 years in favour of growth taking place in Asia and the Pacific (8%), the Middle East (2%) and Africa (1%) (UNWTO, 2016c). Due to the geomorphology of Europe there is a vast coastal region which provides a lot of natural and cultural wealth and this is exploited by T&L and interrelated (e.g. recreation) BE industries.

A European market summary overview is shown below:

- International arrivals reached 1.087 billion people in 2013, a 5% increase from 2012,
- In Europe the leisure travel market grew 5.4%,
- Travel and tourism accounts for 9.5% of global GDP and 1 in 11 jobs worldwide,
- Market growth is forecast to rise 4.2% per annum from 2014–2024,
- Market continues to be driven by high growth in online bookings.

All the 4 MARIBE sea basins are represented in the top ten most visited countries (Figure 9.2). From a sea basin perspective, worldwide the most

Figure 9.2   Most visited tourism destinations in the world relevant to MARIBE basins. Data based on international tourist arrivals per selected countries (UNWTO, 2016c).

visited is the Mediterranean, followed by the North Atlantic, the Caribbean and in last comes the Baltic Sea. Nevertheless, the tourism industry in the Baltic Sea region contributes significantly to the economies of the countries in the region, with a total of 73 million international arrivals in the region. Given that Europe accounts for half of the world's total tourism, the Baltic Sea region accounts for around 13% of the tourism in Europe and 7% of the world's tourism measured by arrivals (Winther and Jensen, 2013).

Tourism has been a major contributor to the European economic recovery. A weaker euro in 2015 boosted the number of visitors from outside the EU and contributed to Europe becoming the fastest tourism growth region. Northern Europe recorded a 7% growth in international arrivals mainly due to double digit growth in Iceland, Ireland and Sweden (North Atlantic basin). During this same period, Central and Eastern Europe returned to growth (+5%), Tajikistan entered the tourist destination map, and Hungary, Romania, Slovakia and Latvia (Baltic Basin) all recorded double digit growth. Southern Europe and the Mediterranean also grew by 5%, which is remarkable given the maturity and size of this popular destination region. Only France and Belgium saw a modest growth in the face of the terrorist attacks. The Caribbean also grew 7% in 2015, driven by Cuba (18%), Aruba (14%), Barbados (14%) and Haiti (11%) (UNWTO, 2016a).

Overall, the 4 European sea basins are in a good position and it is expected that coastal and maritime tourism continues to grow. The main drivers that will affect the market are social demographic changes – age, education, income; individual countries holiday legislation; international politics and economy.

Each individual country (Figure 9.2) will benefit or lose market share depending on: currency fluctuation – in 2015 the weaker pound enabled an inbound growth to the UK whereas the stronger dollar drove USA outbound growth (IPK, 2016); terrorism and perceived safety – Tunisia, Egypt and Turkey lost international visitors, but Spain and Portugal had a strong increase of 5 and 10% respectively (UNWTO, 2016a); international visitors country of origin – Chinese citizens now benefit from a two 7-day paid holidays per year and are going on more sun and beach holidays (IPK, 2016). The destination countries will have to adapt and develop strategies to accommodate these changes, as an example, Turkey is targeting more Muslim visitors by developing halal travel to compensate for the 30% decline in international arrivals following terror attacks and the coup attempts. From a basin perspective, it is crucial to understand the type of holiday that visitors book or would like to experience. In general terms, Europeans are moving away from the sun and beach holidays and are booking more city trips (+15%) and events (+6%). Whereas Asians (+20%) and Americans (+11%) are still very keen on sun and beach holidays. It is expected, however, that the growing population in coastal cities will want to experience a different type of coastal and maritime tourism than the typical sun and beach holidays currently on offer (IPK, 2016).

9.2.2 Major Business Models

The majority of the travel market is influenced by large, multinational, integrated Leisure Tourism Businesses (LTB). For example, the TUI Travel group has a €17bn turnover and 67,000 employees (TUI Travel, 2014). They offer a range of package holiday products in several countries under a range of brands and into a number of consumer segments. The Online Travel Agent (OTA) dominates the current market. It is enabled by companies which have strong online presence with access to tools to help the customer throughout their decision making and transaction process (i.e. travel booking). Hence the customers are able to research and apply their personal criteria prior to booking their holiday experience. Numerous T&L product suppliers provide online access for booking hotel, flights, car rental or combinations of these, including: online tour operators (e.g. Expedia and its subsidiaries), budget airlines (e.g. EasyJet), research enabling (Trip Advisor), and leisure specialist & activity companies (mainly based at destinations).

The OTA market is made of online travel suppliers some of which are subsidiaries of a corporate group. A common strategy for the latter is to invest into individual brands rather than in high profile mergers as consumers typically visit multiple travel sites prior to booking travel. Hence having a multi-brand strategy increases the likelihood that those consumers will visit one or more of a large subsidiary group sites.

Customers are increasingly moving towards online channels to fulfil their travel and accommodation needs. This trend has been noticed in the past 5 years and is expected to continue to be the fastest growing segments in the leisure travel industry, driven by the growth in online bookings and supplemented by strong demand from the emerging markets. Therefore, tour operators who offer package holiday products:

- invest heavily in online presence
- increase participation in social media
- move towards an online-driven company culture

Demand (and growth) for online holiday travel and accommodation is expected to continue for the near future (TUI Travel, 2014).

- Online booking accounts for 45% of total travel sales in Europe (Rheem, 2009) and therefore shows that this has matured in many Western countries. There is seen to be limited scope for expansion as the remaining 55% is taken by a combination of traditional physical travel agent shops and business travel booking handled by specialist corporate-travel agents.
- Business growth in a mature market drives acquisitions of products and revenue streams from other industries. One example is Priceline's purchase of OpenTable, a restaurant-reservation website.
- Globally, the market is growing with online travel gaining share on an estimated global travel market of over $1 trillion.
- These trends are driving the innovation of new technologies (e.g. apps, online booking tools, management software, interface tools).
- During 2012, a monthly average of approximately 60 million unique visitors come to Expedia sites to research, plan and book travel (Expedia, 2013).
- Travel companies need to adopt a holistic approach (ETOA, 2010), they need to follow customers through-out all of the steps in the booking funnel. If not, they will face threats from companies active at other stages of the funnel.

9.2.3 Supply Chain Business Models

Businesses that operate as a supplier to the big corporate travel market offers an opportunity for rental, hotel, activity and restaurant services in tourist destinations to greatly improve return on marketing effort while ensuring a reliable influx of visitors. In this particular case, the tier 1 suppliers are not dealing with customers directly, but sell to a larger corporate, for example: hotel owner sells to tour operator; water sports centre sells to hotel; flight operator sells to tour operator; software/app designer sells to online travel agent. As visitors are also booked well in advance, this allows for further investment and improved revenue predictions. Additionally, it presents a large opportunity for emerging markets as corporations can have more focussed advertising, or redirection of an established existing customer base, into these markets. By selling directly to corporate, visitors will be easier to manage and provide appropriate services for as visitor details and demographics are known prior to arrival. However, reduced cost and risk comes at a price as corporate customers will expect a profit sharing arrangement deal (on price per room, for example).

Package holidays are the most common example of selling directly to corporate tourism. Resurgence in popularity in recent years has resulted in 46% of UK holidaymakers in 2013 opting for package holidays when travelling abroad, increasing to 51% in 2014.

Supply chain trends and future outlooks can be summarised as follows (ABTA, 2014, ABTA, 2015, ETOA, 2010, Euromonitor International, 2013):

- Mobile apps from tour operators and OTA have the potential to increase the success of the corporate travel market to ease of use. Package travel may grow in popularity compared to booking hotels and flights individually.
- Growth in Chinese and Middle Eastern tourist travel to Europe – growing target market for package operators (particularly luxury ones), which may result in an increase in corporate tourism business success.
- Growth in peer-to-peer travel where tourists receive services from individuals and micro businesses facilitated by websites or apps such as Airbnb or Blablacar. This poses a risk to corporate tourism as they cannot match 'authenticity' that these means can provide. This directly disrupts the large corporate supply chain by removing their mark up on local products. However, ease and simplicity of package tours will remain a large draw for many tourists.
- City breaks overtook beach holidays as the most popular type of holiday in 2014 (within the UK). These are facilitated by web based services for low cost flights and accommodation websites. Package holidays can easily adapt to these changing trends as 'weekend breaks' can be advertised to the public, particularly outside of the summer months, however, the public mindset has shifted to last minute and custom travel, so the marketing would need to be appropriate for the package holidays to sell effectively.
- Wellness and spa break holidays are also an increasing trend in 2015, and the market is set to grow.

By selling directly to corporate tourist companies, businesses based in successful destinations can better capitalise on their growing popularity by using the corporation's capital and resources to better advertise the destination. This will grow the client base and reputation of the destination.

9.2.4 Vertical Integration

By providing services at multiple levels within the same supply chain, it is much easier for an organisation to capture a larger portion of the market.

Within the tourism industry, vertical integration usually presents itself in the form of a single organisation that is simultaneously a travel agency, airline and tour operator. The organisation is then better placed to control their pricing and interact with companies at all levels of the chain. This results in more comprehensive package deals to its customers at a more competitive rate (or with higher profit margins). This also serves to reduce the amount of risk that an organisation takes when introducing a new service, as it can use other levels within the same supply chain to steer reliable income towards that area. For example, if an organisation wished to begin a flight service to a new destination, the travel agency owned by that organisation could then begin promoting this to its customers, and any hotels owned by the organisation at that destination could begin offering reduced price deals for prospective visitors.

This example of vertical integration in tourism is most commonly seen with the 'Big Four' of the UK tourism industry, compromising of TUI Group, First Choice, MyTravel and Thomas Cook. Each of these companies owns multiple travel agents, airlines and hotels and as such have significant influence over the UK travel market.

Vertical integration market trends & future outlooks:

- Rise of independent and peer-to-peer travel is a threat to vertical integration, as it gives organisations less of a control over the market as customers cannot be steered into selecting certain deals or offers.
- Websites and apps such as Airbnb provide an unbiased service for accommodation. Smaller companies are then able to capture back some of the market as it is easier to communicate their services to prospective customers.
- While these developments do pose a challenge to vertical integration tourism, the tourism market continues to grow steadily year on year. As vertical integration generally allows for the least expensive and best advertised deals, it is likely that this tourism theme will continue to see widespread use in the coming years.
- Large companies, such as the big UK four, are well placed to expand into new forms of tourism as they emerge. For example, wellness tourism businesses, such as spa breaks or retreats, can be purchased by these large organisations to add to their portfolio of holiday destinations and activities.
- Due to their advertising power, large organisations are also well placed to move into the growing Arabic and Chinese markets.

Source: (Lafferty and van Fossen, 2001); (Tremblay, 1998).

9.2.5 Cruise Tourism

A popular subsector of tourism is Cruise Tourism, with growing numbers of vessels and passengers across Europe. The Cruise Lines International Association publishes an annual European market summary which has the following headlines:

- During 2015, 39 cruise lines operated 123 cruise ships with a capacity of around 149,000 lower berths. Another 73 vessels (100,000 lower berths) were deployed in Europe by 23 non-European lines.
- 6.59 million European residents booked cruises, a 3.1% increase over 2014, 30% of all cruise passengers worldwide.
- 6.12 million passengers embarked on their cruises from a European port, a 4.6% increase over 2014. Of these 5.0 million were European and 1.1 million came from outside Europe.

(CLIA, 2016)

Direct expenditure in 2015 was almost €17bn, with indirect outputs in addition to this. The significant market contribution of these vessels relates to on-board activities for the passengers, and additional income is generated when vessels visit each destination port. The other major contribution to the European economy is in shipbuilding and maintenance. Almost a third of total direct spend (€4.6bn) relates to ship building. The majority of all cruise vessels ordered and under construction worldwide (48 out of 50) are in European shipyards, showing that this is serving internal and export markets (Cruise Lines International Association, 2016). Launched in 2016, Harmony of the Seas was the largest passenger ship in the world, built in Saint-Nazaire, France, with gross tonnage of 227,000 GRT, 6780 passengers and 2100 crew. She was the third vessel built to this design (Royal Carribbean International, 2017). To provide a context for this level of activity, the rapid expansion of the offshore wind sector in Europe lead to 3000MW of new wind turbine installations in 2015 and a similar total investment in that year of €18bn (Pineda et al., 2015).

Italy is the largest market, with the UK second and Germany third, showing strong activity in all European sea basins. The busiest port is Barcelona, with 2.5m passengers visiting, starting or finishing their journey at this port (Cruise Lines International Association, 2016)). Vayá *et al.* (2017) investigated the economic impacts of cruise activity at this port and show that all sectors of the local economy benefit from the cruise activity. However, there are negative impacts related to localised congestion. The rise

Table 9.2 Lifecycle of the Tourism and Leisure industry within the studied basins (see Table 9.1 for definitions)

	T&L Selling Directly to the Customer	T&L Selling Directly to Corporate	T&L Vertical Integration	T&L Selling Direct via Online Marketplaces (Expedia, Airbnb)
North Atlantic	Post mature diversification	Mature	Mature	Growing
Caribbean	Growing	Mature	Mature	New
Baltic	Growing	Growing	Growing	New
Mediterranean	Post mature diversification	Mature	Mature	Growing

of responsible tourism has highlighted some practices of the cruise sector in destination ports, where the vessels charge a significant premium for onshore excursions, reducing the local benefit of the activity (Klein, 2011). There are also concerns about the environmental impact of large numbers of cruise visitors in some ports. The most obvious impact from a large cruise ship with 6000 passengers is the management of waste streams, in particular waste water, air emissions and solid waste streams (Klein, 2011).

9.3 Sector Industry Structure and Lifecycle

In general tourism & leisure is a mature industry aiming at diversification; from a Maribe context only the T&L in the Baltic is not yet considered a matured sector. A summary assessment of lifecycle is given in Table 9.2.

9.4 Working Environment

9.4.1 Employment and Skills

Coastal and maritime tourism is the largest maritime activity in Europe and employs almost 2 million people (Figure 9.3) (European Commission, 2014a). However, the coastal tourism sector is not attracting or maintaining enough skilled personnel due to its seasonality and lack of long term career opportunities which can lead to problems in service quality and hamper competitiveness. The sector needs well qualified professionals who are service minded and multilingual. It also needs dynamic and creative entrepreneurs who can link local enterprises, administration and stakeholders and deal with the following challenges:

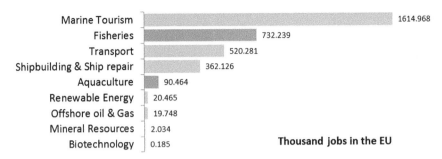

Figure 9.3   Blue Economy Jobs in the EU, thousands (European Commission, 2014a).

1. energy and GHG emissions (particularly relevant for the cruising industry)
2. water consumption (particularly relevant in the Mediterranean and Caribbean)
3. waste and pollution management (also in conflict with other industries, cargo transport, petrol and gas extraction, agriculture)
4. loss of biological diversity; (of special interest in the Mediterranean and Caribbean)
5. effective management of cultural heritage (all basins)
6. aging population (all basins)

From a client's perspective there is a growing consumer population concerned about the ethical labour conditions in the places they are visiting (ILO, 2010). On the other hand, Rheem (2009) argues that less than a third of American travellers indicate a willingness to pay some sort of premium for "green" travel, higher prices (cost premium) being seen as a demand barrier for 67 per cent of respondents. Therefore, visitors' attitudes and perceptions from different countries will affect the visiting countries in different ways, for instance if in the Caribbean the main visitors are from the United States of America there will be less incentive to develop a "green" travel approach, whereas in Europe it might be the opposite case. Australia is leading the way in ecotourism (Dowling, 2002) and includes a system of ecotourism badges for suitably trained tour guides.

9.4.2 Economic Climate for Small Businesses

The economic benefit of tourism does not always benefit those working locally. For example, the cruising industry generates business opportunities for the ports visited, but for coastal regions it is not always easy to capture economic benefits generated by this cruise tourism, due to the margins retained by the vessels, while pressures to invest in port infrastructures and to preserve the environment are increasing. Benefits of Ocean Energy to local businesses, specifically fishing, was studied by Reilly et al. (2015). This concluded that continuing dialogue was the most important factor and that fishermen would be open to re-training to undertake job roles within a different BG sector.

This study and others showed that these problems can be tackled and the EU has invested in two projects to address these specific issues:

1. The SubArcheo project (Euroreso, 2003) to facilitate the re-skilling of fishermen by developing new distance-learning methodologies and materials for the training of Underwater Archaeologists and Guides in coastal areas around the Mediterranean.
2. The Europe-wide online platform TourismLink (2014) connecting small businesses with travel agents and tour operators.

9.4.3 Contribution of Tourism to GVA and Revenues

Coastal and maritime tourism is the second largest maritime activity in Europe generating a total of €183 billion in gross value added and representing over one third of the maritime economy as shown in Figure 9.4. Visitor Exports are an essential component of the value added for tourism. A visitor export is expenditure by international visitors to a country, including expenditure for travel. The tourism industries in the Baltic region accounts for between 1.8 percent (Sweden) and 3.2 percent (Estonia) of the total employment in the countries. Tourism has also established itself as an essential

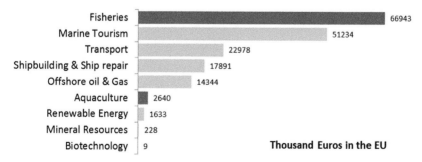

Figure 9.4 The Blue economy in the EU, unit thousand euros (European Commission, 2014b).

activity for many Caribbean economies; the direct contribution of tourism to the GDP of the Caribbean was around 5% of GDP, with indirect and induced effects this rises to 15% of GDP (World Travel and Tourism Council, 2017).

A critical challenge facing mature tourism destinations worldwide is how to maximize the economic impact of tourism spending (Alegre and Cladera, 2006, Kozak and Martin, 2012). Many mature Caribbean and Mediterranean 'sun and sand' holiday destination like Malta, southern Spain and the Balearic Islands, Greek islands and more recently Turkey, are following the well-documented path towards greater product diversification in an attempt to target the most profitable visitors (Bramwell, 2004). Knowing this need for product diversification, The European Union has put in place several funding schemes that should contribute to GVA:

- The European Structural and Investment Funds can co-finance sustainable tourism investments linked to ICT, entrepreneurship, SME's competitiveness, energy efficiency, employment and labour mobility, etc., and can promote the exchange of good practice, transnational networks and clustering.
- Horizon 2020 is the financial instrument for the EU's research and innovation strategy, with BG as one of its focus areas. The COSME framework programme aims to enhance SME competitiveness, increase tourism demand, diversify offer and products, and enhance quality, sustainability, accessibility, skills and innovation.
- The Creative Europe programme supports synergies between cultural and nature tourism, including coastal and maritime heritage.
- The Erasmus+ programme helps in terms of employability, new sector curricula and innovative forms of vocational teaching and training.
- The LIFE+ programme can co-finance innovative projects affecting coastal and maritime tourism and resource efficiency. The Proposal for the 7th EU Environment Action Programme is also linked to coastal and maritime tourism. In addition, the European Investment Bank provides SMEs with financing for investments in tourism and/or in convergence regions.

9.4.4 Changes to Infrastructure and Support Service Requirements

Traditional mass tourism such as "sun-and-sand" resorts has reached a steady growth stage. In contrast, ecotourism, nature, heritage, cultural and "soft adventure" tourism, as well as sub-sectors such as rural and community tourism are taking the lead in tourism markets and are predicted to grow most rapidly over the next two decades. It is estimated that global spending on eco-tourism is increasing by 20 per cent a year, about six times the industry-wide rate of growth. Nature-based tourism is an important economic component of the entire tourism market (TEEB, 2009). Also, a research report conducted by SRI International for the Global Spa & Wellness Summit (GSWS, 2014) found that wellness tourism represents a US$439 billion market, equivalent to 14% of world tourism expenditures. Wellness tourism category will grow 9% annually through 2017, 50% faster than "regular" tourism. Considering these 2 areas it can be foreseen that the following infrastructures and areas will be required:

- sustainable ports (desalination treatment, renewable energy usage, pollution containment or prevention)
- sustainable design of hotels and lodges (water treatment, using renewable energies, eco-friendly materials, energy efficient, aesthetically balanced with the natural environment)
- biotechnological developments to produce cosmetics and smart foods

To provide some co-ordination for these activities, the EU through its funding programs has invested in creating links between the different states, the private sector and universities, also known as the Triple Helix. Consequently, The European Network of Maritime Clusters is a confederation of Clusters aiming to disseminate best practice and a platform for exchange through a website, informal talks and an annual summit during which each country gives a brief presentation of the economic situation of its maritime sector and the recent actions of its national organization. The aim is to establish a framework for future common targeted actions.

9.5 Public Policy Regulatory Framework

The ratification and implementation of the Treaty of Lisbon has brought about an increased role for the European Commission within the area of tourism. EU Policies which are relevant to or have an impact on tourism fall into the following categories:

1. Agriculture and rural development;
2. Development and cooperation.
3. Climate change;
4. Internal market and services;
5. Competition;
6. Justice;
7. Education and culture;

8. Maritime affairs and fisheries;

9. Employment, social affairs and inclusion;

10. Mobility and transport;

11. Energy;

12. Regional policy;

13. Enlargement;

14. Research and innovation;

15. Environment;

16. Taxation;

17. Health and consumer protection;

18. Trade;

19. Home affairs;

20. Communications Networks, Content and Technology;

The pressure from tourism on the environment of the coastal zone is a key issue that requires management. This provides clean bathing waters (as part of the Water Framework Directive) and can also protect wildlife. The Natura 2000 is an EU-wide network of nature protection areas, established under the 1992 Habitats Directive, aimed at ensuring the long-term survival of Europe's most valuable and threatened species and habitats, including in the marine environment. Successful management of the sites allows their enjoyment as part of holiday experiences and their conservation for all future users. Overall, stakeholders have two main visions of EU regulation, positive benefits related to equalised standards and negative concerns of the effects of legislation on profits; these are summarised in Table 9.3 and Table 9.4. Global legislation and guidelines are summarised in Table 9.5.

Table 9.3   The benefits of the EU Directives and Legislation

Tourists Welfare	Reducing Uneven Regional Development	Ongoing Work
Timeshare Directive	Development of regional airports	Roaming charges for mobile phones
European Tourism Quality Principles 2014	Baltic Region Strategy	Services Directive.
Consumer Rights Directive	Danube Region Strategy	VAT systems implementation
Package Travel Directive		
Regulation (EC) No 810/2009 on visa policy		
Enforcement activities on airlines' sales		
Regulations on air passenger travel rights		
Habitats Directive		
Water Quality Framework		

Table 9.4   Profitability concerns of the EU Directives and Legislation

EU/National Regulation Burden on Business	Differences in Implementation of EU Legislation	Differences in Quality Labelling and Standards	Obstacles for Non-EU Inbound Tourists
Emission Trading Scheme, impact on EU airlines	VAT Directive, member state variations	Common regulatory framework impractical	To obtain visas is time consuming
Working Time Directive, Impact on hotel segment	Tour Operators Margin Scheme, member state variations	Costly to introduce	Schengen visa is expensive
	Services Directive, licences implemented in some states	No independent controlling body reduces credibility of labelling schemes	

Table 9.5   Advancing tourism: world policy, programmes, codes and guidelines

	United Nations
Education & Training	UNWTO Capacity programme
	UNWTO.TedQual programme
Climate Change	The Davos Declaration puts tourism in the forefront of the global response to climate change through:
	The Davos Process on Tourism and Climate Change
	Global Sustainable Tourism Criteria
Sustainability	UNWTO Global Code of Ethics for Tourism
	UNWTO Technical Cooperation Programme
	Sustainable Tourism – Eliminating Poverty Initiative (ST-EP)
	Tourism and Poverty Alleviation: Recommendations for Action
	Global Sustainable Tourism Criteria
	Tourism, Microfinance and Poverty Alleviation
	UNWTO Sustainable Development of Tourism Programme
	Lusaka Declaration on Sustainable Tourism Development, Climate Change and Peace
	Kuala Lumpur Declaration on Climate Change and Tourism
	Legazpi Declaration on Tourism's Response to Climate Change
	Djerba Declaration
Biodiversity	UNWTO Recommendations on Tourism and Biodiversity.
	Specialized Unit on Tourism and Biodiversity
	UNWTO Consulting Unit on Tourism and Biodiversity
	Practical Guide for the Development of Biodiversity-Based Tourism Products
Ethics	World Committee on Tourism Ethics
	Global Code of Ethics for Tourism
	UNWTO Ethics & Social Dimensions of Tourism Programme
	Protect Children from Exploitation in Tourism and Travel
	The Responsible Tourist and Traveller
Green Economy	UNEP – Green Economy Initiative
	UNWTO – Journey to Rio+20

9.6 Innovation

Tourism is a mature and resilient sector – in the face of change it adapts. Trends in consumer behaviour change fast and innovation is the key for resilience. Therefore, tourism is investing in two main innovation areas: technology and social trends.

Technology enables tourism companies to offer customers very specific targeted information, including niche marketing, information during a holiday and audio-visual interaction with attractions. Mobile phones and tablets are increasingly being used by tourists to search for and book travel with increasing numbers of mobile application downloads (Expedia: 90m downloads in 1 year; annual report, 2013). To effectively compete on these new channels, companies need to build flexible systems architecture to reach all types of consumers on all types of screens (ETOA, 2010). Effective interaction will increasingly rely on the use of big data to present customers with targeted options, which is in line with travellers' expectation of a more unique and tailored consumer experience. However, social media has strengthened peer-to-peer interaction and travellers routinely share reviews and other content. This enables the sharing economy and consumers are now active in providing holiday apartments, cars, meals and tours.

Consumers are increasingly becoming "always connected" and there is a rising demand for real time services and companies need to provide personalised alerts and information. This is particularly important in times of disruption, so that, for example, updates on delayed flights can be provided. Customer service needs to be available 24/7 and easy to access via mobile devices, and via a number of channels: voice, email and social media. Extension to this is the potential for interactions that are enabled by wearable technologies and connected cars, these can provide customised service and allow on-the-go consumers to find information, book hotels and other services.

Social trends and the visitors' social attributes can be used to innovate towards new tourism products for a better visitor experience. For example, personal attributes such as status, age, gender and political outlook can be reflected in visitors' choice of specialist holidays including cultural tourism, wellness tourism and eco-tourism. Cultural tourism has gone beyond the typical museum or gallery display, or the passive activity of visiting a heritage site; today cultural tourism includes music, food, literature and crafts festivals; and heritage sites are used to show transmedia installations profiting from technology. A related and growing sector is Eco tourism, including the development of natural trails, marine reserves, bird watching, dolphin watching, whale watching and many other value added activities.

Wellness tourism which was mainly restricted to health spas and clinics is becoming more diverse – it now targets people interested in nurturing the body and soul in creative ways. These include innovative ways to exercise, rest, meditate and eat. A research report conducted by SRI International for the Global Spa & Wellness Summit found that wellness tourism represents a US$439 billion market, equivalent to 14 percent of world tourism expenditures. Wellness tourism category will grow nine percent annually through 2017, 50 percent faster than "regular" tourism (GSWS, 2014).

One recent example of innovation aiming to develop new forms of leisure activities are the underwater museums. Cancun was first to offer in 2009 site visits to a monumental contemporary museum of art – MUSA consisting of over 500 underwater sculptures. In Europe the first underwater museum opened in 2015, in Antalya, Turkey (Mediterranean basin) (Side, 2017), and in the beginning of 2017 Lanzarote (Spain) followed the trend and now offers an underwater experience in the Atlantic Museum. BALTACAR – Baltic history Beneath Surface, is an EU funded project to develop the underwater cultural heritage of the Baltic Sea and has ambitions to provide underwater heritage trails by 2019 (Baltacar, 2017).

9.6.1 Cross-Sectoral Opportunities

There are a number of examples where other industries have benefited from the existence of tourism and the combination of sectors has provided additional value to both. One well established example is the primary production of wine, where tourists visit vineyards as part of the holiday. In some countries (e.g. Chile, New Zealand, Portugal) this type of tourism has contributed to the growth of the wine production sector. Fisheries is another example, with the movement of fish markets from industrial warehouses to tourist friendly retail premises in prime waterfront areas. Many companies are combining seafood production, aquaculture and tourism. Three examples are:

- Scotland: Lock Fyne oysters, restaurant and oyster bar
  (http://www.lochfyne.com/)
- Indonesia: Tours to visit a Seaweed Farm near Bali
  (https://www.govoya gin.com/activities/visit-a-seaweed-farm-on-a-serene-island-near-bali/1127)
- Belgium: Maritime Oostend – a joint local cooperation of fishing, farming and tourism enterprises to attract visitors to the area
  (http://translate. google.pt/translate?hl=en&sl=nl&u=http://www.ostendaise.be/&prev=search)

Recreational fishing, also called sport fishing, is the application of the skills of the fishing industry to a tourism sector. In the USA, about 12 million recreational saltwater fishers generate $30 billion in economic impact and support 350,000 jobs. This directly supports:

- manufacture and retail of fishing tackle
- construction of recreational fishing boats
- companies providing fishing boats for charter and guided fishing trips

When aquaculture tourism is included, this also supports:

- manufacturers and retailing of aquaculture apparel (cages, ropes, pumps etc.)
- manufacturers of vessels adapted for aquaculture operations (e.g. mussel harvest)

In all these examples, fishing and aquaculture becomes a (partial) reason for tourists to visit a particular area and, in addition to specialist enterprises, this also benefits the local hotels, restaurants, nightlife and heritage centres (museums, historic houses). It is also possible for the fishing/aquaculture facility to provide additional services such as a conference venue or wedding venue.

In the Baltic and North Sea regions, tourism services and offshore wind energy have already been successfully combined to integrate offshore wind energy into regional tourism development concepts. This includes:

- Boat tours
- Sightseeing flights
- Routes for motor and sailing boats
- Offshore restaurants and merchandising products
- Routes for motor and sailing boats
- Combined offshore and onshore wind energy tour
- Viewing platform with telescopes
- Offshore information centre

Source: (Business LF, 2013); (German Offshore Wind Energy Foundation).

9.7 Investment

Tourism is a profitable and mature sector with the big players listed on the major stock markets. Therefore, well planned tourism developments with a good marketing strategy will be able to raise investment funds. A number of specialist organisations produce annual market reports, which

Table 9.6 Capital Investment 2016 (WTTC, 2017)

	Capital Investment 2016 US$bn	Relative Terms 2016 % of Total Capital Investment	Capital Investment 2016 Growth %
Caribbean	6.8	12.3	2.9
European Union	159.6	4.9	2.0

include details of recent investments. A number of these have been quoted in this chapter (Euromonitor International, 2013) (ABTA, 2015) (IPK, 2016) (Cruise Lines International Association, 2016) (UNWTO, 2016c) (World Travel and Tourism Council, 2017). The overall picture is that these reports show growth in turnover and growth in investments in the sector over recent years. Table 9.6 shows recent capital investments and this highlights that tourism is a significant part of the whole of the economy, not just the BE. For the Caribbean in particular, tourism is a significant part of the whole economy.

9.8 Uncertainties and Concluding Remarks

Coastal and maritime tourism is an important sector worldwide. In general there are no specific uncertainties within the tourism sector; it is a consolidated, mature sector, well-regulated and innovative:

- In Europe, tourism is the biggest maritime sector in terms of gross value added and employment and is expected to grow by 2–3% by 2020.
- In 2012, Cruise tourism alone represented 330,000 jobs and a direct turnover of €15.5 billion and is expected to grow.
- As a consequence tourism has been a major contributor to the European economic recovery.
- Investment is expected to rise in the next 10 years in all sea basins.

However, general uncertainties apply to the coastal tourism sector in the same way as the other sectors within the BE:

- Fluctuation in fossil fuel prices
- Terrorism/War conflicts
- Migration policy
- Food security
- Disease outbreaks
- Supply chain dependencies
- Climate change and extreme weather events, including:

- Ocean level rise
- Tsunamis
- Hurricanes
- Heat waves
- Flooding
- Storms

The combination of different BE sectors and tourism is already a reality and many more are projected to happen in the very near future. Well established joint combinations which are already taking place between tourism and other sectors of the BE include:

1. Fisheries
2. Aquaculture
3. Offshore wind

It is vital that we plan for new approaches to tourism that are complementary and original, better integrated into the host societies and their environment, and that are able to deliver alternatives to the classic resort-based mass tourism as well as taking on board the concept of sustainable development. Politicians, policy makers, CEOs of other industries not directly related to tourism but also researchers often underestimate the importance and impact of tourism on the economy. Previous reports from some EU BG projects also show a lack of vision or deeper understanding of the tourism sector and its importance for other sectors within the BE.

Historically, tourism has been driving important sectors of the economy and has contributed to prevent the loss of traditional production practices (e.g. endemic strains of wines and artisanal fisheries practices). Tourism is humanitarian in nature and provides direct relationships with people from different backgrounds. Tourism can connect people from different countries, re-connect people to the natural environment and through visitor attractions such as an offshore wind visitor centre, can educate and inspire people for a sustainable future.

References

ABTA 2014. Travel Trends Report 2014.
ABTA 2015. Travel Trends Report 2015.
ALEGRE, J. & CLADERA, M. 2006. Repeat visitation in mature sun and sand holiday destinations. *Journal of Travel Research,* 44, 288–297.
BALTACAR. 2017. *Baltic*

History Beneath Surface: Underwater Heritage Trails In Situ and Online [Online]. Available: **http://database.centralbaltic.eu/pr**oject/67 [Accessed 21/6/2017].

BRAMWELL, B. 2004. *Coastal mass tourism: Diversification and sustainable development in Southern Europe*, Channel View Publications.

BUSINESS LF. 2013. *Offshore Wind Farms and Tourism Potentials in Guldborgsund* Municipality [Online]. Available: http://www. southbaltic-offshore.eu/reports-studies/img/OFFSHORE_WIND_FARM S_AND_TOURISM.pdf [Accessed 21/6/2017].

CRUISE LINES INTERNATIONAL ASSOCIATION 2016. Contribution of Cruise Tourism to the Economies of Europe 2015.

DOWLING, R. K. 2002. Australian ecotourism–leading the way. *Journal of Ecotourism,* 1, 89–92.

ECORYS 2013. Study in support of policy measures for maritime and coastal tourism at EU level.

ETOA 2010. Groups mean Business. A Charter For Successful Tourism. European tourism association.

EUROMONITOR INTERNATIONAL 2013. World Travel Market Global Trends Report 2013.

EUROPEAN COMMISSION. 2014a. *A European Strategy for more Growth and Jobs in Coastal and Maritime Tourism* [Online]. Available: http://ec.europa.eu/maritimeaffairs/documentation/publications/docume nts/coastal-and-maritime-tourism_en.pdf [Accessed 21/6/2017].

EUROPEAN COMMISSION. 2014b. *Blue Growth infography* [Online]. Available: http://ec.europa.eu/assets/mare/infographics/ [Accessed 21/6/2017].

EURORESO. 2003. *SubArcheo – Novas Tecnologias na Formação de Arqueologos Subaquáticos* [Online]. Available: http://www.euroreso.eu/index. php/projects/9-sin-categoria/75-subarcheo [Accessed 20/6/2017].

EXPEDIA. 2013. *Annual Report* [Online]. Available: http://files.shareholder. com/downloads/EXPE/65641886x0x750253/48AF365A-F894-4E9C-8 F4A-8AB11FEE8D2A/EXPE_2013_Annual_Report.PDF [Accessed 21/6/2017].

GERMAN OFFSHORE WIND ENERGY FOUNDATION. *The Impact of Offshore Wind Energy on Tourism – Good Practices and Perspectives for the South Baltic Region* [Online]. Available: http://www.southbaltic-offshore.eu/reports-studies-the-impact-of-offshore-wind-energy-on-tour ism.html [Accessed 21/6/2017].

GSWS. 2014. *Top 10 Global Spa and Wellness Trends Forecast-2014 trends report* [Online]. Available: http://www.spafinder.co.uk/newsletter/trend s/2014/2014-trends-report.pdf [Accessed 21/6/2017].

ILO 2010. Global Employment Trends, January 2010. International Labour Office.

IPK 2016. ITB World Travel Trends Report 2016/2017. Berlin: IPK International.

KLEIN, R. A. 2011. Responsible cruise tourism: Issues of cruise tourism and sustainability. *Journal of Hospitality and Tourism Management,* 18, 107–116.

KOZAK, M. & MARTIN, D. 2012. Tourism life cycle and sustainability analysis: Profit-focused strategies for mature destinations. *Tourism Management,* 33, 188–194.

LAFFERTY, G. & VAN FOSSEN, A. 2001. Integrating the tourism industry: problems and strategies. *Tourism Management,* 22, 11–19.

PINEDA, I., RUBY, K., HO, A., MBISTROVA, A. & CORBETTA, G. 2015. The European Offshore Wind Inductry: Key Trends and Statistics 2015. *The European Wind Energy Association (EWEA),* 23.

REILLY, K., O'HAGAN, A. M. & DALTON, G. 2015. Attitudes and perceptions of fishermen on the island of Ireland towards the development of marine renewable energy projects. *Marine Policy,* 58, 88–97.

RHEEM, C. 2009. *PhoCusWright's. Going Green: The Business Impact of Environmental Awareness on Travel, February 2009* [Online]. Available: http://ghaward.ie/ghaward/userfiles/file/PhocusWright%20Nov%20201 0%20-%20going %20Green%20-%20the%20business%20impact.pdf [Accessed 20/6/2017].

ROYAL CARRIBBEAN INTERNATIONAL. 2017. *Harmony of the Seas: By the Numbers* [Online]. Available: http://www.royalcaribbean.com/ connect/harmony-of-the-seas-by-the-numbers/ [Accessed 19/6/2017].

SIDE. 2017. Available: http://www.side-underwatermuseum.com/ [Accessed 21/6/2017].

TEEB 2009. TEEB – The Economics of Ecosystems and Biodiversity for National and International Policy Makers. Summary: Responding to the Value of Nature.

TOURISMLINK. 2014. *TOURISMlink, An EU Initiative, Enters Its Second Phase* [Online]. Available: http://news.gtp.gr/2014/07/24/tourismlink-enters-into-the-second-phase/ [Accessed 20/6/2017].

TREMBLAY, P. 1998. The economic organization of tourism. *Annals of tourism research,* 25, 837–859.

TUI TRAVEL 2014. Annual Report and Accounts.

UNWTO 2016a. Exports from international tourism rise 4% in 2015 [press release]. United Nations World Tourism Organisation.

UNWTO 2016b. *Why Tourism?* [Online]. Available: http://www2.unwto.org/content/why-tourism [Accessed 20/4/2016].

UNWTO 2016c. World Tourism Barometer. United Nations World Tourism Organisation.

VAYÁ, E., GARCIA-SANCHÍS, J. R., MURILLO, J., ROMANÍ, J. & SURIÑACH, J. 2017. *Economic Impact of Cruise Activity: The Port of Barcelona* [Online]. Available: https://papers.ssrn.com/abstract=2884785 [Accessed 21/6/2017].

WINTHER, K. & JENSEN, S. K. 2013. DI Analysis: Tourism in the Baltic Sea Region.: Confederation of Danish Industry.

WORLD TRAVEL AND TOURISM COUNCIL 2017a. Travel and Tourism Economic Impact 2017 Caribbean.

World Travel and Tourism Council (WTTC) 2017b. Travel & Tourism Economic Impact 2017 Region League Table Summary. World Travel and Tourism Council.

PART III

Planning by Sea Basin

10

Regulation and Planning in Sea Basins – NE Atlantic

Anne Marie O'Hagan

MaREI Centre, ERI, University College Cork, Ireland

10.1 Introduction and Geography

The Atlantic Ocean forms the western boundary of the EU and is the second largest of the world's oceans. In an EU context, the Atlantic Basin consists of France, the island of Ireland, Portugal, Spain and, for the immediate future, the UK. Within Britain, the western parts of England and Scotland and all of Wales are included. Additional boundaries occur with the Crown dependencies of Jersey, Guernsey and the Isle of Man. All the Atlantic Basin countries are parties to the UN Law of the Sea Convention and many have large offshore maritime areas. Portugal has one of the largest Economic Exclusive Zones in Europe covering more than 1.7 million km^2, more than 18 times the country's territorial space (ECORYS *et al.*, 2014). Ireland's EEZ is approximately nine times its land mass (Government of Ireland, 2012). The Mid-Atlantic Ridge, extending from Iceland to approximately 58° South, forms a natural East – West boundary dividing the North Atlantic into two large troughs with depths from 3,700–5,500m (Marine Institute and Marine Board – ESF, 2011). The international waters of the Atlantic Ocean stretch westward to the Americas, eastward to Africa and the Indian Ocean, southward to the Southern Ocean and northward to the Arctic Ocean. Between the EU, cooperation with USA and Canada is common in relation to certain specific areas e.g. Atlantic Ocean Research and ocean observation through Galway Statement on Atlantic Ocean Cooperation (Marine Institute, 2013). All EU countries bordering the Atlantic Ocean have extensive coastlines and large populations reside within the coastal zone, except for those on the open Atlantic seaboard which have fewer than 10 inhabitants per km^2 in some

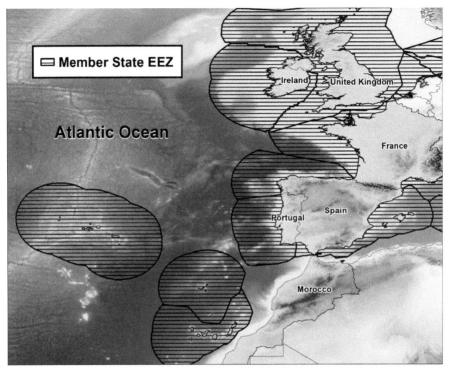

Figure 10.1   The NE Atlantic Basin.

remote areas (Marine Institute and Marine Board – ESF, 2011). Coastal areas vary from indented rocky coastlines, to sandy beaches and sheltered estuarine mudflats.

The North East Atlantic (Figure 10.1) may be one of the richest oceans in the world, but it is also one that is under increasing threat from over-fishing, pollution, abstraction and shipping traffic (EC, 2005). The region is characterised by a wide variety of coastal geologies including small island archipelagos, rocky headlands, cliff formations, salt marshes, sand dunes, bays, estuaries and numerous sandy beaches many of which contain important habitats and species protected under EU nature conservation legislation and site designation. Regional ecology varies widely. Irish and UK waters are categorised as cool-temperate waters with the waters around Atlantic parts of France, Spain and Portugal described as warm-temperate waters (OSPAR, 2010). This means that the Celtic Sea area is at the southern limit of the distribution range for some cold-water species, such as herring and cod, while some warm-water species, such as sea bass and sardine, come up from the south. Ocean depths vary widely: from being fully oceanic at the shelf break to the west of Ireland, through to the relatively shallow semi-enclosed Irish Sea as well as brackish estuarine systems along the west coast of the UK. Almost all parts of the NE Atlantic area support breeding and migratory birds dependent on the sea. The waters to the south and west of Ireland host a number of cetacean species, including common dolphins and bottlenose dolphins. Further to the west of Ireland, around the 200-mile limit, are a number of designated cold water coral (*Lophelia pertusa*) sites. Cold water coral is also found in Portuguese continental shelf waters to the south.

10.1.1 Overview of Key Marine Sectors

Studies conducted on Blue Growth within the Atlantic Sea Basin indicate that the total size of the Atlantic Blue Economy is at least €26.8 billion in gross value added (GVA) and accounts for more than 800,000 jobs, excluding maritime economic activities that could not be quantified (ECORYS *et al.*, 2014). The OSPAR Quality Status Report (2010) found that changes occurring to coastal and marine ecosystems are largely a function of human intervention. Currently coastal and marine tourism, fishing and shipbuilding, maritime transport and shipping are the most established sectors in the Atlantic basin though their relative importance varies between the individual Member States. In the UK and France, for example, shipbuilding is a much more economically significant activity than it is for Ireland. More than a third of the value of the maritime sector in the NE Atlantic is generated by shipping and coastal tourism (OSPAR, 2010). Shipping and maritime transport was the largest contributor in terms of turnover and value added in 2012 to Ireland's ocean economy (Vega *et al.*, 2015). Marine tourism and leisure is the next largest category overall in the Atlantic sea basin and is the largest contributor in terms of employment. At EU level, coastal tourism represents one of the sectors identified in the Blue Growth agenda and currently represents over one third of the maritime economy. Statistics indicate that more than four out of nine nights spent in accommodation establishments in the EU are spent in coastal areas (EC, 2014). In Ireland, Portugal and Spain tourism is showing an increase in GVA and employment contribution despite the economic crisis, though in the latter two countries this is more pertinent to the Mediterranean basin area (ECORYS *et al.*, 2014). Cruise tourism is one of the fastest growing maritime economic activities in Europe, though the Atlantic area is a less strong destination area than the Baltic or the Mediterranean.

Despite this, cruise tourism in the Atlantic is growing with a number of specific ports promoting their respective areas as a cruise destination. In Ireland, cruise tourism has experienced year on year growth since 1990. The number of ships increased from 61 in 1994 to 229 in 2012 and passenger numbers increased by over 200% (Fáilte Ireland, 2012; Vega et al., 2015).

Fisheries is an important employment sector in Spain, Portugal and France though its contributes more economically (GVA) to Portugal, Ireland and the UK (ECORYS *et al.*, 2014). Spain, the United Kingdom and France are the largest producers in terms of volume in the EU (EU, 2016). Over 74% of the total EU catch in 2013 came from the NE Atlantic (EU, 2016). The fishing fleet is large though it consists primarily of small vessels (under 12m). One of the key challenges identified for fisheries in an EU context and specifically for the NE Atlantic relates to stock sustainability and how reform of the CFP will impact upon this through changes to quotas and Total Allowable Catches (ECORYS *et al.*, 2014). Aquaculture is also showing a steady growth rate in Spain and Portugal, but less so in the other Atlantic basin countries. In the UK, for example, aquaculture activity is limited to the west coast of Scotland. In the EU, overall aquaculture output has been largely constant in volume since 2000 (EC, 2013). In four of the five Atlantic countries, fisheries and coastal tourism are the two largest sectors in terms of employment. Hydrocarbon (oil and gas) exploitation is limited in the Atlantic when compared to other sea-basins (e.g. the North Sea area) with activities currently limited to exploration. Short-sea shipping is out-performing deep-sea shipping in terms of tonnage in the Atlantic. Given the geographic conditions of the Atlantic basin, passenger ferries are important for transportation to and between islands belonging to, for example, Spain (Canary Islands) and Portugal (Azores, Madeira). Generally, there is no convergence in terms of one leading maritime sector in the Atlantic area and according to the Blue Growth study this is likely to stay the same for the immediate future (ECORYS *et al.*, 2014).

The governments of various Atlantic countries, as well as some individual companies, have targeted blue biotechnology as an emerging sector. In Ireland there is an active research community working on biomaterials, bioprocessing, food ingredients including functional foods, drugs and other therapeutic products, animal health and agriculture, aquaculture, medical devices, cosmetics and environmental remediation which receives funding from national, EU and international sources (Marine Coordination Group, 2012). Deep-sea mining and exploration activities relating to seabed minerals are under investigation by the UK and France, where exploration contracts have been signed with the International Seabed Authority. Only one of these pertain to the Atlantic basin in an area beyond national jurisdiction in the Mid-Atlantic Ridge.[1] Research organisations in both Spain and Portugal are also interested in this sector. Marine renewable energy (offshore wind, wave and tidal) has also been identified as a promising future sector for Atlantic seaboard countries (OEE, 2016). Ireland and parts of the UK (Scotland and Northern Ireland) have either dedicated marine renewable energy strategies or national strategies that recognise the potential for development of marine renewable energy in their waters (e.g. DCENR, 2014; Scottish Government, 2017; DETI, 2012). The potential represented is not limited to the physical resource but also the potential in terms of supply chain opportunities, job creation etc. In Wales for example, the sector directly supported 36 FTE jobs and 174 person years of employment in 2015 whereas already in 2017, that has risen to 137 FTE jobs and 350 person years of employment (Marine Energy Wales, 2017).

10.1.2 Most Promising Marine Sectors

A list of the most promising maritime economic activities for Member States in the Atlantic Basin is shown in Table 10.1.

Table 10.1 List of promising maritime economic activities at EU Member State level (ECORYS *et al.*, 2014)

France	Ireland	Portugal	Spain	UK
Ocean Energy	Ocean Energy	Blue Biotechnology	Ocean Energy	Blue Biotechnology
Blue Biotechnology	Blue Biotechnology	Ocean Energy	Blue Biotechnology	Offshore wind
Deep-sea mining	Environmental monitoring	Environmental monitoring	Desalination	Ocean Energy
Ship-building	Offshore wind	Offshore wind	Deep-sea mining	Environmental monitoring
Oil and Gas	Yachting and marinas	Deep-sea mining	Offshore wind	Ship-building
Environmental monitoring	Cruise tourism	Oil and gas	Environmental monitoring	Oil and gas
Maritime surveillance	Coastal protection	Ship-building	Maritime surveillance	Cruise tourism

Note: Cells in colour represent maritime economic activities that are prominent in all countries.

---

[1] See https://www.isa.org.jm/deep-seabed-minerals-contractors/overview

Ireland, Spain and parts of the UK appear to have policies and strategies for each maritime sector outlining development objectives for the future. In some cases, these are reflected together in an over-arching marine/oceans policy such as Harnessing Our Ocean Wealth in Ireland (Government of Ireland, 2012), *Grenelle de la Mer* [national strategy for the sea] in France (Ministère de l'écologie, de l'énergie, du développement durable et de la mer, 2009) and *Estratégia Nacional para o Mar* [national strategy for the sea] in Portugal (Governo de Portugal, 2012) each of which focus on specific maritime priorities and sectors for that country. The French strategy contains over one hundred commitments sub-divided into four categories covering energy, fisheries, transport and pollution. In Ireland, the national plan seeks to double the value of ocean wealth to 2.4% of GDP by 2030 and increase the turnover from the ocean economy to exceed € 6.4bn by 2020 (Government of Ireland, 2012). The Portuguese strategy highlights governance, living and non-living resources, observation and other activities as the five priority areas (Governo de Portugal, 2012). The UK has a Marine Policy Statement which basically acts as a framework for developing marine plans and for decisions affecting the marine environment but does not contain any national objectives per se (HM Government, 2011). Regional marine plans focus on maritime sectoral activities in English regions. Elsewhere, maritime spatial plans are in the process of being implemented in Scotland (Scottish Government, 2015) or are at the advanced planning phase (DAERA, 2017; Welsh Government, 2015). The Scottish National Marine Plan contains sectoral objectives for a number of marine sectors (fisheries, aquaculture, oil & gas, marine renewables, tourism, shipping, cables etc.) within a wider marine planning context. The draft Welsh marine plan follows a similar format. It is not yet known what approach will be taken in Northern Ireland.

10.1.3 Key Features Affecting Policy

10.1.3.1 EU membership

National maritime policies in the Atlantic are heavily influenced by membership of the European Union. The EU's Integrated Maritime Policy (COM(2007)575) encouraged the development of a dedicated Maritime Strategy for the Atlantic Ocean Area in 2011 (COM(2011)782). This was later supplemented by an Action Plan in 2013 (COM(2013)279) and has resulted in increased cooperation between Atlantic Arc countries through mechanisms such as the Atlantic Forum and Atlantic Stakeholder Platform as well as motivating a renewed focus on marine sectors within the individual Atlantic

countries. The key objective of the Atlantic Action Plan is to identify investment and research priorities in the sea basin that could be considered for EU
financial support in the new programming period of 2014–2020. This contains
four priorities:

- Promote entrepreneurship and innovation;
- Protect, secure and enhance the marine and coastal environment;
- Improve accessibility and connectivity;
- Create a socially inclusive and sustainable model of regional
  development.

A progress report on the IMP (COM(2012) 491 final) generally found that
through IMP initiatives and Member State actions, many sectors were making
progress but that certain sectors needed a more targeted approach hence the
adoption of the Blue Growth agenda in 2012. In future, the UK's departure
from the EU and the terms it negotiates with the Commission could have
wider policy implications for the Atlantic area.

10.1.3.2 Geography and jurisdiction

All the countries surrounding the Atlantic Ocean have claimed EEZs and
have conducted substantial seabed mapping programmes which have been
instrumental in contributing to knowledge of marine resources. In 2013, as
part of an event entitled "The Atlantic – a shared resource", the Galway
Statement on Atlantic Ocean cooperation was signed by representatives from
the EU, Canada and the USA. This sought to align the respective ocean
observation efforts of all parties so as to promote sustainable management of
the Atlantic's resources. Previous efforts at seabed mapping have also been
used to inform continental shelf claims. The water depth of the Atlantic sea
basin is much deeper than the North Sea and the Baltic and this could limit the
development of offshore wind energy in the immediate future. In the longer
term however, floating installations may be an option. The Atlantic basin also
has opportunities in terms of its position as a gateway for maritime transport
which links Europe to other global markets. Its rich natural resources in wind
and wave energy could be instrumental in future maritime sector growth.

International maritime boundaries remain to be settled between Ireland,
UK, Denmark and Iceland with respect to the Hatton-Rockall area, and the
anticipated hydrocarbon resources within that area. A maritime boundary
between Ireland and the UK in the Rockall area was settled by international agreement in 1988, though this does not extend into the coast. The
Rockall-Hatton area is also claimed by Denmark (on behalf of the Faroe

Islands) and by Iceland. Informal consultations between officials from the four countries concerned had taken place since 2001 but broke down without agreement prior to 2009. Ireland lodged its own submission in 2009 but, as the area is disputed, the Commission on the Limits of the Continental Shelf (CLCS) will not get involved or make any recommendations in relation to the claims received. France, Ireland, Spain and the United Kingdom made a joint submission to CLCS in 2006 in respect of an area to the south of Ireland in the area of the Celtic Sea and the Bay of Biscay. The area concerned is approximately 83,000 km^2 and was accepted by the CLCS in 2009. Ireland had previously made a separate claim relating to the Porcupine Abyssal Plain in 2005 which was accepted by the CLCS in 2007. Extended continental shelf claims, and the fact that EU Atlantic States have huge maritime jurisdictional zones, raise the relevance of oceans governance for both the States concerned and the EU more generally. In 2009 the Commission published a Communication on 'Developing the international dimension of the Integrated Maritime Policy of the European Union' (COM(2009)536); launched a public consultation on international ocean governance in June 2015 (SWD(2016)352 final) and subsequently issued a joint Communication on international ocean governance: an agenda for the future of our oceans in 2016 (JOIN(2016)49 final).

10.1.3.3 Economic and political climate

The countries of the Atlantic basin are politically stable and whilst there have been economic recessions in Ireland, Spain and Portugal it would appear that a new level of economic stability is emerging. Many Atlantic coastal communities are peripheral in nature and are having to cope with declines in traditional maritime sectors such as fisheries and ship-building. Opportunities associated with emerging marine sectors such as ocean energy and offshore aquaculture are viewed as having the potential to address unemployment and declining populations as well as stimulating regional development. The Atlantic is not a source of global insecurity or tension with limited opportunity for conflict. As mature countries, policies tend to be well developed though from the perspective of integrated maritime governance this could be problematic as management has tended to be approached separately rather than cooperatively. Maritime Spatial Planning, for example, is well developed in the UK and Portugal but does not yet exist in Ireland or Spain. The French approach to MSP is highly decentralised.

The EU is in the process of negotiating a trade and investment deal with the USA, the Transatlantic Trade and Investment Partnership (TTIP).

This is said to be the world's biggest bilateral trade and investment deal and will open up the US to EU companies as well as create new rules for exporting, importing and investing overseas. It is thought that an agreement of this nature would stimulate trade and business at a time of ongoing economic crisis, generating growth of businesses and jobs. Economically it is projected that, when fully operational, TTIP could bring benefits to the EU economy worth an additional 0.5% of GDP (Francois *et al.*, 2013). Originally the negotiations were expected to be finalised by the end of 2014 but this did not occur and uncertainty currently reigns over the future of the agreement. Separately the EU is also in the process of negotiating the Comprehensive Economic and Trade Agreement (CETA) with Canada. The European Parliament voted in favour of CETA on 15 February 2017 (but EU national parliaments must approve CETA before it can take full effect). It is intended to have many of the same implications as TTIP; stimulating trade, investment and growth. All EU Atlantic sea basin countries have strong trade links with Canada currently. Canada ranks in the top 20 for all Atlantic countries as the biggest trade partner outside the EU: 5th for UK, 7th for Ireland, 9th for Portugal, 15th for France and 20th for Spain (European Commission, 2017).

10.2 Environmental Policy

10.2.1 Regional Sea Level

The Convention for the Protection of the Marine Environment of the NE Atlantic (OSPAR Convention) applies to the waters of the Atlantic basin which stretch from the Arctic southward to Portugal. All five Atlantic States of the EU are signatories to the OSPAR Convention as is the European Union. The OSPAR area is sub-divided into five areas: Arctic waters; greater North Sea; Celtic Seas; Bay of Biscay and Iberian Coast; and wider Atlantic. The latter three subdivisions incorporate the Atlantic sea basin in the EU sense. The OSPAR Convention has its genesis in addressing dumping and was later expanded to include land-based sources of pollution. In 1998 a new annex on biodiversity and ecosystems was adopted to address non-polluting human activities that can adversely affect the sea. The overall goal of the OSPAR Convention is to conserve marine ecosystems and safeguard human health and, when practicable, restore marine areas that have been adversely affected by preventing and eliminating pollution and by protecting the maritime area against the adverse effects of human activities (Article 2(1)). This is taken forward through a dedicated NE Atlantic Environment Strategy (OSPAR, 2010) supported by five thematic strategies with a Joint Assessment and

Monitoring programme – Biodiversity and Ecosystems; Eutrophication; Hazardous Substances; Offshore Industry; Radioactive Substances. The strategic objective of the biodiversity and ecosystems strategy is to halt and prevent further loss of biodiversity in the OSPAR maritime area; to protect and conserve ecosystems; and to restore, where practicable, marine areas which have been adversely affected by 2020. This, therefore, links directly with the EU's Birds and Habitats Directives, Biodiversity Strategy and the Marine Strategy Framework Directive.

One of the key aims of the OSPAR Convention is the creation of an ecologically coherent network of Marine Protected Areas (MPAs) in the NE Atlantic that is well managed by 2016 (OSPAR Recommendation 2003/3, para.2.1, as amended). The Commission operates an online MPA Map tool[2] which contains spatial and non-spatial data from OSPAR Contracting Parties on their respective MPAs. Any SPAs and SACs designated under the Birds and Habitats Directives that are partly or wholly in the OSPAR maritime area should also consider reporting that area as a component of the OSPAR Network of Marine Protected Areas (OSPAR Recommendation 2003/3, para.3.3(a), as amended). OSPAR has also agreed to include MPAs in areas of the NE Atlantic that are outside the jurisdiction of the Contracting Parties (ABNJ areas). The majority of MPAs are in territorial waters (23.59%), significantly less in the EEZs (3.06%) and slightly over 6% of MPAs are in areas beyond national jurisdiction (i.e. the High Seas, the Area and the extended continental shelf areas) (OSPAR, 2015). Table 10.2 shows

Table 10.2   Breakdown of OSPAR MPAs in Atlantic sea basin countries (adapted from OSPAR, 2015)

Country	No. of OSPAR MPAs	MPA Coverage (km^2)			
		Territorial Sea	EEZ	ABNJ	TOTAL
France	39	15,821	6,283	n/a	22,104
Ireland	19	1,594	2,542	n/a	4,135
Portugal	8[3]	1,022	4,656	22	5,700
Spain	13	7,277	12,985	n/a	20,262
United Kingdom	244[4]	28,239	98,155	17,158	143,522

---

[2]See http://mpa.ospar.org/home_ospar

[3]Portugal has nominated a total of 12 MPAs to OSPAR. Four of these MPAs occur on an extended shelf claim area, submitted to the UN CLCS.

[4]The United Kingdom has nominated a total of 244 MPAs to OSPAR. Two of the 244 MPAs occur on the extended continental shelf of the UK. The North West Rockall Bank SAC straddles the UK EEZ and the extended continental shelf of the UK.

the breakdown of OSPAR MPAs for Atlantic sea basin countries, noting that some parts of these countries fall into other OSPAR subdivisions e.g. part of the UK is within the greater North Sea region. Most of the OSPAR MPAs under national jurisdiction are subject to the management provisions of the Birds and Habitats Directives. In areas beyond national jurisdiction, the OSPAR Commission has already agreed on Recommendations for the management for each of these areas which guide OSPAR Contracting Parties in their actions and in the adoption of measures to achieve the site objectives (OSPAR, 2015). To complement this approach, a "Collective Arrangement between competent international organisations on cooperation and coordination regarding selected areas in areas beyond national jurisdiction in the NE Atlantic" was agreed between the OSPAR Commission and NE Atlantic Fisheries Commission (NEAFC) with respect to fisheries in 2014 (OSPAR Agreement 2014-09).

Under the 'Human Activities' work area, the OSPAR Commission works on marine renewable energy, marine litter, underwater noise, mariculture and fishing, shipping, dredging and dumping, and conventional munitions. OSPAR is primarily concerned with the impacts of marine activities on the receiving environment. The OSPAR Commission has developed guidance on environmental considerations for the development of offshore wind farms, the purpose of which is to assist in the identification and consideration of some of the issues associated with determining the environmental effects of offshore wind farm developments (OSPAR, 2008). In relation to fisheries and aquaculture, OSPAR's Eutrophication; Hazardous Substances; and Biodiversity and Ecosystems strategies all contain measures to monitor, assess and regulate the impacts of mariculture and fisheries. Aside from human activities, OSPAR works on specific anthropogenic issues such as noise, marine litter and ballast water. Whilst none of the OSPAR Commission's documentation refers explicitly to multi-use platforms or combined activities, its guiding principles and cross-cutting work on topics such as Maritime Spatial Planning, risk assessment and implementation of the Marine Strategy Framework Directive has the potential to influence coexistence of marine activities in future.

10.2.2 EU Level

10.2.2.1 Sea basin strategy

A Maritime Strategy for the Atlantic Ocean Area was produced by the Commission in 2011 (COM(2011)782) as outlined briefly in Section 10.1.3. The Strategy highlighted the need to take an ecosystem approach to the management of the Atlantic area as well as the activities going on within the area. The Atlantic Action Plan (COM(2013)279) published in 2013 sought to deliver on the areas contained in the strategy. One of the priorities in the Action Plan is to protect, secure and enhance the marine and coastal environment. There are four specific objectives under this priority – improving maritime safety and security; exploring and protecting marine waters and coastal zones; sustainable management of marine resources; and exploitation of the renewable energy potential of the Atlantic area's marine and coastal environment. Under the specific objective on exploring and protecting marine and coastal areas, the focus is primarily on the creation and development of observation systems and capabilities. The Action Plan also seeks to contribute to the development of tools and strategies to address global climate change issues incorporating assessment of impacts and sharing best practices. The Action Plan complements on-going work under the MSFD to achieve Good Environmental Status (GES) by agreeing on good practices, evaluation processes, encouraging coordination and facilitating integrated monitoring programmes as envisaged under OSPAR. It also seeks to support Member States in their implementation of Integrated Coastal Management and Maritime Spatial Planning. Almost all of these objectives are complementary to other areas of EU activity through existing legislative or policy instruments. In terms of implementation, the Action Plan is not overly specific but outlines possibilities for funding, collaboration and support. Cooperation and implementation of the Action Plan is voluntary. Regular communications on the implementation of the Action Plan are facilitated through annual conferences and the Support Team for the Atlantic Action Plan (See Section 10.5.1).

10.2.2.2 Marine Strategy Framework Directive (MSFD)

At EU level, the Marine Strategy Framework Directive requires Member States to produce marine strategies for their marine waters so that Good Environmental Status can be achieved. In order to achieve its goal, the Directive establishes European marine regions and sub-regions on the basis of geographical and environmental criteria. The NE Atlantic Ocean is one of these regions. All five Atlantic countries have now transposed the requirements of the Directive into national law, assigned their competent authorities, completed their initial assessment, determined what Good Environmental Status (GES) means for their marine waters, identified their environmental targets and associated indicators, submitted their monitoring programmes and, with the exception of the UK, agreed their programmes of measures.[5] The Commission's progress report on MSFD implementation states that overall Member States' definition of GES and the path out to achieve it lacks coherence across the EU, even between neighbouring countries within the same marine region (COM(2014)97, p.2). Coherence is, however, identified as being strongest within the NE Atlantic. Cooperation within the OSPAR framework was stronger for the initial assessment and the definition of GES than for the establishment of the environmental targets and indicators (Milieu Ltd., 2014).

10.3 Regulatory Regimes

10.3.1 Overview

All five Atlantic countries are subject to the legal requirements deriving from international conventions and treaties to which they are party or to which the European Union is a party. EU legislation listed in Annex 1 is also applicable in the Atlantic sea basin and to the associated countries.

10.4 Spatial Impact and Planning

10.4.1 Spatial Considerations

In terms of the MARIBE combinations identified for the Atlantic Sea Basin, a number of the combinations will be dependent on specific physical and geographic characteristics in order to be realised. Oil and gas, wind and wave installations are all resource dependent. Generally, the spatial requirements for wave and tidal energy farms are not yet well-established and will depend on the device chosen. Ireland (SEAI, 2010), England (DECC, 2011), Scotland (The Scottish Government, 2007) and Northern Ireland (AECOM and Metoc, 2009) have all conducted Strategic Environmental Assessments for marine renewables generally whilst Spain has conducted a similar exercise specifically for offshore wind (Ministerio de Industria, Energía y Turismo, 2009). This helps to identify potential areas for the development of commercial scale projects. In Portugal, the national Maritime Spatial Plan reflects current ocean energy test sites and it is anticipated that such sites could host

---

[5]See http://ec.europa.eu/environment/marine/eu-coast-and-marine-policy/implementation/ scoreboard_en.htm (information dated 27 March 2017, accessed 5 April 2017).

commercial scale deployments in future (O'Hagan, 2016). The UK is the only country with dedicated Renewable Energy Zones and leasing rounds for marine renewable energy (O'Hagan, 2018). France also operates a leasing round type call for projects (tidal and offshore wind) (e.g. Ministère de l'Environnement, de l'Énergie et de la Mer, 2017). Trade association work in Ireland, through the Marine Renewables Industry Association, has previously identified suitable zones for marine renewables in Irish waters but as yet the Irish Government has not conducted leasing rounds for any technology (MRIA, 2010).

Aquaculture planning along the Atlantic basin is approached in a non-systematic way with farmers proposing areas for operation in their licence applications e.g. in Ireland and France (O'Hagan *et al.*, 2017). Scotland implements the approach advocated by the UN FAO involving zoning, site selection and area management (Aguilar-Manjarrez *et al.*, 2017). Zones are designated along the west coast, western isles and northern isles primarily for salmon production but also shellfish (mussel) production. Locally these are subdivided into farm management areas, decided in association with industry representatives, which involves synchronised approaches to fallowing and treatments. This is supplemented with disease management areas for the control of notifiable diseases. Marine Scotland, as the consenting authority, is responsible for strategic aquaculture policy and have published locational guidelines (carrying capacity) for the sector (Marine Scotland Science, 2015). All along the Atlantic coast there is a perceived lack of available sites for expansion of aquaculture which, it is hoped, will be addressed through the recent marine planning system in Scotland. Aquaculture does not take place along English or Welsh coastlines. The EU Directive on Shellfish Waters (2006/113/EC) requires Member States to designate waters that need protection in order to support shellfish life and growth. In France planning of aquaculture installations depends on whether it will be situated in marine waters or inland (FAO, 2017). The main type of mariculture in France is shellfish farming, which represents 80 percent of the total aquaculture production (European Commission/DG MARE, 2017). In Spain the applicable legal framework for aquaculture development is the responsibility of the Autonomous Communities, who apply their own norms for authorisations or leases (MAGRAMA, 2014). This also varies according to whether the aquaculture is inland or marine and in public or private areas. Only Spanish citizens or entities may hold concessions or authorisations for marine cultures under Law No 23/1984. Portugal is the lowest aquaculture producer in the Atlantic Arc, which can be attributed to a range of factors including a complex licensing process, the small number of optimal sites explored to date, the lack of a dedicated aquaculture zoning system and the situation of existing installations within protected areas where certain new technologies for production may not be permitted (Ministério da Agricultura e do Mar, 2014).

10.4.2 Maritime Spatial Planning

Maritime Spatial Planning can help to create better conditions for proceeding with particular developments since synergies and conflicts between different sectors and other sea uses are usually addressed during development of the plan. An overview of the status of MSP in the different Atlantic sea basin Member States is presented in Table 10.3. This situation will evolve over the coming years as Member States are now required to develop Maritime Spatial Plans at the latest by 2021 under the EU's MSP Directive.

10.5 Related Strategies

10.5.1 Atlantic Strategy

The EC's Communication on Developing a Maritime Strategy for the Atlantic Ocean Area (COM(2011)782) identified five key themes of relevance to the Atlantic sea basin – Implementing the ecosystem approach; Reducing Europe's carbon footprint; Sustainable exploitation of the Atlantic seafloor's natural resources; Responding to threats and emergencies; Socially inclusive growth. Subsequently the EC adopted an Action Plan for a Maritime Strategy in the Atlantic area: delivering smart, sustainable and inclusive growth (COM(2013)279) in 2013 as outlined briefly above. The Action Plan sets out priorities for research and investment to advance the 'blue economy' in the Atlantic area. The Action Plan was developed through consultations conducted in the Atlantic Forum which consisted of representations from each of the five Atlantic Member States, the European Parliament, regional and local authorities, civil society and industry. The Action Plan states that to be effective it needs to be supported by targeted investment, increased research capacity and higher skills. The priority areas identified in the Action Plan are complemented by a range of specific objectives but each of these pertain to specific marine and coastal sectors or

Table 10.3 Information on MSP in Atlantic basin countries

COUNTRY	MSP in Place?	Activities Covered	Comments	Links
France	No	Unknown	National Maritime and Coastal Strategy adopted in February 2017. Legislation transposing MSP Directive enacted in May 2017. Documents Stratégique de Façade, created under the National strategy will be developed for the four French sea basins and will implement both the MSFD and MSP.	http://www.dirm-memn.developpement-durable.gouv.fr/gestion-integree-de-la-mer-et-du-littoral-giml-r266.html https://www.legifrance.gouv.fr/eli/decret/2017/5/3/DEVH163206D/jo/texte
Ireland	No	Unknown	Ireland has an Integrated Marine Plan called Harnessing Our Ocean Wealth. This recognised the need for MSP and the Marine Coordination Group has produced reports on an appropriate framework for MSP in Ireland which will be taken forward in due course	http://www.ouroceanwealth.ie/ https://www.ouroceanwealth.ie/publications http://www.housing.gov.ie/planning/maritime-spatial-planning/maritime-spatial-planning-directive/maritime-spatial-planning
Portugal	Yes	All	All maritime activities were identified, mapped and studied including their conflicts and interaction with local communities	http://www.dgpm.mam.gov.pt/Pages/POEM_PlanoDeOrdenamentoDoEspacoMarinho.aspx
Spain	No	Unknown	MSP could be progressed as part of the implementation of the MSFD according to Spanish law.	*None available*
UK: England	Yes	All	Marine plans for each marine region	https://www.gov.uk/government/collections/marine-planning-in-england
Wales	No	Unknown	Under development	http://gov.wales/topics/environmentcountryside/marineandfisheries/marine/marine-planning/?lang=en
Scotland	Yes	All	National plan with regional plans and sectoral plans	http://www.gov.scot/Topics/marine/seamanagement and links to additional resources
N. Ireland	No	Unknown	Under development	https://www.daera-ni.gov.uk/articles/marine-plan-northern-ireland

cross-cutting issues. There is no mention of coexistence or multiple use of space.

The *Support Team for the Atlantic Action Plan* provides guidance and proactive support for public and private organisations, research institutions and universities, institutional and private investors from the Atlantic region who wish to engage in the implementation of the Atlantic Action Plan. The Support Team are represented in each Atlantic Member State by a specific focal point who can provide interested parties with updated information, networking opportunities, funding and project ideas so as to advance the Action Plan priorities. These national points are coordinated by a central office in Brussels. Funding for projects comes through the European Maritime and Fisheries Fund as well as through programmes such as INTERREG Atlantic Area.

INTERREG's Atlantic Area programme forms part of the EU's Cohesion Policy, supporting transnational cooperation projects in 37 regions across the five Atlantic countries, in recognition of the primary features which are common between the regions: environmental heritage and maritime dimension as well as a territorial and urban development common pattern based on a majority of intermediate rural areas and a limited number of large metropolitan areas (INTERREG Atlantic Area, 2015). Projects funded under this programme should contribute to the achievement of economic, social and territorial cohesion in the areas of innovation, resource efficiency, environment and cultural assets, and supporting regional development and sustainable development. The current Programme focuses on four main priorities axes and specific objectives:

- Priority 1: Stimulating innovation and competitiveness;
- Priority 2: Fostering resource efficiency;
- Priority 3: Strengthening the territory's resilience to risks of natural, climate and human origin; and
- Priority 4: Enhancing biodiversity and the natural and cultural assets.

These complement the priorities of the Atlantic Action Plan quite strongly and also the over-arching Europe 2020 strategy for growth.

10.5.2 Existing Maritime Clusters

Clusters are considered important for the progress of the Blue Growth strategy as the development and growth of maritime sectors are often dependent on collaboration and cooperation between local players.

Table 10.4 Maritime Clusters in the Atlantic Sea basin

Country	Region	Cluster	
France	Bretagne	Brest	Defence, blue biotechnology, shipbuilding, fisheries, ocean renewable energy
	Aquitaine	Bordeaux	Yacht building and repair
Ireland	Cork	IMERC	Marine Energy, Shipping, Logistics and Transport; Maritime Safety and Security, Marine Recreation
	Galway	MI	Offshore energy; blue biotechnology, aquaculture; deep sea technologies
Portugal	Norte	Porto	Deep and short-sea shipping; coastal, nautical and cruise tourism; marine minerals mining
	Norte	Aveiro	Industrial fisheries, aquaculture, fish processing; nautical tourism; R&D
	Lisboa	Lisboa	(Industrial) fisheries, marine biotechnology, metallic and non-metallic minerals, freight transport, marine aquaculture, cruise tourism
	Algarve		(Coastal/nautical and cruise) tourism; transhipment; (industrial) fisheries; aquaculture and marine biotechnology
	Azores		Aquaculture, R&D
	Madeira		Tourism, fisheries, R&D
Spain	Galicia	Cluster Maritimo de Malaga	Coastal tourism, Maritime transport (deep and short-sea shipping), Fisheries, Aquaculture, Offshore renewable energy, Shipbuilding
	Basque Country		Shipbuilding, Marine energy, Coastline tourism, Maritime transport (port of Bilbao).
	Canarias	Cluster Maritime de Canarias	Shipbuilding and ship repairs; port services; fishing and aquaculture; and auxiliary industries
UK	South West England	Dorset and Somerset	Fisheries/Aquaculture, Biotechnology, Renewable Energy, Minerals and Aggregates, Coastal protection, yachting/marinas, Ship/(leisure) boat building
	Scottish West Coast	Highlands and Islands, SW Scotland	Offshore wind, marine aquaculture, fisheries, ocean renewable energy, blue biotech

Source: Country fiches, 2014.

The INTERREG Atlantic Programme states that there is a need to "encourage more clustering and cooperation mechanisms between complementary sectors and between research and economic actors in a transnational context". The view from INTERREG is that promotion of cooperation through clustering assists in the transfer of knowledge and technology to industry and also contributes to the "free movement" of knowledge in the Atlantic Area (INTERREG Atlantic Area, 2015).

10.6 Supporting Blue Growth

The Support Team for the Atlantic maintain a list of current funding opportunities covering both national and European sources. The EU funding sources that can be used to fund Blue Growth activities are listed in Table 10.5.

In some circumstances specific sectors may receive governmental support in the form of State-aid but this can be subject to very strict conditions. With respect to shipbuilding for example, the European Commission has created three exemptions for the shipbuilding industry that are considered not to disturb the internal market and competition between companies and countries.

1. Regional aid – if the investment is used for upgrading or modernising existing yards and is not used to restructure the yard financially
2. Innovation aid – for innovation in existing shipbuilding, ship repair or ship conversion yards provided that it relates to the industrial application of innovative products and processes
3. Export credits – ship owners may be granted State-supported credit facilities for new buildings or vessel conversions (EC, 2003).

In 2011, these rules expanded the scope of the current rules to include inland waterway vessels, as well as floating and moving offshore structures and further provided refined rules on innovation aid (EC, 2011). These rules might, therefore, have implications for multi-use platforms in future.

10.7 Key Lessons

10.7.1 Key Messages and Relevant Research Notes

There is a strong law and policy basis for Blue Growth activities at EU level through the Atlantic Action Plan and at national level through Member State strategic plans. One potential problem here is the fact that strategic plans tend to be sectoral in nature making it difficult to reconcile objectives and future development plans. The rural character of the Atlantic area's coastal regions mean it has strong traditional knowledge of maritime sectors that could now be in decline but those same skillsets could be harnessed

Table 10.5 Possible sources of EU funding for Blue Growth activities

Funding Name	Purpose/Type of Activity Covered	Links
2014–2020 EU financial framework	Partnership contracts between national governments and the Commission, operational programmes for regional development & work programmes for research.	http://ec.europa.eu/budget/mff/index_en.cfm
European Agricultural Fund for Rural Development (EAFRD)	Sustainable management of natural resources and climate action and the balanced territorial development of rural areas.	http://ec.europa.eu/agriculture/rural-development-2014-2020/financial-instruments/index_en.htm
European Maritime and Fisheries Fund (EMFF)	Maritime and fisheries related activities including sea-basins such as the Atlantic. Aims at achieving the objectives of the reformed CFP and IMP.	http://ec.europa.eu/fisheries/cfp/emff/index_en.htm
European Social Fund (ESF)	Main financial instrument for investing in people. Seeks to increase employment opportunities and promote education.	http://ec.europa.eu/regional-policy/en/funding/social-fund/
European Regional Development Fund (ERDF)	Aims to strengthen economic and social cohesion in the European Union by correcting imbalances between its regions.	http://ec.europa.eu/regional-policy/en/funding/erdf/
Cohesion Fund	Helps Member States with a GNI per inhabitant of less than 90% of the EU-27 average to invest in TEN-T transport networks and the environment.	http://ec.europa.eu/regional-policy/en/funding/cohesion-fund/
European Territorial Cooperation Fund	Provides a framework for the exchanges of experience between national, regional and local actors from different Member States as well as joint action to find common solutions to shared problems.	http://ec.europa.eu/regional-policy/en/policy/cooperation/european-territorial/
European Groupings of Territorial Cooperation (EGTCs)	Designed to help specific countries/regions overcome complicated differences between national rules and regulations.	https://portal.cor.europa.eu/egtc/Pages/welcome.aspx
Connecting Europe Facility	A new, integrated instrument for investing in EU infrastructure priorities in transport, energy and telecoms.	https://ec.europa.eu/inea/en/connecting-europe-facility
Programme for the Competitiveness of Enterprises and small and medium-sized enterprises (COSME)	Aims to strengthen the competitiveness and sustainability of the Union's enterprises and encourage an entrepreneurial culture by promoting the creation and growth of SMEs.	http://ec.europa.eu/growth/access-to-finance/cosme-financial-instruments/index_en.htm
Horizon 2020	Research and innovation funding for various types of research project.	http://ec.europa.eu/programmes/horizon2020/
LIFE+	Covers the environment, biodiversity, resource efficiency, governance and all aspects of climate change.	http://ec.europa.eu/environment/life/funding/lifeplus.htm

for the development and implementation of Blue Growth activities. This could be a unique selling point for the region in future. There are already a significant number of maritime clusters along the Atlantic region but it has been recognised that these require stronger local and regional involvement (ECORYS *et al.*, 2012), as well as greater support and recognition at the EU level. Access to finance is important for shipyards and marine equipment suppliers as their investments are capital intensive and uncertain. Similarly, new and developing SMEs also need access to finance as characteristically they do not lend themselves to having the funds necessary for large scale development and deployment e.g. wave and tidal energy developers. Studies conducted by the EC on Blue Growth in the Atlantic sea basin suggests that trends over time are more diverse across the Atlantic Arc (ECORYS *et al.*, 2014). Certain sectors already recognise that lack of suitable maritime space is a key limiting factor for expansion (e.g. aquaculture). Situations such as this, however, have not yet stimulated a move towards combining uses on multi-use platforms. Co-existence may take on a more prominent role or warrant additional consideration at Member State level as States respond to the EU's Maritime Spatial Planning Directive. This can already be seen in the UK where the evidence reports involve stakeholders in deciding if and where specific marine activities can co-exist.

One of the factors that may limit the region's growth in future is the north south gap, clearly evidenced on aspects such as demography, accessibility, higher education, early school leaving, economic development, competitiveness and Innovation regional performance (Innovation Union Scoreboard), share of Natura 2000 sites and capacity to adapt to climate change (INTERREG, 2015). Parts of the Atlantic region are more industrialised that others, hence the region could face challenges with respect to cumulative impacts of marine activities within the context of OSPAR and achieving Good Environmental Status of marine waters by 2020. Many of the Atlantic region countries have huge maritime territories which also present challenges for enforcement, compliance, security and surveillance. Moving closer to shore, where marine activities predominate and where there are already some examples of competition for space, there is still a prevalence of sectoral planning systems over more integrated planning systems at Member State level. Institutional structures also follow this trend with multiple entities often involved in planning, decision-making, implementation, monitoring and decommissioning of marine activities and structures. These realities have the potential to impede development of multi-use platforms. Permissions, authorisation, licences and leases are very much based on single sector activities. This could make it difficult to licence a platform which would host multiple activities in a shared space. The novelties presented by multiple-use platforms raise many unresolved legal issues such as those relating to liability and insurance, which need to be further explored from a legal perspective.

Given the Atlantic region's geographic position, it could be considered a gateway to continental Europe. There are strong maritime transport links and a captive market for food and energy products which could be derived from increased offshore aquaculture activities as well as the deployment of offshore wind, wave and tidal devices. The latter, however, will require grid integration with continental Europe which also necessitates dedicated funding and high levels of both political and public support. From a policy perspective there appears to be considerable support for Blue Growth and the Atlantic strategy generally from many government actors, industry and wider stakeholder groups. The implications of Brexit are as yet unknown but will more than likely create both opportunities and threats to the region as a whole. In terms of multi-use platforms there are no policies exploring or advocating this approach as a way forward but from this analysis it would seem that there is nothing, in theory, limiting developments of that nature.

References

AECOM and Metoc. (2009). Offshore Wind and Marine Renewable Energy in Northern Ireland Strategic Environmental Assessment (SEA) Non-Technical Summary (NTS). Cheshire: AECOM Limited.

Aguilar-Manjarrez, J., Soto, D. and Brummett, R. (2017). Aquaculture zoning, site selection and area management under the ecosystem approach to aquaculture. Full document. Report ACS113536. Rome, FAO, and World Bank Group, Washington, DC. 395 pp.

Department of Agriculture, Environment and Rural Affairs. (2017). Marine Plan for Northern Ireland: Stakeholder Newsletter February 2017. [online] Belfast: DAERA, 4pp. https://www.daera-ni.gov.uk/sites/default/files/publications/daera/Marine%20Plan%20January%202017%20newsletter%20%28with%20logo%29.pdf [Accessed 3 April 2017].

Department of Communications, Energy and Natural Resources (2014). Offshore Renewable Energy Development Plan. [online] Dublin: DCENR, 60pp. http://www.seai.ie/Renewables/Ocean-Energy/The-Offshore-Renewable-Energy-Development-Plan-OREDP/ [Accessed 14 Feb. 2014].

Department of Energy and Climate Change (DECC). (2011). UK Offshore Energy Strategic Environmental Assessment. OESEA2 Non-Technical

Summary Future Leasing/Licensing for Offshore Renewable Energy, Offshore Oil & Gas, Hydrocarbon Gas and Carbon Dioxide Storage and Associated Infrastructure. [online] DECC, London, 28pp. Available from: https://www.gov.uk/government/publications/uk-offshore-energy-strategic-environmental-assessment-2-environmental-report [Accessed 10 April 2017].

Department of Enterprise, Trade and Investment Northern Ireland. (2012). Offshore Renewable Energy Strategic Action Plan 2012–2020. Belfast: DETI, 45pp.

ECORYS, Deltares, Oceanique Development, (2012). Blue Growth: Scenarios and drivers for sustainable growth from the oceans, seas and coasts. Third Interim Report. [online] Rotterdam/Brussels: ECORYS, 126pp. Available at: https://ec.europa.eu/maritimeaffairs/sites/maritimeaffairs/files/docs/publications/blue_growth_third_interim_report_en.pdf [Accessed 26 March 2012].

ECORYS, S-PRO and MRAG. (2014). Study on Deepening Understanding of Potential Blue Growth in the EU Member States on Europe's Atlantic Arc. Sea basin report – FWC MARE/2012/06 – SC C1/2013/02 for DG Maritime Affairs and Fisheries. [online] Rotterdam/Brussels: ECORYS, 186pp. Available at: https://webgate.ec.europa.eu/maritimeforum/sites/maritimeforum/files/Blue%20Growth%20Atlantic_Seabasin%20report%20FINAL%2007Mar14.pdf [Accessed 2 Nov. 2015].

European Commission. (2003). Framework on State-aid to Shipbuilding (2003/C 317/06). Official Journal of the European Union, C317/11, 30 Dec. 2003.

European Commission. (2011). Framework on State-aid to Shipbuilding (2011/C 364/06). Official Journal of the European Union, C364/9, 14 Dec. 2011.

European Commission. (2013). Aquaculture. http://ec.europa.eu/fisheries/cfp/aquaculture/.

European Commission. (2014). Communication from the Commission to the European Parliament, the Council, the European Economic and Social Committee and the Committee of the Regions. A European Strategy for more Growth and Jobs in Coastal and Maritime Tourism COM(2014) 86 final. [online] Brussels: EC. http://eur-lex.europa.eu/legal-content/EN/TXT/?qid=1479224038190&uri=CELEX:52014DC 0086 [Accessed 24 Feb. 2014].

European Union, (2016). Facts and figures on the Common Fisheries Policy: Basic statistical data 2016 Edition. [online] Brussels: European

Union. ISBN 9789279609725. doi:10.2771/607841. https://ec.europa.eu/fisheries/sites/fisheries/files/docs/body/pcp_en.pdf [Accessed 01 April 2017].

European Commission. (2017). CETA in your town – Country infographics for France, Ireland, Portugal, Spain and the UK. [online] Available at: http://ec.europa.eu/trade/policy/in-focus/ceta/ceta-in-your-town/ [Accessed 5 April 2017].

European Commission/DG MARE. (2017). France: Multiannual national plan for the development of sustainable aquaculture an overview. [online] Available at: https://ec.europa.eu/fisheries/sites/fisheries/files/docs/body/summary-aquaculture-strategic-plan-france_en.pdf [Accessed 10 April 2017].

Fáilte Ireland. (2012). Cruise Tourism to Ireland Research Report – 2010. [online] Dublin: Bord Fáilte. Available at: http://www.failteireland.ie/FailteIreland/media/WebsiteStructure/Documents/3_Research_Insights/1_Sectoral_SurveysReports/ReportCruiseTourismIreland.pdf?ext=.pdf [Accessed 03 April 2017].

FAO. (2017). National Aquaculture Legislation Overview: France. [online] Available at http://www.fao.org/fishery/legalframework/nalo_france/en [Accessed 10 April 2017].

Francois, J., Manchin, M., Norberg, H., Pindyuk, O. and Tomberger, P. (2013). Reducing Transatlantic Barriers to Trade and Investment: An Economic Assessment. Final Project Report, prepared under implementing Framework Contract TRADE10/A2/A16 for the European Commission. [online] London: Centre for Economic Policy Research. Available at: http://trade.ec.europa.eu/doclib/docs/2013/march/tradoc_150737.pdf [Accessed 01 April 2017].

Government of Ireland. (2012) Harnessing Our Ocean Wealth – an Integrated Marine Plan for Ireland. Dublin: Government of Ireland.

Governo de Portugal. (2012). Estratégia Nacional para o MAR 2013–2020. [online] Lisboa: Governo de Portugal, 73pp. Available at: http://www.dgpm.mam.gov.pt/Documents/ENM.pdf [Accessed 03 April 2017].

HM Government. (2011). Marine Policy Statement. [online] London: HMSO, 51pp. Available at: https://www.gov.uk/government/uploads/system/uploads/attachment_data/file/69322/pb3654-marine-policy-statement-110316.pdf [Accessed 5 April 2011].

INTERREG Atlantic Area. (2015). Atlantic Area Programme 2014–2020: Approved Cooperation Programme. [online] Available at: https://light.

ccdr-n.pt/index.php?data=d967d3135a6e92ebbcd443b1ad9b3d7103da7 81463063390097ce7311d8cb819dcaccabbc4731c5d7a34ebb3b112294a [Accessed 10 April 2017].

Marine Institute and Marine Board – ESF. (2011). A Draft Marine Research Plan for the European Atlantic Sea Basin Discussion Document. WP 6 – Task 6.1 SEAS-ERA project. [online] Dublin: marine Institute, 46pp. Available at: https://ciencias.ulisboa.pt/sites/default/files/fcul/investigac ao/A%20Draft%20Marine%20Research%20Plan%20for%20the%20eu ropean%20atlantic%20sea%20basin%20(October%202011).pdf [Accessed 28 March 2017].

Marine Coordination Group. (2012). Our Ocean Wealth: Towards an Integrated Marine Plan for Ireland, Background Briefing Documents Part II: Sectoral Briefs.

Marine Energy Wales. (2017). Marine Energy Wales: Investment Jobs Supply Chain 2017. Pembrokeshire: Marine Energy Wales. [online] Available at: http://www.marineenergywales.co.uk/wp-content/uploads/2017/03/ Marine-Energy-Wales-Investment-Jobs-Supply-Chain-2017.pdf [Accessed 31 March 2017].

Marine Institute. (2013). Galway Statement on Atlantic Ocean Cooperation. [online] https://www.marine.ie/Home/sites/default/files/MIFiles/Docs_ Comms/SignedGalwayStatement24MAY2013.pdf [Accessed 03 April 2017].

Marine Renewables Industry Association (MRIA). (2010). Initial Development Zones to Focus On Realizing Ireland's Ocean Energy Potential White Paper. [online] Dublin: MRIA. Available at: http://www.mria.ie/ documents/19e0290049ea2b4cb770f70fc.pdf [Accessed 1 Aug. 2010].

Marine Scotland Science. (2015). Locational Guidelines for the Authorisation of Marine Fish Farms in Scottish Waters. Edinburgh: The Scottish Government, 27pp.

Milieu Ltd. (2014). Article 12 Technical Assessment of the MSFD 2012 obligations NE Atlantic Ocean. Final version 7 February 2014. Brussels: Milieu Ltd., 43pp.

Ministère de l'écologie, de l'énergie, du développement durable et de la mer. (2009). Le Livre Bleu des engagements du Grenelle de la Mer. 10 et 15 juillet 2009. [online] Paris: Ministère de l'écologie, de l'énergie, du développement durable et de la mer, 71pp. http://www.la documentationfrancaise.fr/var/storage/rapports-publics/094000356.pdf [Accessed 03 April 2017].

Ministrère de L'écologie, du Développment Durable et de L'énergie. (2014). Plan Stratégique National: Développement des aquacultures durables 2020. French version. Paris: Ministrére de L'écologie, du Développment Durable et de L'énergie.

Ministère de l'Environnement, de l'Énergie et de la Mer. (2017). A l'occasion des 4èmes Assises nationales des énergies renouvelables en mer, Ségolène Royal annonce de nouvelles actions en faveur des énergies marines. Press release 22 March 2017. Available at: http://www.deve loppement-durable.gouv.fr/sites/default/files/2017.03.22_cp_assises_na tionales_enr_en_mer.pdf [Accessed 10 April 2017].

Ministério da Agricultura e do Mar. (2014). Plano Estratégico para a Aqui-cultura Portuguesa 2014–2020. 86pp. (in Portuguese). [online] Lisboa: Ministério da Agricultura e do Mar. Available at: https://www.dgrm.mm. gov.pt/xportal/xmain?xpid=dgrm&xpgid=genericPageV2&conteudo Detalhe_v2=4829995 [Accessed 18 Jan. 2017].

Ministerio de Agricultura, Alimentación y Medio Ambiente (MAGRAMA). (2014). Plan Estratégico Plurianual de la Acuicultura Española 2014–2020. [online] Madrid: MAGRAMA. Available at: http://www.mapama. gob.es/es/pesca/temas/acuicultura/plan_estrategico_6_julio_tcm7-3890 36.pdf [Accessed 15 January 2017].

Ministerio de Industria, Energía y Turismo. (2009). Estudio Estratégico Ambiental del Litoral Español para la Instalación de Parques Eólicos Marinos. [online] Madrid: Ministerio de Industria, Energía y Turismo. Available at: http://www.aeeolica.org/uploads/documents/562-estudio-estrategico-ambiental-del-litoral-espanol-para-la-instalacion-de-parques-eolicos-marinos_mityc.pdf

Ocean Energy Europe. (2016). Ocean Energy Strategic Roadmap 2016, building ocean energy for Europe. [online] Brussels: OEE. Available at https://webgate.ec.europa.eu/maritimeforum/sites/maritimeforum/files/ OceanEnergyForum_Roadmap_Online_Version_08Nov2016.pdf [Accessed 11 Nov. 2016].

O'Hagan, A.M., Corner, R.A., Aguilar-Manjarrez, J. Gault, J., Ferreira, R.G., Ferreira, J.G., O'Higgins, T., Soto, D., Massa, F., Bacher, K., Chapela, R. and D. Fezzardi. (2017). Regional review of Policy-Management Issues in Marine and Freshwater Aquaculture. Report produced as part of the Horizon 2020 AquaSpace project. 170pp.

O'Hagan, A.M. (2016). Consenting Processes for Ocean Energy – a Report prepared on behalf of the IEA Technology Collaboration Programme for Ocean Energy Systems (OES); www.ocean-energy-systems.org

O'Hagan, A.M. (2018). Chapter 10: Consenting and Legal Aspects. In: Greaves, D. and Iglesias, G. (eds.) Wave and Tidal Energy. London: Wiley.

OSPAR Commission. (2003). Recommendation 2003/3 on a network of marine protected areas. [online] Available at http://www.ospar.org/docu ments?d=32867 [Accessed 15 Dec. 2016].

OSPAR (2008). OSPAR Guidance on Environmental Considerations for Offshore Wind Farm Development (Reference number: 2008-3). [online] London: OSPAR Secretariat. Available at http://www.ospar.org/docu ments?d=32631 [Accessed 15 Dec. 2016].

OSPAR (2010). The NE Atlantic Environment Strategy of the OSPAR Commission for the Protection of the Marine Environment of the NE Atlantic 2010–2020 (OSPAR Agreement 2010-3). [online] London: OSPAR Secretariat. Available at http://www.ospar. org/site/assets/files/1200/ospar_strategy.pdf [Accessed 6 June 2012].

OSPAR Commission. (2015). Status Report on the OSPAR Network of Marine Protected Areas. [online] London: OSPAR Commission, 64pp. Available at http://www.ospar.org/documents?d=33572 [Accessed 05 April 2017].

Scottish Government. (2017). Scottish Energy Strategy: The future of energy in Scotland. [online] Edinburgh: Scottish Government. 82pp. Available at http://www.gov.scot/Resource/0051/00513466.pdf [Accessed 7 Feb. 2017].

Sustainable Energy Authority of Ireland (SEAI). (2010). Strategic Environmental Assessment (SEA) of the Offshore Renewable Energy Development Plan (OREDP) in the Republic of Ireland. Environmental Report. Produced for SEAI by AECOM, Metoc and CMRC. [online] Available from: http://www.seai.ie/Renewables/Ocean_Energy/Offshore_Rene wable_SEA

The Scottish Government. (2007). Strategic Environmental Assessment for Wave and Tidal Energy. Report prepared for the Scottish Executive by Faber Maunsell and Metoc PLC. Available at: http://www.gov.scot/Publications/2007/03/seawave

The Scottish Government. (2015). Scotland's National Marine Plan: A Single Framework for Managing Our Seas. Edinburgh: The Scottish Government, 144pp.

Vega, A., Hynes, S. and O'Toole, E. (2015). Ireland's Ocean Economy, Reference Year 2012, SEMRU Report Series [Online] Galway: SEMRU,

62pp. http://www.nuigalway.ie/semru/documents/semru_irelands_ocean _economy_web_final.pdf [Accessed 14 July 2015].

Welsh Government. (2015). The Welsh National Marine Plan Initial draft (November 2015). [online] Cardiff: Welsh Government, 230pp. http://gov.wales/docs/drah/publications/151130-welsh-national-marine-plan-initial-draft-november-2015-en.pdf [Accessed 9 Jan. 2017].

11

Regulation and Planning in Sea Basins – North and Baltic Seas

Hester Whyte* and Shona Paterson

Future Earth Coasts, MaREI Centre, ERI, University College Cork,
Ireland
*Corresponding Author

11.1 Introduction and Geography

The Baltic Sea is an almost completely enclosed sea basin (Figure 11.1).
Located in the North-Eastern part of Europe, and it covers a surface area
of 377,000 km^2. The basin includes, amongst others, the Gulf of Bothnia,
Gulf of Finland, Gulf of Riga, and the Kattegat. In general, Baltic Sea is a
shallow, with an average depth is 55m (Walday and Kroglund, 2002a). Due
to, among other biophysical characteristics, its low salinity levels, the marine
environment is very vulnerable. In particular, eutrophication, the build-up of
nutrients from urban waste water, coastal agriculture, industrial pollution and
atmospheric deposition, poses a major threat to this basin (Ferreira et al.,
2010; HELCOM, 2014). The Baltic Sea region encompasses eight countries (Sweden, Denmark, Estonia, Finland, Germany, Latvia, Lithuania and
Poland) with a collective 85 million inhabitants (17% of the EU population)
that share common features and challenges.

The North Sea is a semi-closed sea basin, adjacent to the Atlantic Ocean.
The sea basin is more than 970 km long and 580 km wide, with an area of
around 570,000 km^2. The North Sea is mostly shallow, less than 200m, with
an average depth of 95m (European Environment Agency, 2012). However,
in the northern part of the sea, off the coast of Norway, there is a deep trench
measuring up to 700m in depth. The southern part of the North Sea is generally the shallowest but also the most congested in terms of human activities.

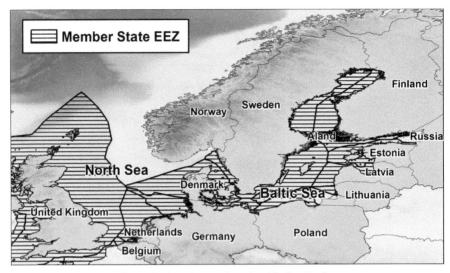

Figure 11.1   The North and Baltic Sea Basins.

Extreme weather conditions that have a direct impact on hydrography and strong tides are characteristic of this sea (Walday and Kroglund, 2002b). The North Sea region is bordered by the United Kingdom (England and Scotland), France, Belgium, Netherlands, Germany, Denmark, Sweden and Norway (non-EU member but a member of the European Economic Area).

11.1.1 Overview of Key Marine Sectors

Both sea basins not only support extensive marine industry but also face extensive expansion in human activities and potential impacts and conflicts associated with those activities (Table 11.1). While, traditional and novel marine-dependent industries are an important source of jobs, they also represent the cornerstones of a competitive maritime economy which has become a key policy focus for the EU and neighbouring countries (Piante and Ody, 2015; EUNETMAR, 2013). Economic uses such as shipping (The Nautical Institute and World Ocean Council, 2013), recreational uses associated with coastal tourism and nature conservation areas (Papageorgiou, 2016), and industrial infrastructure such as cables and pipelines (Directorate General for Energy, 2010), are sectors that are all expected to grow rapidly (Zaucha et al., 2014). Increased shipping is being/will be further facilitated by expansion and upgrading of ports and harbours in both sea basins and emerging fields

Table 11.1 Key marine sectors in Baltic/North Sea countries

Country	Key Marine Sectors
Sweden	Coastal tourism, cruise tourism, passenger ferry services, marine aquaculture, offshore wind, short-sea shipping (incl. Roll on-Roll off)
Denmark	Offshore wind, marine aquaculture, fish for human consumption, short-sea shipping (incl. RO-RO), passenger ferry services, coastal tourism
Estonia	Short-sea shipping (incl. RO-RO), deep-sea shipping, yachting and marinas, water projects, shipbuilding (excl. leisure boats) and ship repair, fish for human consumption
Finland	Coastal tourism, passenger ferry services, short-sea shipping (incl. RO-RO), shipbuilding (excl. leisure boats) and ship repair, yachting and marinas, offshore wind, water projects
Germany (North Sea (north-west coastline) and the Baltic Sea (north-east coastline))	Offshore wind, coastal tourism, yachting and marinas, short-sea shipping (incl. RO-RO), cruise tourism, shipbuilding (excl. leisure boats) and ship repair, blue biotechnology
Latvia	Shipping and port services, tourism, fisheries
Lithuania	Short-sea shipping (incl. RO-RO), shipbuilding (excl. leisure boats) and ship repair, fish for human consumption, water projects, coastal tourism
Poland	Offshore wind, shipbuilding (excl. leisure boats) and ship repair, coastal tourism, offshore oil and gas, yachting and marinas (leisure boat building), protection of habitats/marine aquaculture/environmental monitoring
Belgium	Offshore wind, construction of water projects (incl. protection against flooding), deep-sea shipping, short-sea shipping, inland waterway transport, cruise tourism, blue biotechnology
The Netherlands	Coastal tourism, offshore oil and gas, yachting and marinas, catching fish for human consumption, inland waterway transport, short-sea shipping (incl. RO-RO), deep-sea shipping
United Kingdom (East Coast)	Offshore oil and gas, coastal tourism, passenger ferry services, short-sea shipping, shipbuilding and repair, fisheries for human consumption, deep-sea shipping
Norway	Oil and gas, aquaculture, shipping (oil and gas related deep-sea shipping)
Russia (St. Petersburg, the Leningrad region, and the Kaliningrad region)	Shipbuilding

Source: European Commission, 2017c.

such as offshore renewables, together with the necessary grid infrastructure, are projected to impact existing landscapes (Konstantelos et al., 2017). The cost and technological constraints are likely to restrict new offshore renewable installations to areas that are already covered by a number of other uses (e.g., shipping, fishing), increasing the potential for conflicts between uses and neighbouring zones (Jansen et al., 2015; Hammar et al., 2017). In contrast, military usage, aggregate extraction, oil and gas extraction, and fisheries are expected to stay stable or decline over time (Zaucha et al., 2014).

11.1.2 Key Features Affecting Maritime Policy

Public policies are influenced by a variety of factors including public opinion, economic conditions, new scientific findings, technological change, interest groups, NGOs, business lobbying, and political activity.

Baltic Sea

The European Union Strategy for the Baltic Sea Region (EUSBSR) is the first macro-regional strategy in Europe (European Commission, 2017a). EUSBSR is organised according to three overall objectives: saving the sea, connecting the region and increasing prosperity. Each objective relates to a wide range of policies and has an impact on the other objectives. EUSBSR provides an integrated framework for improving the environmental condition of the sea, transport bottlenecks and energy interconnections as well as facilitating the development of competitive markets across borders and common networks for research and innovation. EUSBSR also aims to reinforce cooperation within this large region in order to face several challenges by working together as well as promoting more balanced development in the area. Cooperation with neighbouring countries such as Russia, Iceland, Norway and Belarus has also encouraged by the EUSBSR.

North Sea

The North Sea Region 2020 (NSR2020) strategy paper is designed to treat the North Sea region as a territorial cooperation area (CPMR North Sea Commission, 2016). NSR2020 indicates that the North Sea Region has the potential to act as an engine for growth in Europe and as a centre of excellence for wider EU issues through developing existing cooperation efforts, improving policy efficiency and value for public money. The strategic focus of NSR2020 is, therefore, on major challenges and common characteristics where there is added value in transnational action and collaboration.

Closely linked to the EU2020 objectives (European Commission, 2017b), the aims of the NSR2020 are four-fold: i) to help the Region improve performance as a competitive, attractive and sustainable area of Europe, ii) to more efficiently address common transnational challenges and exploit opportunities related to sustainable economic growth, climate, energy, accessibility and management of the maritime space, iii) to ensure a better governed region through cross-sectorial coordination and multi-level governance, and iv) to provide a potential pilot for a macro-regional strategy that is different to the EU strategies for the Baltic Sea and Danube areas (CPMR North Sea Commission, 2016).

Important policies, strategies and organisations to ensure coordination

There are a number of overlapping and complimentary policies and directives that are important to consider and ensure cross coordination (Qiu and Jones, 2013) including:

- The Europe 2020 Strategy is the key EU commitment to jobs and smart, sustainable, inclusive growth (European Commission, 2017b). It has five headline targets: promoting employment; improving the conditions for innovation, research and development; meeting climate change and energy objectives; improving education levels; and promoting social inclusion, in particular through the reduction of poverty, and addressing the challenges of ageing.
- Coordination with National Strategic Reference Programmes (European Commission, 2007) existing in all of the EU member states.
- Consistency with EU legislation and policies is at the core of the NSR2020. Therefore examples like the Strategy for the Single Market (Single Market Act) and the Digital Agenda, TEN-T (currently under revision), White Paper on Transport, energy (TEN-E) networks, the EU post-2010 biodiversity target and strategy, the Water Framework Directive, the Integrated Maritime Policy, the Marine Strategy Framework Directive; the Common Fisheries Policy, the Renewable Energy Directive, the Communication on Offshore Wind Energy all resonate with the NSR2020.
- Other international cooperation such as OSPAR, IMO, Trilateral Wadden Sea Cooperation.

- Interreg IV programmes: Interreg IV B North Sea Region Programme, Two Seas Programme (Belgium, France; Netherlands, UK) – ØKS/ KASK (Denmark, Norway and Sweden).

11.2 Environmental Policy

For EU Member States, the EU Marine Strategy Framework Directive (MSFD, 2008/56/EC) establishes a framework within which Member States shall take the necessary measures to achieve or maintain good environmental status of the marine environment by the year 2020 at the latest (Article 1). Member States are required to follow a common approach which involves reiterative six-year cycles. This approach includes a number of targets within the framework that must be achieved:

1. Assessing the current state of the marine environment (Art. 8 MSFD)
2. Determining good environmental status (Art. 9 MSFD)
3. Establishing environmental targets to guide progress towards achieving good environmental status (Art. 10 MSFD)
4. Establishing monitoring programmes for ongoing assessment and regular updating of targets (Art. 11 MSFD)
5. Developing programmes of measures to achieve or maintain good environmental status (Art. 13 MSFD)

Baltic Sea

The Helsinki Commission (HELCOM) is the current governing body of the Convention on the Protection of the Marine Environment of the Baltic Sea Area. The HELCOM Baltic Sea Action plan is a strategy designed to tackle major environmental problems with the aim of restoring good ecological status by 2021 (HELCOM, 2007). In addition, HELCOM proposes the designation of more protected habitats as more ecological and biophysical data become available over time with the view of developing a network of well managed areas (HELCOM Recommendation 15/5).

North Sea

OSPAR is the mechanism by which fifteen Governments of the western coasts and catchments of Europe, together with the European Community, cooperate to protect the marine environment of the North-East Atlantic. OSPAR extends beyond the North Sea basin extending westwards to the east coast of

Greenland, eastwards to the continental North Sea coast, south to the Strait of Gibraltar and northwards to the North Pole. OSPAR's mission is focused on the need to conserve marine ecosystems and safeguard human health in the North-East Atlantic by preventing and eliminating pollution; by protecting the marine environment from the adverse effects of human activities; and by contributing to the sustainable use of the seas (OSPAR, 2006).

11.3 Regulatory Regimes

A detailed list of the most important treaties and legislation is provided in Appendix I. In addition, the main regulatory bodies for the Baltic basin are: Baltic Development Forum; Baltic Sea Parliamentary Conference; Baltic Sea States sub-regional cooperation; Council of the Baltic Sea States; VASAB; Helsinki Commission (Helcom).

11.4 Spatial Impact and Planning

The North Sea is one of the most heavily used sea basins in Europe. It supports a large number of traditional activities, such as fishing, shipping & trade, energy, sand mining, defence and recreation. Increased economic activity in the sea, such as growing shipping traffic and the development of offshore wind farms, has led to increased competition for space (CPMR North Sea Commission, 2016). In this sense, North Sea countries have quite well established MSP processes, and several conflicts between offshore renewables and other sea uses are currently being addressed through on-going or completed projects like C-SCOPE (Smith et al., 2012), PISCES (PISCES, 2012), BLAST and NorthSEE.

In the same way, Baltic Sea countries are already taking steps to improve MSP processes. Within the region, MSP is highly promoted by both the Helsinki Commission (HELCOM) and the Vision and Strategies around the Baltic Sea (VASAB). The Horizontal Action (HELCOM, 2007), clearly encourages "... *the use of Maritime Spatial Planning in all Member States around the Baltic Sea and develop a common approach for cross-border cooperation*". The EU Strategy for the Baltic Sea Region ensures HELCOM and VASAB a prominent role in promoting MSP in the region together with other stakeholders (European Commission, 2017b). However, a large expansion of offshore wind energy will require more attention through MSP to find adequate space (Kyriazi et al., 2016; Hammar et al., 2017).

Table 11.2   Characteristics of MSP in the Baltic/North Sea

North Sea	MSP is relatively advanced in the majority of the North Sea countries, including use of GIS-based planning tools to map and visualize the spatial uses and pressures. Conflict patterns are fairly well understood and the majority of North Sea countries have made efforts to designate suitable areas for offshore renewable energy (ORE), giving the industry and other uses more certainty in where ORE are likely to develop. However, one can see quite different spatial priorities, for example in the Dogger Bank area. This area spreads over four national sea basins – Denmark, Germany, the Netherlands and UK.
Baltic Sea basin	In most of the countries, a comprehensive MSP legal framework has yet to be developed. However, the Baltic Sea is the basin with the largest number of non-binding cross-border regional cooperation initiatives related to MSP, energy and grid (i.e. VASAB-HELCOM, BaltSeaPlan, the EU Strategy for the Baltic Sea Region and the Baltic Sea Region Energy Cooperation). Even though the planning in the EEZ has been reformed to a more integrated approach there is still a need to co-ordinate the different planning competencies in an overarching informal institution.

Source: Payne et al., 2011.

In particular, opportunities will need to be sought for co-existence or multi-use, such as using the spaces between adjacent wind farms to reduce turbulence and regenerate the wind resource, and for other sea use functions like fishing or lower frequency shipping lanes (Rodríguez-Rodríguez et al., 2016; Astariz and Iglesias, 2016). This opportunity for co-existence will become more relevant as future offshore wind farms are developed in large clusters. Furthermore, sea uses which are not location sensitive, or can be relocated or decreased in size without undue impact, should ideally be investigated to find space for additional low cost offshore renewables. Table 11.2 summaries the main characteristics of both basins in relation to MSP and Appendix II further details the different MSP initiatives adopted by specific countries.

11.5 Related Strategies

11.5.1 Maritime Clusters

The concept of a cluster is defined as a "geographically proximate group of interconnected companies and associated institutions in a particular field, including product producers, service providers, suppliers, universities, and trade associations, from where linkages or externalities among industries result" (Porter, 1998). To fulfil the mandate of ensuring cooperation with EU partners, it is important for regional clusters to establish collaborative networks (Salvador et al., 2015). Networking with clusters and across complementary clusters is an important factor for their successful development. Maritime clusters across the Baltic and North Sea Basins are discussed below.

11.5.1.1 North Sea
Belgium

Four maritime clusters were identified in Belgium; these are Antwerp, Oostende, Zeebrugge and Ghent. Zeebrugge is the only cluster identified as being in the "growing" stage and focused on shipping exclusively. Antwerp is the largest of the clusters and while it can rely heavily on its strategic location its weakness is the lack of flexibility and strong attachment to established activities such as oil and gas or shipping. These can provide a hindrance in opening up towards new activities and innovative ventures.

Denmark

In April 2013, the Ministry of Science, Innovation and Higher Education adopted the Strategy for Denmark's Cluster Policy. The overall objective of the strategy is for cluster establishment and cluster development to promote enterprise competitiveness, export growth, investment promotion, employment and productivity in Danish businesses, via innovation, innovative solutions to society's problems and research and competence development. There are cluster organisations located in every part of Denmark. The Danish regions are working on second-generation development strategies, focusing on growth and job creation through the use of cluster organisations. The regions are giving priority to the most significant regional clusters, many of which have main activities that go beyond the regional borders. Every year, more than 6,000 Danish enterprises participate in activities launched by local, regional or national cluster and network organisations. More than 80% are small and medium-size enterprises. There are around 75 clusters in Denmark. The five most relevant ones are: Det Blå Danmark (The Blue Denmark), Europas Maritime Udviklingscenter (Maritime Development Centre of Europe), Offshore Centre Denmark (Offshoreenergy.dk), Energiklyngecenter Sjælland (Energy Cluster Zealand) & Konsumfisk (Edible Fish).

Germany

The maritime economy in Germany is concentrated in the coastal regions and can be described as a cluster alliance with five integrated sub-clusters. The cluster area accounts for almost 160,000 employees in 4,000 enterprises.

- Ems-Axis: region in the north-west of Germany along the river Ems (quite unknown but many shipping companies; important shipyards and fast growing offshore wind industry);
- Metropolitan Region Bremen-Oldenburg: region between and surrounding Bremen and Oldenburg;
- Metropolitan Region Hamburg: Hamburg and its hinterland (most important sub-cluster including Hamburg and the surrounding regions of Schleswig-Holstein and Lower Saxony);
- Schleswig-Holstein: NUTS-3 regions of Schleswig-Holstein not included in Metropolitan Region Hamburg, almost belonging to Germany's Baltic Sea region;
- Mecklenburg-Vorpommern (NUTS-2 region in the north east).

The Netherlands

Three clusters have been identified in the Netherlands, two of which are located in the port regions of Rotterdam and Amsterdam and one 'shipbuilding cluster' in the Northern Netherlands. The port clusters consist of port activities, shipping and inland shipping activities, maritime services and ship repair activities. Rotterdam is the largest port cluster, with over 70,000 persons employed directly or indirectly in the maritime economy. Amsterdam is also a cluster of substantial size, with 40,000 persons employed directly or indirectly in the maritime economy. Shipbuilding is the core of the cluster in the Northern part of the Netherlands, suppliers and ship operators are included in the cluster. All three clusters have a mature status and contain similar or interlinking activities. This provides considerable concentration of expertise. At the same time, weaknesses such as access to skilled labour are also present.

United Kingdom

Clusters in the UK are located in multiple sea basins: the Atlantic and the North Sea. The cluster relevant to the North Sea is the North east Scotland cluster (Aberdeen). The main maritime economic activity in the cluster is the Oil and Gas industry. Various international oil companies and independent oil companies and the Institute of Energy University of Aberdeen are associated with this cluster. The strengths of this cluster are the offshore engineering and hydrocarbon exploration and production expertise.

11.5.1.2 Baltic Sea
Sweden

The Västra Götaland cluster is Sweden's most advanced maritime cluster and with the broadest coverage. It gathers public agencies (Swedish Agency for Marine and Water Management, Swedish Institute for the Marine Environment), universities (Chalmers University of Technology, University of Gothenburg), research centres (Ocean Energy Centre, Centre for Marine Research, Lighthouse Maritime Competence Centre), industry organisations (Swedish Ship-owners Association), the Port of Gothenburg and large companies (STENA Line, Volvo Penta, SK, SAAB, among others). The sea is one of West Sweden's core assets and entrepreneurship linked to marine environment goes back a long time. The Maritime Forum, based in Stockholm, is a member of the European Network of Maritime Clusters and focuses predominantly on the shipping industry.

Norway

Three main clusters all focusing on different maritime industries. The cluster of Stavanger is a centre for offshore oil and gas activities and is a global leader in terms of industry expertise in deep sea and subsea production. Western Norway hosts aquaculture clusters which build especially on salmon production and have strong research potential for the inclusion of new species. The sectors' continuous growth is sparked by strong demand growth in BRIC-countries. The third main cluster in Norway is the shipping cluster located on the west coast of Norway, which is linked closely to the oil and gas sector and as such it is highly dependent on the willingness by oil and gas companies to continue search and production.

- Oil and gas cluster (Stavanger) – Norway, UK
- Aquaculture cluster Western Norway – Norway, Scotland
- Shipping cluster – Norway, UK, Denmark (Oil and gas related deep sea shipping)

Finland

Five different clusters related to maritime areas have been identified in Finland: Helsinki, Turku, KotkaHamina, Vaasa and Meridiem (covering all of

Finland). The first four don't have a legal framework but in each case, these clusters gather different sectors and types of organisations (public/private) in a specific geographical area. Helsinki, the largest city of Finland, hosts many maritime activities. It is firstly a tourist site, welcoming passengers from ferries and cruise ships; it is also an industrial centre with shipping and ship-building activities. Helsinki shipyard, Arctech, focuses on Arctic shipbuilding technology (icebreakers and other Arctic offshore and special vessels). It started its activities in 2011 and is a joint venture between STX Finland Oy and Russian United Shipbuilding Corporation. The strength of the Helsinki cluster is also the fact that it is a capital city, which means the presence of a large university, headquarters of Finnish companies and public institutions. Turku (165 km west of Helsinki) is not as touristic as Helsinki but is one of the Finnish entrances for ferry passengers and maritime freight. Turku port is a frontrunner in Finland on the implementation of liquefied natural gas (LNG) terminal, the distribution of LNG could begin by the end of 2015 with a EUR 60 million investment (Gasum and Port of Turku signed a letter of intent in 2012). In the Vaasa area, located on the Bothnian Gulf (420 km North of Helsinki), there is a large concentration of SME companies focused on leisure ship construction: design, production, etc. In addition, the Energy Institute is located in Vaasa (research, consultancy, projects, education), which focuses, among other areas, on wind energy. Kotka-Hamina area (130 km east of Helsinki) has developed two major activities related to Marine Economic Activities (MEAs): maritime transport and wind energy. Wind energy is not specifically oriented to offshore wind as at present this MEA is still very limited in Finland, but there are spillover effects from onshore wind research and development on offshore wind activity. The most notable companies involved are Winwind Ltd, TuuliWatti and Cursor Oy. Kotka-Hamina is one of the main seaports of Finland and its location on the way to Russia allows the development of maintenance services for ships. Meridiem is supported by the Maritime Cluster Programme (OSKE). Contrary to the other Finnish clusters mentioned, it has a legal framework and its geographic scope is large, the whole of Finland. It is a networking and coordination organisation for the maritime economy, particularly shipbuilding.

Estonia

It is characteristic of Estonia that cluster creation or cluster building are not taking place in the maritime field, but different maritime actors are involved in clusters of other economic sectors. Estonian ports are mainly transit transportation ports and are integrated with railway transport, road transport and other transit transport servicing companies in the logistics cluster. The Estonian Logistics cluster is a joint initiative dedicated to the international marketing of the members' services, introduction of the logistic advantages of Estonia to the target markets, research and development and logistics education. Passenger transport between Tallinn-Helsinki and Tallinn-Stockholm along with cruise shipping is tied to the activities of the tourism cluster (tourism companies, accommodation, commerce). The best example of smaller, local clusters is the Saaremaa Small Craft Cluster. Saaremaa small craft construction is characterised by a diverse production range: output varies from renovating old wooden boats to building modern high-end yachts and workboats. Saaremaa boat builders and subcontractors have formed a Small Craft cluster that represents the core of the Association of Estonian Shipyards, a member of the European Boating Industry. The Small Craft Competence Centre looks for cooperation and mutual business opportunities with foreign universities, research institutions and companies. For companies, the Competence Centre provides product development and trial manufacturing opportunities in cooperation with the Competence Centre and local companies. Other clusters are under development (Estonian Wind power cluster, Estonian Cruising Association).

Lithuania

At the time of publication, there are no officially registered marine related clusters in Lithuania. However, there are several initiatives and processes on-going to fill the gap. Several maritime business associations are working actively, with the most promising one being "Baltic valley", which unites maritime business and science. Maritime activities in Lithuania are concentrated around the city of Klaipeda, which has given its name to the Klaipeda Maritime Cluster. It has a national scope, including the whole Lithuanian maritime sector, but the country's small size and the geographical concentration of activities gives it a strong regional character. The main industrial focal points are shipping, shipbuilding and fishing. In general, the Klaipeda Maritime Cluster has strong research capabilities in the marine environment, which is an area where the industry is rather weak. Conversely, the industrial activity is more pronounced in areas relating to maritime technology, where the research capabilities are on the weak end of the spectrum. Within the Klaipeda Maritime Cluster, several organisations work with facilitating and stimulating cluster development, in order to increase its innovation capacity and promote economic development.

Latvia

Latvian maritime clusters are not defined in any official policy document. Creation of clusters is not taking place in the maritime sector, but different maritime sectors are involved in clusters of other sectors. This is due to a lack of some key elements like critical mass, e.g. in the shipping, shipbuilding and maritime equipment sectors, and because cooperation among sectors is generally weak. The port sector and related sectors are almost independent from the shipping sector and other maritime sectors. Shipbuilding is also not linked to the shipping sector as shipbuilding focuses on ship repair activities and there is weak demand on the national market. Cluster type networks of enterprises can be observed around ports, where shipping companies and cargo handling companies have cooperative relationships with the land transportation sector. Vertical cooperation within the value chain is common for Latvian maritime companies. For example, in the maritime logistics sector ports, shipping companies and cargo handling companies have rather close cooperation, and in the maritime industry field shipyards, design and engineering companies have well-functioning networks. In general, there is very weak tendency to cooperate between ports. There are also networks fostering maritime development activities, such as the Association of Latvian major ports (Rīga, Vetspils and Liepāja) and the Association of small ports.

Poland

According to the European Cluster Observatory (ECO), maritime clusters exist in Poland in two seaside NUTS-2 regions: Pomorskie and Zachodniopomorskie. The Zachodnipomorskie maritime cluster is a relatively small cluster (in terms of employment size, specialisation and focus), with a total employment of 4,139 people. The Pomorskie maritime cluster is a larger cluster, employing 7,305 people. Both clusters are mature but as the employment levels are falling they can actually be qualified as declining. They are specialised in traditional maritime activities: fishing, processing of fish and shipbuilding. There are no very large and specialised clusters present in Poland and no other potential maritime clusters. It is worth noting, however, that according to ECO in 2011 the maritime sector (more narrowly defined that in the present study) in Poland employed a total of 32,500 people and was represented by 7,952 enterprises. The number of enterprises is increasing while the employment levels are decreasing. The two identified clusters thus account only for roughly 1/3 of the national potential. The result is highly surprising and seems to be biased, which puts the ECO estimation into question.

11.6 Supporting Blue Growth

Table 11.3 shows the main important funding and supporting schemes to which Baltic/North Sea countries can apply. In addition, Annex 11.1 details the most important support schemes on a country-specific basis.

Table 11.3 European support schemes

Eurostars Programme http://www.eurostars-eureka.eu/	The Eurostars Programme ('Eurostars') is a European innovation programme. Its purpose is to provide funding for market-oriented research and development with the active participation of specifically research and development performing small and medium-sized enterprises (R&D-performing SMEs).
Bonus: Joint Baltic Sea System Research Programme http://www.bonusportal.org/	To integrate the Baltic Sea System research into a durable cooperative, interdisciplinary, well integrated and focused multinational programme to support the region's sustainable development.
Nordic Environment Finance Corporation (NEFCO) http://www.nefco.org/	NEFCO is an international financial institution established by five Nordic countries: Denmark, Finland, Iceland, Norway and Sweden. NEFCO finances investments and projects primarily in Russia, Ukraine, Estonia, Latvia, Lithuania and Belarus, in order to generate positive environmental effects of interest to the Nordic region.
European Investment Bank's (EIB) Loans http://www.eib.org/products/loans/index.htm	Within the EU the EIB has 6 priority objectives for its lending activity: • Cohesion and Convergence; • Support for small and medium-sized enterprises (SMEs);

(Continued)

Table 11.3   Continued

	• Environmental sustainability; • Implementation of the Innovation 2010 Initiative (i2i); • Development of Trans-European Networks of transport and energy (TENs); • Sustainable, competitive and secure energy.
JASPERS (Joint Assistance in Supporting Projects in European Regions) TA fund if fields of TENs networks, transport, environmental remediation, waste management, renewable energy, water and sanitation services, etc. http://www.jaspers-europa-info.org/	
ELENA (European Local Energy Assistance) http://www.eib.org/products/technical_assistance/ elena/index.htm	
Nordic Investment Bank's (NIB) Loans http://www.nib.int/loans/loan_characteristics	NIB focuses in particular on four sectors: • energy; • environment; • transport, logistics and communications; • innovation. The proceeds of NIB loans can be used to cover any part of projects costs.
Baltic Sea Environment (BASE) Lending Facility	The Baltic Sea Environment (BASE) lending facility is established to operate as the financing source for projects with a positive effect on the Baltic Sea. The facility is aimed at assisting in the implementation of the Baltic Sea Action Plan adopted by the Baltic Marine Environmental Protection Commission—HELCOM with the purpose of restoring the ecological status of the Baltic marine environment by 2021.

Table 11.3 Continued

Climate Change, Energy Efficiency and Renewable Energy (CLEERE) Lending Facility	The lending facility supports actions for combating and adapting to climate change around the world. Under the facility, NIB finances projects: • in renewable energy; • in energy efficiency; • using cleaner production technologies that reduce greenhouse gas emissions in industries; • dealing with the adaptation of power networks and infrastructure to climate change, such as extreme weather conditions.
European Bank for Reconstruction and Development http://www.ebrd.com/index.htm	
Swedish International Development Cooperation Agency Baltic Sea Unit http://www.sida.se/balticseaunit	The work of the SIDA Baltic Sea Unit seeks to develop cooperation between actors in the Baltic area. It has a special assignment from the Government based on Swedish interests to support activities in the fields of the Environment & Energy, Social and Health issues and Civil Security.
Northern Dimension http://www.ndphs.org/?about_nd	The Northern Dimension aims at providing a common platform for promoting dialogue and concrete cooperation as well as strengthening stability and promoting economic integration, competitiveness and sustainable development in Northern Europe.

11.7 Key Lessons

In both North Sea and Baltic Sea basins, the need and importance of an integrated way of planning limited maritime space is the agreed way forward to secure Blue Growth. However, the policy, legislation and planning mechanisms are not fully in place. A major constraint is the implementation cost. The Sustainable Blue Growth Agenda for the Baltic Sea Region, adopted by the European Commission in 2014 highlights the potential for development of the maritime economy in the Baltic. An extensive stakeholder dialogue in the region was undertaken by the European Commission in 2016 to identify the main drivers and challenges of Blue Growth and work towards a desired vision for 2030. Shipping, blue bioeconomy (incl. aquaculture), coastal and maritime tourism and environmental and monitoring technology were identified as the main thematic areas for growth. Although the Baltic Sea Region is a good example of transnational cooperation much still appears to happen within one single sector and increased understanding of other sectors is needed. In addition, a robust funding strategy is needed to enable smaller companies to access technical advice and support services on marketing and market research, risk assessment as well as investor readiness. Clusters could play a key role here (European Commission, 2017d).

> *"The expected intensification in the use of the North Sea, which is partly the result of an increase in the number of designated uses, demands responsible use of the limited available space. Increasing use is exerting pressure on the marine ecosystem. Policy is a prerequisite for harmonising the various designated uses of the North Sea and ensuring a healthy ecosystem."* (Dutch Ministries of I&E and EA, 2015, p. 8).

A policy document on the North Sea for 2016–2012 published by the Dutch Government sets out the desired policy for the use of space, within the limits of the marine ecosystem. It sets the spatial frameworks, allowing the use of space in the North Sea to develop in an efficient and sustainable way. Multiple use of space is considered an important principle in this regard, offering balanced opportunities for all uses of the North Sea within the European frameworks (Water Framework Directive, Marine Strategy Framework Directive, Birds Directive, Habitats Directive and the Malta Convention). A number of actions have been set out in regards to renewable energy namely drawing up a North Sea Energy Master Plan 2030–2050 and more research into combined energy farms to ensure this is implemented wherever possible. The North Sea 2050 Spatial Plan specifically emphasizes that energy areas at sea in which electricity is generated using different techniques is the vision of the North Sea in 2050. The spatial agenda shows that such energy farms, combining wind, tidal and wave energy, are promising, but that the combination of aquaculture and/or mariculture with wind farms is less obvious, unless the wind farms were to be located close to the coast (Dutch Ministries of I&E and EA, 2015).

As administrative and political division of responsibilities especially for the territorial waters differ per country, international cooperation and aligned spatial strategies are key to facilitate sustainable Blue Growth.

Annex 11.1 – National support schemes in Baltic & North Sea

Latvia	Operational Programme "Entrepreneurship and Innovations" www.esfondi.lv	The Programme aims to contribute to improved innovation and the use of knowledge, high value-added production, and enhanced export capacity among the existing enterprises, as well as to encourage the formation of new knowledge-based and technology intensive enterprises.
Finland	European Fisheries Fund www.mmm.fi	Priority 1: Adaptation of the EU fishing fleet; Priority 2: Aquaculture, inland fishing, processing and marketing of fishery and aquaculture products; Priority 3: Measures of common interest; Priority 4: Sustainable development of fisheries areas.
Lithuania	Operational Programme 'Economic Growth' www.esparama.lt	The programme is dedicated to increase business productivity especially by creating a favourable environment for innovations and SMEs, promote R&D, increase efficiency of transport and energy infrastructure.
Estonia	Operational Programme for the European Fisheries Fund www.agri.ee	The main goal of the programme is to restructure the fisheries sector in order to ensure sustainable management in the fisheries sector and an increase of the income of the persons engaged in fishery.
Sweden	Operational Programme for the Swedish Fisheries Sector www.fiskeriverket.se	The programme aims at promoting an ecologically, economically and socially sustainable fisheries sector in Sweden by creating a balance between fish resources and fleet capacity, increasing profitability in the fisheries sector, promoting employment in rural areas in relation to the fisheries sector, decreasing the negative environmental effects brought about by the fisheries sector and ensuring the sustainability of both the environment and natural fish stocks.

(Continued)

Annex 11.1 Continued

Denmark	• European Fisheries Fund www.ferv.fvm.dk/ Fiskeriudvikling • Operational Programme "Innovation and Knowledge" www.ebst.dk	• Core targets of the programme refer to four "Growth Drivers" seen as crucial to promoting growth (innovation, entrepreneurship, new technology, human resources)

Source: http://www.balticsea-region.eu/funding-sources

References

Astariz, S. & Iglesias, G. 2016. Co-located wind and wave energy farms: Uniformly distributed arrays. *Energy,* 113, 497–508.

CPMR North Sea Commission 2016. North Sea Region 2020: North Sea Commission Strategy – Contributing to the Europe 2020 Gothenburg, Sweden: CPMR North Sea Commission.

Directorate General for Energy 2010. Energy Infrastructure Priorities for 2020 and Beyond – A Blueprint for an Integrated European Energy Network.

EUNETMAR 2013. Study on Blue Growth, Maritime Policy and the EU Strategy for the Baltic Sea Region.

European Commission 2007. Cohesion Policy 2007–13. National Strategic Reference Frameworks. Luxembourg: Office for Official Publications of the European Communities, 2007.

European Commission 2017a. Commission Staff Working Document: European Union Strategy for the Baltic Sea Region Action Plan {COM(2009) 248}. Brussels.

European Commission 2017b. *Europe 2020 Strategy* [Online]. Available: https://ec.europa.eu/info/strategy/european-semester/framework/europe-2020-strategy_en [Accessed 07/04/2017].

European Commission 2017c. *European MSP Platform* [Online]. Available: http://msp-platform.eu/msp-practice/countries [Accessed 07/04/2017 2017].

European Commision 2017d. Directorate-general for Maritime Affairs and Fisheries. Towards and implementation strategy for the sustainable blue growth agenda for the Baltic Sea region. ISBN 978-92-79-69371-7

European Environment Agency. 2012. *North Sea physiography (depth distribution and main currents)* [Online]. Available: http://www.eea.europa.eu/data-and-maps/figures/north-sea-physiography-depth-distribution-and-main-currents [Accessed 7/4/2017].

Ferreira, J. G., Andersen, J. H., Borja, A., Bricker, S. B., Camp, J. & Cardoso Da Silva, M. 2010. Marine strategy framework directive, Task Group 5 Report, Eutrophication. JRC 58102, EUR 24338 EN. ISBN 978 92 79 15651 9. ISSN 1018-5593. Luxembourg: Office for Official Publications of the European Communities, European Union and ICES.

Hammar, L., Gullström, M., Dahlgren, T. G., Asplund, M. E., Goncalves, I. B. & Molander, S. 2017. Introducing ocean energy industries to a busy marine environment. *Renewable and Sustainable Energy Reviews,* 74, 178–185.

HELCOM 2007. HELCOM Baltic Sea Action Plan. Krakow, Poland.

HELCOM 2014. Eutrophication Status of the Baltic Sea 2007–2011 – A Concise Thematic Assessment. Baltic Sea Environment Proceedings, No. 143.

Jansen, J., Van Der Welle, A., Kraan, C., Nieuwenhout, F. & Veum, K. 2015. Sharing benefits and costs of integrated offshore grid structures. NorthSeaGrid Policy Brief.

Konstantelos, I., Pudjianto, D., Strbac, G., DE Decker, J., Joseph, P., Flament, A., Kreutzkamp, P., Genoese, F., Rehfeldt, L., Wallasch, A.-K., Gerdes, G., Jafar, M., Yang, Y., Tidemand, N., Jansen, J., Nieuwenhout, F., Van Der Welle, A. & Veum, K. 2017. Integrated North Sea grids: The costs, the benefits and their distribution between countries. *Energy Policy,* 101, 28–41.

Kyriazi, Z., Maes, F. & Degraer, S. 2016. Coexistence dilemmas in European marine spatial planning practices. The case of marine renewables and marine protected areas. *Energy Policy,* 97, 391–399.

OSPAR 2006. Convention for the Protection of the Marine Environment of the North-East Atlantic.

Papageorgiou, M. 2016. Coastal and marine tourism: A challenging factor in Marine Spatial Planning. *Ocean & Coastal Management,* 129, 44–48.

Payne, I., Tindall, C., Hodgson, S. & Harris, C. 2011. Comparison of national Maritime Spatial Planning (MSP) regimes across the EU. Seanergy 2020.

Piante, C. & Ody, D. 2015. Blue Growth in the Mediterranean Sea: the Challenge of Good Environmental Status. MedTrends Project. WWF-France.

PISCES 2012. PISCES: Partnerships Involving Stakeholders in the Celtic Sea Ecosystem. Towards Sustainability in the Celtic Sea.

Porter, M. 1998. "Clusters and the New Economics of Competition". *Harvard Business Review* 76, 77–90.

Qiu, W. & Jones, P. J. S. 2013. The emerging policy landscape for marine spatial planning in Europe. *Marine Policy,* 39, 182–190.

Rodríguez-Rodríguez, D., Malak, D. A., Soukissian, T. & Sánchez-espinosa, A. 2016. Achieving Blue Growth through maritime spatial planning: Offshore wind energy optimization and biodiversity conservation in Spain. *Marine Policy,* 73, 8–14.

Salvador, R., Simoes, A. & Soares, C. G. 2015. Features of the European Maritime Clusters. 55th Congress of the European Regional Science Association: "World Renaissance: Changing roles for people and places", 25–28 August 2015, Lisbon, Portugal.

Smith, N., Belpaeme, K., Maelfait, H., Vanhooren, S. & Buchan, K. 2012. Why one size won't fit all: Marine spatial planning in Belgium & Dorset.

The Dutch Ministry of Infrastructure and the Environment & The Dutch Ministry of Economic Affairs, 2015. Policy Document on the North Sea 2016–2021. www.noordzeeloket.nl

The Dutch Ministry of Infrastructure and the Environment & The Dutch Ministry of Economic Affairs, 2014. North Sea 2050 Spatial Agenda.

The Nautical Institute & World Ocean Council 2013. The shipping industry and Marine Spatial Planning: A professional approach.

Walday, M. & Kroglund, T. 2002a. The Baltic Sea. *In:* Pinborg, U. & Larsson, T.-B. (eds.) *Europe's biodiversity – biogeographical regions and seas.* UNEP/GRID Warsaw.

Walday, M. & Kroglund, T. 2002b. The North Sea. *In:* Pinborg, U. & Larsson, T.-B. (eds.) *Europe's biodiversity – biogeographical regions and seas.* UNEP/GRID Warsaw.

Zaucha, J., Vision & Secretariat, S. A. T. B. S. 2014. *The Key to Governing the Fragile Baltic Sea: Maritime Spatial Planning in the Baltic Sea Region and Way Forward*, VASAB Secretariat.

Additional Web-Based Resources on Blue Growth and Maritime Policy

- http://ec.europa.eu/maritimeaffairs/atlas/seabasins/balticsea/long/index_en.htm
- https://www.havochvatten.se/4.732980de143b1b1de53286b.html
- https://webgate.ec.europa.eu/maritimeforum/en/node/3550
- http://www.unesco-ioc-marinesp.be/spatial_management_practice/belgium
- http://www.mermaidproject.eu/sharepoint/func-startdown/355/

- http://oceana.org/
- http://ec.europa.eu/maritimeaffairs/policy/maritime_spatial_planning/index_en.htm
- http://cur-lcx.curopa.cu/LcxUriScrv/LcxUriScrv.do?uri=COM:2013:0133:FIN:EN:PDF
- http://ec.europa.eu/maritimeaffairs/policy/maritime_spatial_planning/index_en.htm
- http://eur-lex.europa.eu/LexUriServ/LexUriServ.do?uri=COM:2013:0133:FIN:EN:PDF

12

Regulation and Planning in the Mediterranean Sea

Christine Röckmann[1,*], Tomás Vega Fernández[2,3] and Carlo Pipitone[2]

[1]Wageningen Marine Research, Stichting Wagening Research, WUR, P.O. Box 68 1970 AB IJmuiden, The Netherlands
[2]Consiglio Nazionale delle Ricerche, Via Giovanni da Verrazzano 17, 91014 Castellammare del Golfo, Italy
[3]Stazione Zoologica Anton Dohrn, Villa Comunale, 80121 Naples, Italy
*Corresponding Author

12.1 Introduction and Geography

The Mediterranean Sea (Figure 12.1) is the largest of the European regional seas, covering an approximate area of 2.5 million km^2 (Suárez de Vivero and Rodríguez Mateos 2015). It is situated between three continents: Europe to the north, Africa to the south, and Asia to the east and is bordered by more than 20 coastal states, with 11 countries in Europe, 5 in Africa, and 5 in Asia. This is the largest number of coastal countries among European seas (Suárez de Vivero and Rodríguez Mateos 2015). The Mediterranean has an average depth of 1,500 m, with a maximum depth of 5,150 m along the southern coast of Greece. Its coastline is ca. 45,000 km long in total including more than 5,000 islands and islets, with Greece, Italy, Croatia and Turkey accounting for 75% of this length. This semi-enclosed sea has only two communication waterways with outside oceans: the 14 km wide and 300 m deep Strait of Gibraltar to the west and the few-meter wide artificial Suez Canal to the south-east. As a result, water turnover time is estimated to be very low, about one century (Robinson et al. 2001). The main source of replenishment is the continuous inflow of surface water from the Atlantic Ocean through the Strait of Gibraltar. The scarce inflow and low precipitation, coupled with high evaporation, makes the Mediterranean more saline than the Atlantic Ocean.

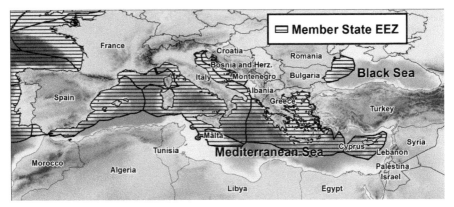

Figure 12.1   Mediterranean and Black Sea Basin.

The Mediterranean is a generally oligotrophic sea with a more productive western basin (D'Ortenzio and Ribera d'Alcalà 2009). Higher productivity occurs in upwelling areas such as the Ligurian Sea, the Alboran Sea and the Strait of Sicily, and in areas characterized by high organic input from natural or human origin such as the north western Adriatic Sea, the Gulf of Lion and the northern Aegean Sea (Barausse and Palmeri 2014). Although its biota was greatly impoverished after a salinity crisis in the late Holocene, the Mediterranean Sea represents a biodiversity hotspot, comprising temperate as well as subtropical species, with about 20% of endemic biota (Sala 2004; Bianchi et al. 2012). Its rocky reefs, seagrass meadows, and upwelling areas support enormous biodiversity. Seagrasses protect the seashore from erosion and maintain water quality, particularly through oxygen production and sediment burial (Salomidi et al. 2012). Mediterranean ecosystems provide suitable habitats for the endangered Mediterranean monk seal *Monachus monachus* as well as for endangered seabirds, endemic fish, invertebrates and communities (e.g. coralligenous). The Mediterranean Sea is also a hotspot of cumulative human pressures (Halpern et al. 2008, Micheli et al. 2013) that pose serious threats to its biodiversity (Coll et al. 2010) and facilitate biological invasions (Galil 2006). The already increasing rate of biological invasions is expected to grow even faster with the foreseen enlargement of the Suez Canal in 2017, since this artificial waterway links the relatively small Mediterranean biota with the huge species pool of the Indo-Pacific Ocean (Galil et al. 2015). The Mediterranean is also highly exposed to geo-hazards, such as earthquakes and land-slides (Urgeles and Camerlenghi, 2013).

The oligotrophy, heavy human pressure, high rate of biological invasions, and exposure to geo-hazards in a semi-enclosed sea with a relatively small water body with respect to the long coastline provide the picture of a diverse but fragile system with low physical inertia and limited ecological resilience. Blue growth activities should be therefore carefully planned in order to attain the adequate balance between benefits and trade-offs, and possibly meet adequate environmental standards (e.g. ISO 14001:2015 Environmental management systems; ISO 31000:2009 Risk management; EU Directives 2011/92/EU (Environmental Impact Assessment) and 2001/42/EC (Strategic Environmental Assessment)).

The Mediterranean region may well be considered one of the cradles of human civilization (Lopes 2014); its borders encompass a hugely diverse cultural, legislative and socio-economic landscape that creates a rich, dynamic and complex human environment, making it more complex to consider a global strategy at basin or sub-basin level (EUNETMAR 2014b). To make things even more complicated, the Mediterranean Sea does not form a clear, unitary geographical entity in the way other European seas do. According to the International Hydrographic Organization (IHO 2002), the Mediterranean Sea is subdivided into a number of smaller water bodies, each with its own designation. These water bodies can be clustered into three sub-regions (although other different subdivisions exist):

Western Mediterranean:

- Strait of Gibraltar;
- Alboran Sea, between Spain and Morocco;
- Balearic Sea, between mainland Spain and the Balearic Islands. In other nomenclatures, the Balearic Sea is part of the Algerian Basin;
- Ligurian Sea between Corsica and Liguria (Italy);
- Tyrrhenian Sea enclosed by Sardinia, the Italian peninsula and Sicily;

Central Mediterranean:

A submarine ridge between Sicily and Tunisia, corresponding to the Strait of Sicily, divides the Western and Eastern basins. Two main water bodies are considered:

- Ionian Sea between southeastern Italy, Sicily, Albania and Greece;
- Adriatic Sea between Italy, Slovenia, Croatia, Bosnia and Herzegovina, Montenegro and Albania.

Eastern Mediterranean:

- Aegean Sea between Greece and Turkey;
- Levantine Basin (or Levant Sea), that is the easternmost part of the Mediterranean.

12.1.1 Overview of Key Marine Sectors

The Mediterranean Sea is amongst the world's busiest waterways (Maritime Forum 2010): Currently, about 30% of the world ship traffic pass the Gibraltar Strait (several hundred ships daily, VT Explorer website) and the Suez Canal (49 transiting vessels daily at present) (Abdulla and Linden, 2008). The latter figure is expected to double with the imminent enlargement of the Suez Canal to a daily average of 97 transiting vessels by the year 2023 (Suez Canal Authority 2017). In addition to long distance shipping of goods, also some of the busiest intercontinental submarine cables for telecommunications pass through the Mediterranean (Bilsky 2009).

A study of twelve Mediterranean countries[1] identified costal tourism and shipping (both deep sea and short-sea shipping) as key marine sectors in the Mediterranean Sea, generating 73% of the total gross value added by maritime economic activities in these countries; this exceeded EUR 63 billion in 2010 (EUNETMAR 2014b). Coastal tourism and maritime transport are significant economic activities not only in these countries, but also in European Neighbourhood Policy's partner countries. In terms of projects and initiatives related to blue growth and IMP in the Mediterranean, more than 80% EU-driven and -funded topics concern primarily environmental monitoring, coastal tourism and maritime transport. In programmes funded by the European Neighbourhood and Partnership Instrument (ENPI), coastal tourism is by far the principal maritime economic activity covered.

Fisheries is another principal marine economic sector, with an estimated regional economic impact of almost EUR 9 billion (GFCM, 2016). Several EU-, GFCM- and country-driven initiatives have been implemented to mitigate overfishing and foster fish stock recovery, from fishing license buyback schemes to spatial-based fisheries management (e.g., EC 2006; cf. Table 12.3).

---

[1]The twelve countries are the EU member states Spain, France, Italy, Slovenia, Malta, Greece, Cyprus, Croatia, the candidate countries Bosnia and Herzegovina, Montenegro, and Albania, and the potential EU candidate Turkey.

In a study that looked at future business trends, marine aquaculture and coastal tourism were identified as "the most promising and relevant" maritime economic activities in almost all of the twelve EUNETMAR countries studied, followed by short-sea shipping and cruise tourism (EUNETMAR 2014a). Tourism is often a pivotal activity and the potential for growth remains significant. Other activities such as oil and gas extraction and aquaculture are also considered as promising in some countries. A further eight activities are important in more than two but fewer than five of the 12 studied countries: Passenger ferry services, yachting and marinas, deep-sea shipping, offshore oil and gas, shipbuilding and ship repair, water projects, inland waterway transport, fishing for human consumption. The remaining potential industry areas, such as blue biotechnology, offshore wind, protection of habitats, securing fresh water supply, maritime monitoring and surveillance were not considered important for those 12 Mediterranean countries (EUNETMAR 2014b, p. 8).

To summarise, Table 12.1 presents an inventory of the main current and future activities of the 21 Mediterranean countries. Actual data for the EU member states was gathered via the new European MSP Platform (http://www.msp-platform.eu/), which is considered "the central exchange forum for the rich knowledge generated in past, current and upcoming MSP processes and projects". For the non-EU Mediterranean countries, no central data exchange point is available, hence, the inventory relies on the data published until 2011 in a research project report for the European Commission (PRC 2011). Our recent inventory comprising all Mediterranean countries slightly modifies the picture based on EUNETMAR (2014a,b): Tourism and aquaculture do represent important current and future maritime activities in many Mediterranean countries. However, in addition to these, also renewable energy production and environmental protection (MPAs) appear repeatedly in the list of future activities (Table 12.1).

12.1.2 Key Features Affecting Maritime Policy

The Mediterranean policy is affected by the particularities of the Mediterranean geography (see above) and by the diversity of the jurisdictional, political, and economic factors (see below) in the coastal states (Suárez de Vivero and Rodríguez Mateos 2015).

Jurisdictional features

The Mediterranean countries have declared different width of their territorial sea (3, 6, 12 nautical miles), resulting in national jurisdictional heterogeneity.

Table 12.1   Inventory of current and future marine/maritime activities, based on msp-platform.eu for all EU member states (except France), and on PRC (2011)

Country	Current Activities	Future Activities
Albania	Predominant focus: fisheries, **aquaculture.** Other activities: coastal **tourism**, offshore wind, oil and gas research, nature protection (1 MPA). Potential regional competition between: tourism, fisheries, onshore energy. Harbour transport, fishing, urban pressure.	**offshore wind farm**; oil and gas exploitation, **nature protection (more MPAs planned)**
Algeria	Fisheries, marine protection. Maritime transport routes, desalination of sea water, marine aggregate (sand) exploitation	**offshore wind/wave energy**
Bosnia and Herze-govina	fishing, **aquaculture**, seashells production. High environmental pressure from fisheries and mariculture.	plan for harbour construction (no MPAs yet, no plans for offshore wind/wave energy yet)
Croatia	**tourism** (coastal + cruise), shipping, fisheries, shipbuilding, ship repair, water projects, passenger ferry services, marine **aquaculture**	**tourism, aquaculture**, shipping, passenger ferry services, yachting, marinas
Cyprus	**tourism**, fishing, shipping, water projects, securing fresh water supply	shipping, **tourism, aquaculture**, oil and gas, securing fresh water supply
Egypt	Maritime traffic; offshore hydrocarbon activities; fishing, **aquaculture**	**nature protection (MPAs planned)**
France	maritime traffic, marine protection,	**fixed offshore wind farms planned**; no recent update available
Greece	nature conservation, shipping, ports, fisheries, **aquaculture, tourism,** under water cultural heritage, oil and gas, submarine cables and pipelines, military; coastal industries (cement, desalination)	**offshore renewable energy production, dive parks, underwater tourism**
Israel	maritime transport, ports/marinas, fisheries/**mariculture**, sea water desalination, oil and gas, MPAs. Economic growth concerns; Perceived conflicts among uses	new and emerging uses, e.g., **aquaculture;** effects of climate change (e.g., sea-level rise, coast cliffs erosion) [winds near Israeli shore not sufficiently strong for offshore wind farms]

Table 12.1 Continued

Italy	Coastal **tourism**; Fishing for human consumption; Short sea shipping; Cruise **tourism**; Shipbuilding and repair; Passenger ferry services; Deep-sea shipping	Short sea shipping; Passenger ferry services; Marine **aquaculture; Protection of habitats**; Coastal **tourism**; Cruise **tourism**
Lebanon	maritime transport, fishery, oil and gas, marine protection (1 MPA)	?
Libya	fisheries management, offshore hydrocarbon exploitation/oil and gas, maritime transport	challenges: **Biodiversity protection/creation of coastal protected areas**; Pollution of coastal waters by municipal, industrial and ship-generated waste; Lack of public awareness and participation; Participation in international agreements
Malta	shipping, ports, **tourism**, fisheries, **aquaculture**, oil and gas, submarine cables and pipelines	**offshore renewable energy production**, submarine cables and pipelines, scientific research
Monaco	navigation, ports, pollution, sustainable development, environmental protection (1 MPA)	?
Montenegro	maritime transport, ports, passenger ferry services, fisheries, **mariculture**, oil and gas, coastal **tourism**, nature protection (MPAs)	coastal **tourism**, nautical **tourism**/marinas, passenger ferry services, oil and gas, **nature protection**
Morocco	maritime transport/shipping, land-based water discharges/pollution, **tourism**, overfishing, sand extractions; high marine biodiversity	?
Palestine/ Gaza strip	natural gas fields; environmental pollution: solid wastes, construction debris, rubble, waste water, overfishing/excessive fishing methods, beach recreation, war	?
Slovenia	**tourism**, fisheries, shipping; water projects, shipbuilding and repair	shipping, **tourism, aquaculture**; biotechnology
Spain	**tourism**, fisheries, shipping/maritime transport, **mari/aquaculture**, ports, hydrocarbons/mineral extraction, offshore renewable energy production/energy corridors	**offshore renewable energy production**
Syria	oil and gas terminals and ports	?

(Continued)

Table 12.1   Continued

Country	Current Activities	Future Activities
Tunisia	hydrocarbons extraction, maritime transport, fishing; coastal **tourism**, nature protection	?
Turkey	shipping/maritime transport, fisheries, **aquaculture**, marine protection	?

Note: Cells with "?" denotes no information found.

A wide range of jurisdictional regimes applies. The waters beyond national jurisdiction fall under the UNCLOS high seas regime, which implies free access to the water for all states, including non-coastal states. The seabed and the underlying subsoil, however, are part of the national jurisdiction down to the continental shelf border, including the slope. Most Mediterranean countries have renounced claims of sovereignty far beyond their territorial sea in order to avoid tensions with their neighbouring states. Roughly one third of the Mediterranean Sea has not been formally/officially claimed (yet) as national territory or EEZ (cf. Figure 2a in Cinnirella et al. 2014). In 2012, the maritime jurisdictions in the Mediterranean Sea were distributed as follows (Suárez de Vivero and Rodríguez Mateos 2015): High Seas 29%; EEZ 26%; Territorial Sea 19 %; Inland waters 7%; Fisheries protection zone 9%; Ecological protection zone 8%; Ecological and fisheries protection zone 1%; Other 1%. This picture is evolving, though, due to several sovereignty disputes (Table 12.2). For example, the borders of territorial seas off the coasts of Syria, Lebanon, Israel, Gaza, and Cyprus are under discussion following the discovery of substantial oil and gas deposits in the marine subsoil (US Geological Survey 2010).

Political and economic features

The Mediterranean region is currently focal point of conflict, crisis, terrorism and mass movements of people (e.g. Albahari 2015, Abbasi et al. 2015, Taghizadeh Moghaddam et al. 2017, Tardif 2017). The political and economic situations in the different coastal countries is very diverse (e.g. Coscieme et al. 2017, Cirer-Costa 2017). The Northern and Southern shores are separated by "one of the most marked economic divides on the planet and at their eastern end are home to one of the most intricate and dangerous geopolitical conflicts in modern international relations" (Suárez de Vivero and Rodríguez Mateos 2015). On the other hand, the Mediterranean countries share an enormous wealth of physical and economic assets, which, according to the European

Table 12.2 Overview of EEZ status of countries bordering the Mediterranean Sea, as of March 2017

COUNTRY	EEZ status (March 2017)
Albania	no
Algeria	No EEZ, but exclusive fishing zone established.
Bosnia and Herzegovina	no
Croatia	provisions for EEZ in Maritime Code of Croatia (1994), but not established. Instead: "ecological and fishery protection zone" beyond the territorial waters of Croatia.
Cyprus	yes
Egypt	yes
France	in progress
Greece	not yet
Israel	established in 2011, but not yet declared according to UNCLOS. Both northern and southern territorial boundaries are in dispute: N with Lebanon, S with Egypt (and Gaza) (Portman 2015)
Italy	no
Lebanon	yes (Decree No. 6433 - Delineation of the boundaries of the exclusive economic zone of Lebanon, 2011)
Libya	yes: EEZ declared in 2009 (General People's Committee Decision No. 260 of A.J. 1377)
Malta	no
Monaco	no
Montenegro	no
Morocco	EEZ established but not enforced in the Mediterranean
Palestine/ Gaza strip	no. 3nm Gaza Marine Activity Zone (1994)
Slovenia	no
Spain	yes: EEZ established in North-west Mediterranean Sea (Royal Decree 236/2013)
Syria	yes: EEZ claimed although Syria has not ratified UNCLOS.
Tunisia	EEZ area defined, but not enforced yet
Turkey	no

Note: EU member states printed yellow on blue (white on black). Cells highlighted in green indicate countries that have established an EEZ. Orange highlight indicates that the a process is ongoing, or that there are still disputes to settle between bordering countries and the entire EEZ has not been established yet. See "Links" in Table 12.4 for references.

Commissioner for Maritime Affairs and Fisheries, Karmenu Vella, need to be further explored to unlock the potential of the Mediterranean blue economy (EU 2009, Vella 2015): 450 ports and terminals; over 400 UNESCO world heritage sites; 236 Marine Protected Areas; 30% of the global sea-borne trade by volume; a quarter of worldwide sea-borne oil traffic; the world's leading tourism destination with one third of total arrivals worldwide; rapidly developing cruise tourism; and a coastal population of 150 million people which more than doubles during tourist season.

12.2 Environmental Policy

Two major governance processes can be distinguished in the Mediterranean Sea (Cinnirella et al. 2014, Suárez de Vivero and Rodríguez Mateos 2015):

1. Regional Sea level: the United Nations Mediterranean Action Plan (MAP) and the Barcelona Convention, along with the regionalization of management;
2. EU level: development and implementation of the EU's Integrated Maritime Policy and marine policies/legislation relating to fisheries, environment, coastal management, maritime spatial planning, the EU strategy for the Mediterranean Sea basin.

Both, sea basin and EU level policies trigger national action, such as the national strategic action plans (NSAP), and the national implementation of the EU framework directives.

12.2.1 Regional Sea level

The Mediterranean Action Plan (MAP), a regional environmental protection initiative, has been adopted in 1975 by sixteen Mediterranean countries and the European Community as the first-ever Regional Seas Programme under United Nations Environment Programme (UNEP) umbrella. Its four main objectives focus on: assessment and control of marine pollution, formulation of national environmental policies, improvement of governance to identify alternative development paths, and optimization of resources allocation. Seven legal protocols complete the MAP's legal framework, specifically as regards pollution control and management: Dumping Protocol from ships and aircraft; Prevention and Emergency Protocol (pollution from ships and emergency situations); Land-based Sources (LBS) and Activities Protocol; Specially Protected Areas (SPA) and Biological Diversity Protocol; Offshore Protocol (pollution from exploration and exploitation); Hazardous Wastes Protocol; Protocol on Integrated Coastal Zone Management (ICZM).

There are several programs and regional activity centres (RAC) to implement the MAP protocols. Also, to address land-based pollution, Strategic Action Plans (SAPs) have been developed since 1993, and countries have prepared and formally endorsed National Action Plans (NAPs) that describe the policy and actions each country intends to undertake to reduce pollution, in line with SAP targets. They incorporate mechanisms for information exchange, technology transfer, and promotion of cleaner technology, public participation and sustainable financing. Their fundamental goal is to develop and implement concrete pollution reduction projects that mobilize both stakeholders and resources, to become a cyclical process on which to build upon, to be mainstreamed into relevant institutional, budgetary and policy frameworks, and to incorporate lessons learnt in the process.

The NAP implementation process is expected to greatly enhance economic, technological and social development at the local level and to contribute to sustainable development. The Mediterranean Ecosystem Approach Strategy (EcAp) was proposed in 2005 and launched in 2009, aiming to achieve a Healthy Environment status of the Mediterranean Sea by 2020.

Today, the European Union and twenty-one countries around the Mediterranean are Contracting Parties of the MAP. In 2016 all parties reconfirmed their commitment "to implement the UNEP/MAP Mid-Term Strategy 2016–2021" to achieve "a healthy Mediterranean with marine and coastal ecosystems that are productive and biologically diverse contributing to sustainable development for the benefit of present and future generations" (UNEP/MAP Mid-Term Strategy 2016–2021 and UNEP Athens Declaration 2016). The MAP is legally binding.

The same MAP parties adopted the Convention for the Protection of the Mediterranean Sea Against Pollution (Barcelona Convention) in 1976. Both instruments, MAP and Barcelona Convention, were amended and adopted again by the Contracting Parties in 1995 (UNEP 1995a, 1995b). Initially the Barcelona Convention was fully focused on nature conservation and protection. Such focus successively shifted towards sustainable development, aiming at meeting the challenges of environmental degradation in the sea and to harmonize sustainable resource management with socio-economic development.

12.2.2 EU Level

Specifically for the Mediterranean region, the basin strategy of the EU's Integrated Maritime Policy (IMP) emphasizes the need for improving cooperation between the more than 20 Mediterranean countries. The IMP's Mediterranean basin strategy is "to improve cooperation and governance while also encouraging sustainable growth" (EU DG MARE website). Cooperation of the Mediterranean countries and among the many marine and maritime Mediterranean sectors is necessary in order to manage maritime activities, protect the marine environment and maritime heritage, prevent and combat pollution, improve safety and security at sea, and promote blue growth and job creation.

There is also a strategic research agenda for the EU Marine Strategy Framework Directive (MSFD) (SEAS ERA Med 2012), with the goal for Mediterranean marine science to "be able to contribute with New Knowledge to efficient Policy Making and sustainable growth of Maritime Economy in response to the societal challenges for Food, Energy, Wellbeing, and a Healthy marine environment following the principles of Ecosystem Approach to Management of Natural Resources" by 2020. The research agenda also includes a focus on support to sustainable economic growth in the region.

12.3 Regulatory Regimes

Appendix 12.1 addresses the main global regulations (treaties and legislation) of relevance for Blue growth sectors. In addition, Table 12.3 shows specific policy frameworks in relation to the sector combinations considered for the Mediterranean. The MARIBE project identified these sectors as relevant for the Mediterranean Sea [Chapter 14 of this book].

Table 12.3 Mediterranean sector specific policies

Sector	Policy/Agreement
Aquaculture	• Directive on Animal Health Requirements (2006/88/EC) • Common Fisheries Policy: Regulation (EC) No 1434/98 • Regulation (EU) No 1380/2013 of the European Parliament and of the Council of 11 December 2013 on the Common Fisheries Policy. • Communication to the European Parliament, the Council, the European Economic and Social Committee and the Committee of the Regions of 29 April 2013 on Strategic Guidelines for the sustainable development of EU aquaculture (COM/2013/229). • Regulation (EU) No 304/2011 of the European Parliament and of the Council of 9 March 2011 concerning use of alien and locally absent species in aquaculture. • Council Regulation (EC) No 708/2007 concerning use of alien and locally absent species in aquaculture. • Commission Regulation (EC) No 710/2009 of 5 August 2009, as regards laying down detailed rules on organic aquaculture animal and seaweed production. • Regulation (EC) No 889/2008 laying down detailed rules for the implementation of Council Regulation (EC) No 834/2007 on organic production and labelling of organic products with regard to organic production, labelling and control. • Council Regulation (EC) No 834/2007 of 28 June 2007 on organic production and labelling of organic products and repealing Regulation (EEC) No 2092/91. • Communication from the Commission to the European Parliament and the Council – Building a sustainable future for aquaculture, A new impetus for the Strategy for the Sustainable Development of European Aquaculture (COM/2009/0162 final).

Table 12.3 Continued

Fisheries	• GFCM Agreement • ICCAT Convention • Common Fisheries Policy: Regulation (EU) 2015/812 of the European Parliament and of the Council of 20 May 2015, as regards the landing obligation. • Regulation (EU) No 1380/2013 of the European Parliament and of the Council of 11 December 2013 on the Common Fisheries Policy. • COUNCIL REGULATION (EC) No 1967/2006 of 21 December 2006 concerning management measures for the sustainable exploitation of fishery resources in the Mediterranean Sea.
Offshore Wind fixed , floating; offshore fixed terminal	• Renewable Energy Directive (2009/28/EC) • Communication from the Commission to the European Parliament and the Council, the European Economic and Social Committee and the Committee of the Regions – Blue Growth, opportunities for marine and maritime sustainable growth (COM/2012/494 final). • Communication from the Commission to the European Parliament and the Council, the European Economic and Social Committee and the Committee of the Regions – Blue Energy Action needed to deliver on the potential of ocean energy in European seas and oceans by 2020 and beyond (COM/2014/08 final).
Tourism	• No specific policy at regional level. The following policies and instruments affect tourism in the Mediterranean at different levels: • UN Agenda 21, is the only policy covering the entire region. • Cotonou Agreement (2000) and its successive amendments allow for cooperation in development between the EU and African countries through the European Development Fund. • At EU level: Communication from the Commission to the Council and the European Parliament – A European Strategy for more Growth and Jobs in Coastal and Maritime Tourism (COM/2014/086 final). • At sub-regional level: Communication from the Commission to the Council and the European Parliament, the Council, the European Economic and Social Committee and the Committee of the Regions concerning the European Union Strategy for the Adriatic and Ionian Region (COM/2014/0357 final).
Oil and Gas	• Industrial Emissions Directive (2010/75/EU) • Directive on safety of offshore oil and gas operations (2013/30/EU) • Directive concerning common rules for the internal market in natural gas (2009/73/EC) • Directive imposing an obligation on Member States to maintain minimum stocks of crude oil and/or petroleum products (2009/119/EC)

(Continued)

Table 12.3　Continued

Sector	Policy/Agreement
	• Directive on the conditions for granting and using authorisations for the prospection, exploration and production of hydrocarbons (94/22/EC) • Communication from the Commission to the Council and the European Parliament – Blue Growth, opportunities for marine and maritime sustainable growth (COM/2012/494 final).
Seabed Mining Offshore	• Directive on environmental liability with regard to the prevention and remedying of environmental damage (2004/35/EC) • Communication from the Commission to the Council and the European Parliament – Blue Growth, opportunities for marine and maritime sustainable growth (COM/2012/494 final).
Biotechnology	• There is currently no overarching regional strategy or plan specifically focusing on marine biotechnology research and development. • General marine science issues are considered by organisations such as CIESM and projects such as the SEAS-ERA Project (www.marinebiotech.eu) • At EU level: Communication from the Commission to the Council and the European Parliament – Blue Growth, opportunities for marine and maritime sustainable growth (COM/2012/494 final).
Nature conservation	• Convention for the Protection of the Mediterranean Sea Against Pollution (Barcelona Convention) • Mediterranean Ecosystem Approach Strategy (EcAp) • Agreement on the Conservation of Cetaceans of the Mediterranean and the Black Sea and contiguous Atlantic Area (ACCOBAMS) • Directive 2008/56/EC of the European Parliament and of the Council of 17 June 2008 establishing a framework for community action in the field of marine environmental policy (Marine Strategy Framework Directive)

12.4 Spatial Impact and Planning

A protocol on Integrated Coastal Zone Management (ICZM) entered into force in 2011 (UNEP 2008), signed by 15 and currently ratified by 10 parties (Albania, Algeria, Croatia, France, Greece, Israel, Italy, Malta, Monaco, Montenegro, Morocco, Slovenia, Spain, Syria, Tunisia and the EU). The main goal of the ICZM Protocol is to allow the Mediterranean countries to better manage and protect their coastal zones, as well as to deal with the emerging coastal environmental challenges (e.g. climate change). The Protocol puts pressures on science and technology to improve practices of MSP and ICZM. The Action Plan for the implementation of the ICZM

Protocol is ongoing (2012–2019). The Protocol is part of EU law (EU 2008) and has binding effects. Furthermore, the EU FP7 research project PEGASO[2] has developed novel approaches to support integrated policies for the coastal, marine and maritime realms of the Mediterranean (and Black) Sea. Building on existing capacities, the approaches are consistent with and relevant to the implementation of the ICZM Protocol for the Mediterranean.

Table 12.4 provides a status overview by country of the existing implemented marine spatial plans and/or any existing related MSP legislation. The overview was constructed on the basis of data from the European MSP Platform, PRC (2011) and the UNESCO country reports (UNESCO website). The EU MSP directive provides a framework for MSP for EU Member States. In order to comply with the MSP Directive EU Member States needed to implement the required laws, regulations and administrative provisions by September 2016. The maritime spatial plans should be implemented as soon as possible, and at the latest by March 2021. Plans will be reviewed at least every 10 years. Of the Mediterranean EU Member States only Croatia, Malta and Slovenia have successfully implemented national MSPs so far. There are no international MSP obligations for the Non-EU states.

12.5 Related Strategies

12.5.1 Mediterranean Cooperation Projects and Initiatives

The EUNETMAR (2014b) project identified 149 cooperation projects and initiatives related to blue growth and integrated maritime policy. About a third of those are specific to the Adriatic and Ionian basins. Table 12.5 gives an overview of international/supranational, EU and national cooperation projects for maritime sectors as well as overarching for the Mediterranean Sea region.

12.5.2 Maritime Clusters

Additionally, clusters are considered important for the progress of the Blue Growth strategy as the development and growth of maritime sectors are often dependent on collaboration and cooperation between local players. A cluster is defined "as a geographically proximate group of interconnected companies and associated institutions in a particular field, linked by commonalities and complementarities (external economies). External economies that occur within a cluster are the economic and financial inter-sector relations, a common knowledge and technology base,

---

[2]http://www.vliz.be/projects/pegaso/

Table 12.4 Overview by country of existing implemented marine spatial plans or related MSP legislation in the Mediterranean Sea, based on msp-platform.eu for all EU member states (except France), PRC (2011) and UNESCO website.

Country	MSP in Place	Existing Legislation	Activities Covered	Comments	Links
Albania	no	Coastal Zone Management plan (2004): National ICZM plan (but no national ICZM law)			PRC 2011
Algeria	no	Coastal Area Management Programme (CAMP) for the Algerian Coastal Zone (2001)		No EEZ, but exclusive fishing zone established.	PRC 2011
Bosnia and Herzegovina	no	Federal Law on Spatial Planning. No legislative instruments/mechanisms/procedures for coastal management; no bodies/agencies for integrated management of the coastal area		Territorial sea is entirely surrounded by Croatia's internal waters. Treaty on maritime borders signed in 1999; however not ratified.	PRC 2011
Croatia	Yes	MSP on regional level the legally binding "Zadar county integrated sea use and management plan". On national level the Physical Planning Act (2013) provides for MSP, but not legally binding, legally binding plan focusing on mariculture and with links to MSP: Zadar county integrated sea use and management plan.	focus on mariculture	Maritime Code of Croatia (1994) contains several EEZ provisions, but has not in reality been established	msp-platform.eu
Cyprus	no	EEZ established. No specific, single legislative act for maritime spatial planning yet. Pilot plans exist			msp-platform.eu

Table 12.4 Continued

Egypt	no	EEZ established. National Committee for Integrated Coastal Zone Management (1994) but clear mandate and authority lacking. Aim to develop ICZM strategy in compliance with the ICZM protocol.		PRC 2011
France*	ongoing	EEZ establishment in progress. MSP: Plan d'action pour le littoral méditerranéen; Marine Protected Areas Strategy, suitable for MSP. National Strategy for the sea and oceans (Livre Bleu, 2009); currently developing a new maritime policy	ecology, pollution, management of new risks, control	PRC 2011
Greece	no	Several sector specific legislative acts, strategic documents and spatial planning frameworks (e.g. aquaculture, tourism, industry and renewable energy)		no specific legislation dealing with MSP yet — msp-platform.eu
Israel	yes	Israel Marine Plan (Technion IMP (TIMP)) completed (by academic initiative), but not yet approved. The southernmost part of the territorial sea is adjacent to the Gaza Strip and therefore closed to Israel civilian uses.	marine mining; fishing; marine protected areas;	Competition likely between: port development-MPAs-channel dredging-recreation-military use; fishing-deep sea MPAs-desalination projects-recreation; — TMP (ongoing); Portman 2015; UNESCO website 2017

(Continued)

Table 12.4 Continued

Country	MSP in Place	Existing Legislation	Activities Covered	Comments	Links
			marine conservation, biodiversity concerns; new and emerging uses	cliffs erosion-nature protection/archaeology; sand supply/port construction-nature protection-sand drift	
Italy	no	EU MSP Directive was transposed in Italian legislation with the Legislative Decree 17 October 2016, n. 201. No binding maritime spatial plan has yet been officially elaborated/adopted		potentially three MSP plans to be elaborated, for (i) Western Mediterranean, (ii) Adriatic Sea, (iii) Ionian Sea and Central Mediterranean	msp-platform.eu
Lebanon	no	EEZ established. No binding MSP legislation, but several ICZM strategies and projects; Framework Law on the Protection of the Environment (444/2002, article 29–34); creation of the National Council of the Environment to coordinate ICZM actions foreseen, but not yet set up.		no maritime activities at border with Israel due to political conflicts. Laws and regulations re planning and environment are overlapping, contradictory and not well-applied, resulting in inter-ministerial disputes related to jurisdiction and mandate.	PRC 2011

Table 12.4 Continued

			controlling development on land	Important and well-preserved beaches and near-shore marine areas (sea turtles nesting areas);	PRC 2011
Libya	no	EEZ established. N\o comprehensive ICZM legalisation for the entire coastline; weak environmental legislation. Existing legislation (1969) and National Spatial Planning Strategy focus on controlling development on land.			
Malta	yes	MSP: Strategic Plan for Environment and Development (SPED, 2015) = overarching document for planning issues on land and at sea in an integrated manner. It constitutes the national Maritime Spatial Plan. Main legislative act: Development Planning Act of 2016 (incl. development at sea); Strategic Plan for Environment and Development (SPED, 2015) = overarching document for planning issues on land and at sea in an integrated manner. It constitutes the national Maritime Spatial Plan.	maritime transport; short sea shipping; port; cruise tourism; yachting, fishing, aquaculture, offshore hydrocarbon exploration; communication cables		Plan: http://www.pa.org.mt/sped; Country MSP info: msp-platform.eu

(Continued)

Table 12.4 Continued

Country	MSP in Place	Existing Legislation	Activities Covered	Comments	Links
Monaco	no				PRC 2011
Montenegro	yes	MSP: Spatial Plan for the coastal zone/Public Maritime Domain as a Special Purpose Area (2007). Draft National Strategy on Integrated Coastal Area Management (NS ICAM)		pollution hotspots and sensitive areas in the coastal sea; planning system in general is well-developed, comprehensive and integrated, but lack of plans/adequate solutions for the coastal/marine area issues	PRC 2011
Morocco	no	EEZ established but not enforced in the Mediterranean. Environmental legislation exists; sub-national coastal management plan developed.			PRC 2011
Palestine/ Gaza strip	no	no information		natural gas fields; environmental pollution: solid wastes, construction debris, rubble, waste water, overfishing/excessive fishing methods, beach recreation, war	FT 2013; Al-Dameer 2009
Slovenia	yes	MSP: National Spatial Plan for the integrated spatial development	Shipping, Ports,		msp-platform.eu

Table 12.4 Continued

		of the port for international traffic at Koper (no new legislation needed for implementation of MSP directive). National Spatial Plan for the integrated spatial development of the port for international traffic at Koper	Tourism (incl. recreation and sports), Nature protection, Military		
Spain	no	Spanish law 41/2010 for planning the marine environment (MSFD implementation); an instrument (Royal Decree) for MSP is being prepared.	coastal and marine tourism, fisheries, maritime transport and mariculture	No MSP covering the entire coastal zone; Act on Protection of the Marine Environment (2010) ICZM strategy. No legal delimitation of the coastal zone.	msp-platform.eu
Syria	no	EEZ claimed although Syria has not ratified UNCLOS. Strategic integrated approach to spatial planning of the coastal area is lacking.		High salinity, low biodiversity, high pollution	PRC 2011
Tunisia	no	EEZ established. Existing maritime law; national MPA programme		ICZM considered high priority. Challenges: high urbanisation, pollution, vulnerability of the marine environment	PRC 2011
Turkey	no	legal framework for ICZM and institutional structure not yet established; wide scope ICZM law provided by current legal framework; uncoordinated management as regards coastal planning		uncoordinated management as regards coastal planning: More than twenty institutions are in charge of the sea and coastal areas resulting in responsibility overlaps and gaps.	PRC 2011

Table 12.5 Maritime clusters per cooperation project/sector in the Mediterranean Sea Region

Cooperation Project	Name and Brief Description	Source/Link
Cooperation EU-non EU	EUROMED (Euro-Mediterranean Partnership) • a portal for news and information about EU cooperation with its Southern neighbours • established by the Barcelona Convention: Union for the Mediterranean (UfM).	http://enpi-info.eu/indexmed.php http://eeas.europa.eu/euromed/index_en.htm
Regional and local cooperation	Euro-Mediterranean Regional and Local Assembly (ARLEM) (EU-Committee of the Regions) • To support the process of decentralisation and promote the "territorialisation" of the UfM's policies, programmes and projects • To strengthen the institutional capacity of local and regional authorities to manage public policies and highlight the role of local and regional authorities as strategic partners for good governance and successful development outcomes • For the implementation of a cohesion policy in the Southern and Eastern Mediterranean region and the adoption of a macro-regional approach. • To bring Euro-Mediterranean cooperation closer to the citizens, therefore producing tangible results in people's daily lives.	http://cor.europa.eu/en/activities/arlem/Pages/arlem.aspx
Regional support	CPMR Inter Mediterranean Commission to express the shared interests of Mediterranean • Regions in important European negotiations • Defending the interests of the Mediterranean Regions in key EU policies • Incorporating the territorial concept and the role of the regional authorities in the Barcelona process and the Mediterranean Union • Undertaking strategic "pilot" projects on key themes with a forceful territorial impact 9 member countries, comprising 40 member regions	http://www.medregions.com/index.php?act=1,1
Regional fisheries management organization (RFMO)	• GCFM (General Fisheries Commission for the Mediterranean) under the FAO • to promote the development, conservation, rational management and best utilization of	http://www.fao.org/gfcm/en/

Table 12.5 Continued

	living marine resources as well as the sustainable development of aquaculture in the Mediterranean, the Black Sea and connecting waters 24 members (23 countries + EU)	
Marine cetaceans: Conservation Monitoring, research Capacity building information	ACCOBAMS (Agreement on the Conservation of Cetaceans in the Black Sea, Mediterranean Sea and Contiguous Atlantic Area) • cooperative tool for the conservation of marine biodiversity in the Mediterranean and Black Seas • to reduce threats to cetaceans in Mediterranean and Black Sea waters and improve our knowledge of these animals	http://www.accobams.org/
Science/Research	CIESM (Mediterranean Science Commission) • support a network of marine researchers • applying the latest scientific tools to better understand, monitor and protect a fast-changing, highly impacted Mediterranean Sea • Structured in six committees and various taskforces, CIESM runs expert workshops, collaborative programs and regular congresses, delivering authoritative, independent advice to national and international agencies. • 23 member states	http://www.ciesm.org/
Coastal nature protection, FR-IT	Accord RAMOGE (Agreement for the prevention and combat against pollution in the marine environment and the littoral of the PACA Region (FR), the Principality of Monaco and Liguria (IT)) • Management and protection of the coast and marine biodiversity, fight against pollution of the marine environment	http://www.ramoge.org/fr/default.aspx
Coastal protection, collaboration	PIM (Petites Iles de Méditerranée) Mediterranean small islands Initiative, Coastal Protection Agency • setting-up practical measures for conservation management and protection of these microcosms	http://www.initiative-pim.org/en
	• facilitating exchange of information and experience between site managers (administrators) and experts from across the Mediterranean Basin.	

(Continued)

Table 12.5 Continued

Cooperation Project	Name and Brief Description	Source/Link
Regionalisation	e.g.: EU Strategy for the Adriatic and Ionian region and its action plan Developing sub-regional strategies to exploit the strengths and address the weaknesses of particular maritime regions	http://ec.europa.eu/maritimeaffairs/policy/sea_basins/adriatic_ionian/index_en.htm
Coast guard: Maritime safety, Surveillance, Monitoring. . .	ECGFF (Mediterranean Coast Guard Functions Forum) • self-governing, non-binding, voluntary, independent and non-political body • brings together administrations, institutions and agencies working on coast guard issues from all Mediterranean countries • network • DG MARE funded	http://www.ecgff.eu/
Virtual knowledge centre Coordination, Cooperation	Virtual Knowledge Centre for marine and maritime affairs in the Mediterranean – part of IMP-MED • instigated by the European Union, European Investment Bank and International Maritime Organization • to facilitate coordination and cooperation, consolidate and share general, technical and sectoral information; to improve synergies across initiatives and projects, promote investment and innovation, and support maritime businesses.	http://www.vkc-med.eu/
Communication platform	Maritime Forum • a common communication platform for EU maritime policy stakeholders to improve their communication to publish events, documents and follow developments in their areas of interest • share information amongst a closed community or published openly	https://webgate.ec.europa.eu/maritimeforum/

and a shared labour market" (PRC 2008). A European Network of Maritime Clusters has been established "as a best practices dissemination and exchange platform", its "aim is to establish a framework for future common targeted actions" (ENMC website). Table 12.6 describes existing clusters in a few Mediterranean countries.

Table 12.6   Maritime Clusters in the Mediterranean Sea basin

Country	Name	Cluster Description
France		Aquitaine, Bordeaux
		Maritime cluster (especially shipping) is active and strong in many niches (e.g. maritime research and technological services and yacht building) (source: CMF, 2008): Yacht building and repair
		Mer PACA = regional competitiveness poles (source: Pôles de Compéitivité, 2008)
Greece		No formal organisation representing sea-related sectors is (yet) established
Italy	Federazione del Sistema Marittimo Italiano	Shipbuilding, Marine equipment, Seaports, Shipping AIDIM (diritto marittimo), ANCIP (lavoro portuale), ANIA (assicurazione), ASSOPORTI (amministrazione portuale), ASSONAVE (cantieristica navale), ASSORIMORCHIATORI (rimorchio portuale), COLLEGIO CAPITANI (stato maggiore marittimo), CETENA (ricerca navale), CONFITARMA (navigazione mercantile), FEDERAGENTI (agenzia e intermediazione marittime), FEDEPILOTI (pilotaggio), FEDERPESCA (navigazione peschereccia), FEDESPEDI (trasporti internazionali), INAIL/exIPSEMA (previdenza marittima), RINA (certificazione e classificazione) e UCINA (nautica da diporto).
Malta		traditional maritime sectors with an employment of 7 600 or 5% of all Maltese employed
Slovenia		Employment in coastal tourism and fisheries
Spain	Cluster Maritimo Espanol (SMC) several regional cluster organisations Cluster Maritimo de Canarias	Fisheries, Shipbuilding, coastal tourism, offshore supply, recreational boating Canaries: Shipbuilding and ship repairs; port services; fishing and aquaculture; and auxiliary industries Spanish maritime cluster excels in the sectors fisheries and coastal tourism and their supporting services

Source: PRC 2008.

12.6 Supporting Blue Growth

International European funding sources for Blue Growth activities are listed in Table 12.7.

Table 12.7 International funding schemes, purpose and link

Funding Name	Purpose/Type of Activity Covered	Links
2014–2020 EU financial framework	Partnership contracts between national governments and the Commission, operational programmes for regional development and work programmes for research	http://ec.europa.eu /budget/mff/index_en. cfm
European Agricultural Fund for Rural Development (EAFRD)	Sustainable management of natural resources and climate action and the balanced territorial development of rural areas.	http://ec.europa.eu/ agriculture/rural-development-2014-2020/financial-instruments/index_en. htm
European Maritime and Fisheries Fund (EMFF)	Maritime and fisheries related activities including sea-basins such as the Atlantic. Aims at achieving the objectives of the reformed CFP and IMP.	http://ec.europa.eu/ fisheries/cfp/emff/index_ en.htm
European Social Fund (ESF)	Main financial instrument for investing in people. Seeks to increase employment opportunities and promote education.	http://ec.europa.eu/ regional_policy/en/ funding/social-fund/
European Regional Development Fund (ERDF)	Aims to strengthen economic and social cohesion in the European Union by correcting imbalances between its regions.	http://ec.europa.eu/ regional_policy/en/ funding/erdf/
Cohesion Fund	Helps Member States with a GNI per inhabitant of less than 90% of the EU-27 average to invest in TEN-T transport networks and the environment.	http://ec.europa.eu/ regional_policy/en/ funding/cohesion-fund/
European Territorial Cooperation Fund	Provides a framework for the exchanges of experience between national, regional and local actors from different Member States as well as joint action to find common solutions to shared problems.	http://ec.europa.eu/ regional_policy/en/ policy/cooperation/ european-territorial/
European Groupings of Territorial Cooperation (EGTCs)	Designed to help specific countries/regions overcome complicated differences between national rules and regulations.	https://portal.cor.europa. eu/egtc/Pages/welcome. aspx
Connecting Europe Facility	A new, integrated instrument for investing in EU infrastructure priorities in transport, energy and telecoms.	https://ec.europa.eu/ inea/en/connecting-europe-facility
Programme for the Competitiveness of Enterprises and small and	Aims to strengthen the competitiveness and sustainability of the Union's enterprises and encourage an entrepreneurial culture by promoting the creation and growth of SMEs.	http://ec.europa.eu/ growth/access-to-finance/cosme-financial-instruments/index_en.

Table 12.7 Continued

Funding Name	Purpose/Type of Activity Covered	Links
medium-sized enterprises (COSME)		htm
Horizon 2020	Research and innovation funding for various types of research project.	http://ec.europa.eu/ programmes/horizon 2020/
LIFE+	Covers the environment, biodiversity, resource efficiency, governance and all aspects of climate change.	http://ec.europa.eu/ environment/life/funding /lifeplus.htm
European Investment Bank (EIB)	Innovation and skills, access to finance for small businesses, environment and climate, infrastructure	http://www.eib.org/ about/index.htm

In some circumstances specific sectors may receive governmental support in the form of State aid but this can be subject to very strict conditions. Accordingly, shipbuilding is one such sector, exemptions are discussed in Section 10.6.

The EUNETMAR (2014b) project also identified potential funding opportunities, i.e. support schemes, related to blue growth and integrated maritime policy in the Mediterranean (Table 12.8).

12.7 Key Considerations

Due to the many anthropogenic as well as natural challenges (oligotrophy, heavy human pressure, high rate of biological invasions, exposure to geo-hazards, semi-enclosed sea with a very low turnover time) the Mediterranean Sea can be characterized as a diverse but fragile system with low physical inertia and limited ecological resilience. Before initiating any type of Blue Growth activity or the combinations thereof, the potential ecological impacts therefore need to be carefully investigated. Due to the socio-economic and cultural diversity of the Mediterranean societies, also social impact studies need to be carried out. An adequate balance between benefits and trade-offs of innovative Blue Growth activities needs to be carefully planned. In light of the geo-political situation the major challenge for the Mediterranean region is to create a safer, peaceful and more prosperous region.

Stability and a common governance framework for the entire Mediterranean Sea region are crucial for Blue Growth – a key conclusion resulting from the international public consultation on Ocean Governance (DG MARE

Table 12.8 Support schemes/potential funding opportunities relevant for projects in the Mediterranean Sea

Cluster	Name and Brief Description
Regional integration and cohesion	UfM (Union for the Mediterranean) http://ufmsecretariat.org/ • multilateral partnership aiming at increasing the potential for regional integration and cohesion among Euro-Mediterranean countries
Blue Growth	UfM Blue Growth call for 2016 and 2017 focussing on several maritime and marine challenges of the BLUEMED Initiative. http://ufmsecretariat.org/within-the-framework-of-its-global-sustainable-development-strategy-the-ufm-launches-a-new-blue-economy-cooperation-initiative-in-the-mediterranean/
International, regional, and sectoral integration	ENPI (European Neighbourhood & Partnership Instrument) • supports the European Neighbourhood Policy (ENP): 16 ENP countries, to achieve the closest possible political association and the greatest possible degree of economic integration IMP-MED (Project on Integrated Maritime Policy in the Mediterranean) designed to help the southern Neighbourhood countries develop integrated approaches to marine and maritime affairs. → ENPI finances actions in the various sectors, including: more equitable development, energy, transport, information society, environmental sustainability, research and innovation. http://ec.europa.eu/europeaid/funding/european-neighbourhood-and-partnership-instrument-enpi_en
Cooperation EU-non EU Maritime transport	EUROMED Transport Programme: Mediterranean Motorways of the Sea – Maritime transport connections http://www.enpi-info.eu/mainmed.php?id_type=10andid=41 • improving transport connections between the EU and its Mediterranean neighbours and to promote the Motorways of the Sea (MoS) concept and assisting the partner countries in further implementing the maritime transport and port operations actions as adopted in the Regional Transport Action Plan (RTAP), a road map for transport cooperation adopted in 2007 (covering 2007–2013).
Fisheries Cooperation	FISHERIES – FARNET (Charter for Mediterranean FLAG Cooperation) https://webgate.ec.europa.eu/fpfis/cms/farnet/charter-mediterranean-flag-cooperation • FR, ES, GR, CY • 2011 • to further projects that contribute to the development of Mediterranean fisheries areas • environmental and educational activities • promotional actions for local fisheries products and fisheries-related tourism

Table 12.8 Continued

Cluster	Name and Brief Description
Fisheries	EMFF (European Maritime and Fisheries Fund) http://ec.europa.eu/fisheries/cfp/emff/index_en.htm • supports the setting up of a network of Maritime Clusters in the Mediterranean by over half a million euro
Marine Protected Areas (MPAs) managers	MEDPAN (network of Marine Protected Areas managers in the Mediterranean) http://www.medpan.org/ • partnership approach to promote the sustainability and operation of a network of Marine Protected Areas in the Mediterranean (ecologically representative, connected, managed effectively) to reduce rate of marine biodiversity loss. • a network for knowledge, information, anticipation and synthesis • >90 MPAs in 18 Mediterranean countries
Conservation Biodiversity management Sustainable development	IUCN – Med Programme, IUCN Centre for Mediterranean Cooperation http://www.iucn.org/about/union/secretariat/offices/iucnmed/iucn_med_programme/ • since 2011 • Make knowledge, information and experience available regarding the conservation and management of Mediterranean biodiversity and natural resources for sustainable-use and rehabilitation efforts. • Strengthen and support IUCN members and Commissions in the region to mainstream social, economic and environmental dimensions in policy-making, management, and the conservation of biodiversity and natural resources. • Promote, both globally and regionally, Mediterranean policies on conservation and sustainable development, and supporting mechanisms for their implementation.
Stability, security, economic opportunity	EU Emergency Trust Fund for stability http://europa.eu/rapid/press-release_MEMO-15-6056_en.htm • so far 1.9 billion euros dedicated to address root causes of irregular migration and promote economic opportunities, including on the Southern coast of the Mediterranean.
Jobs, growth, investment	EU Infrastructure Investment Plan ("Junker Plan") http://ec.europa.eu/priorities/jobs-growth-investment/plan/index_en.htm >300 billion euros • To support connectivity needs in the region through financing, e.g. international energy grids or telecommunications networks.

2015). Investors need a stable governance framework, which ensures business certainty. Considering the huge differences among Mediterranean countries in economic, cultural, societal and legislative setup, these aims are very ambitious and hard to reach. To create new opportunities while keeping focused on the common goals, the challenges need to be tackled collectively by countries, businesses, and citizens. The Mediterranean governance framework would improve through (1) ensuring legitimacy of the institutions involved in management actions; (2) enhancing coordination of maritime affairs inside countries (between ministries and institutions as well as between European, national and regional administrations); (3) establishing coordination schemes among countries (at bilateral and multilateral level); and (4) ensuring cross-sectorial coordination of maritime policies through existing regional organisations, projects and activities.

The MAP and Barcelona Convention together with the specific Mediterranean EU policies, regulations and strategies do represent an impressive governance framework, focusing on protection/restoration of the Mediterranean Sea ecosystem as well as on fostering sustainable development, a sustainable blue economy and blue growth (Cinnirella 2014). The implementation of MSP and ICZM is critical for the preservation of biodiversity and the co-location of different maritime activities. However, policy goals on paper are still far from being met in reality: Environmental problems in the Mediterranean sea are aggravating instead of improving and Mediterranean marine ecosystem services are degrading (Coll et al. 2010, 2012). For example, 93% of the assessed fish stocks in the Mediterranean are not sustainably fished.

Fisheries should be managed in more efficient ways, tackling the overfishing problem and improving the critical state of Mediterranean fish stocks in closer strategic cooperation with partner countries (Vella 2015). Also, marine aquaculture and biotechnologies need to develop further; marketing and communication are needed to allow for the economic viability of the exploitation of fish and seafood products; innovative, high-quality tourist offers should be developed to ameliorate negative impacts from mass tourism; and technology transfer (e.g. traceability in the food industry, eco-labelling of products, fuel efficiency, eco-tourism, security of water supply through desalination, etc.) should be enhanced to warrant the ecological sustainability of economic activities.

The full delimitation of maritime zones in the Mediterranean Sea can also contribute to improving the governance framework. Disputes around contended borders of EEZs need to be settled at sea-basin level. The benefits expected from establishing full EEZs in the Mediterranean considerably exceed the costs, likely offering "synergy and costs saving efficiencies with regard to control and possibly monitoring and data collection" – under the crucial prerequisite that there is political will amongst countries and their neighbours (MRAG et al. 2013, p. 219).

Institutional support, long-term political vision and continuous engagement of stakeholders are still lacking at the regional scale (Cinnirella 2014).

This is a priority for the future development of strategic sectors such as tourism, energy, blue biotechnologies and fisheries. The Mediterranean Sea holds diverse ecosystems providing substantial goods (like food supply) and services (like protection from coastal erosion) to coastal societies from ancient times. However, on-going environmental and ecological degradation erode the potential of the Mediterranean region for Blue Growth because pristine environments and healthy ecosystems constitute important assets for some of the most promising economic activities: Tourism, which contributes most to GDP at regional level, has traditionally taken advantage of the particularly long history of human occupation, huge cultural heritage, mild climate and scenic landscapes of the Mediterranean region. Yet, increasing human pressures are leading to significant degradation of the Mediterranean ecosystems, ultimately putting at risk the continuity of those assets, which traditional tourism is based upon. Innovative eco-tourism approaches, offering distinct cultural and traditional experiences, depend on an intact, sustainably managed and diverse environment. Also sectors such as biotechnology and fisheries heavily rely on the continuity of the delivery of natural goods and the maintenance of the processes supporting them. Some Mediterranean areas appear particularly well suited for the production of renewable energy, such as wind, tidal, or solar. Raising awareness, education and training across all sectors is necessary in order to solve these issues in the medium and long-term. All in all, a picture emerges that shows substantial potential for positive synergies among environmental protection, ecological integrity, cultural diversity and economic growth under a shared political vision for the Mediterranean region.

References

Abbasi, K., Patel, K. and Godlee, F. (2015). Europe's refugee crisis: an urgent call for moral leadership. BMJ: British Medical Journal 351.

Abdulla A., Linden O. (2008) Maritime traffic effects on biodiversity in the Mediterranean Sea: Review of impacts, priority areas and mitigation measures. IUCN Centre for Mediterranean Cooperation, Malaga, 184 pp.

Al-Dameer Association for Human Rights (2009). A special report on marine environment in the Gaza Strip. Environmental Report Series 03. http://dico.pourlapalestine.be/docs/ocha_opt_wash_cluster_specialreport_3_alDameer_June2009-20090715-160238.pdf

Albahari, M. (2015). Europe's refugee crisis. Anthropology Today 31, 1–2.

Barausse A., Palmeri L. (2014). A Comparative Analysis of Trophic Structure and Functioning in Large-Scale Mediterranean Marine Ecosystems. In: Goffredo S., Dubinsky Z. (eds) The Mediterranean Sea. Its history and present challenges. Springer, Dordrecht: 421–434.

Bianchi C.N., Morri C., Chiantore M., Montefalcone M., Parravicini V., Rovere A. (2012). Mediterranean Sea Biodiversity Between The Legacy from the Past and A Future of Change. In: Stambler N. (ed) Life in the Mediterranean Sea: A Look at Habitat Changes. Nova Science Publishers, New York: 1–55.

Bilsky, T. (2009). Disaster's impact on internet performance – Case study. In: International Conference on Computer Networks. Springer, Berlin, Heidelberg,: 210–217.

Camillo (2009). Mediterranean-yachting.com. yachting: The Countries bordering the Mediterranean. http://www.mediterranean-yachting.com/Countries.htm

CIBRA (2005). The Mediterranean Sea. Università degli Studi di Pavia. Centro Interdisciplinare di Bioacustica e Ricerche Ambientali. http://www-3.unipv.it/webcib/edu_Mediterraneo_uk.html

Cinnirella, S., Sardà, R., Suárez de Vivero, J. L., Brennan, R., Barausse, A., Icely, J., Luisetti, T., March, D., Murciano, C., Newton, A., O'Higgins, T., Palmeri, L., Palmieri, M.G., Raux, P., Rees, S., Albaigés, J., Pirrone, N. and Turner, K. (2014). Steps toward a shared governance response for achieving Good Environmental Status in the Mediterranean Sea. Ecology and Society 19.

Cirer-Costa, J.C. (2017). Turbulence in Mediterranean tourism. Tourism Management Perspectives 22, 27–33.

Claus S, De Hauwere N, Vanhoorne B, Deckers P, Souza Dias F, Hernandez F, Mees J (2014). Marine Regions: Towards a global standard for georeferenced marine names and boundaries. Marine Geodesy 37(2): 99–125.

Coll M, Piroddi C, Steenbeek J, Kaschner K, Lasram FBR, Aguzzi J, Ballesteros E, Bianchi CN, Corbera J, Dailianis T, Danovaro R, Estrada M, Froglia C, Galil BS, Gasol JM, Gertwagen R, Gil J, Guilhaumon F, Kesner-Reyes K, Kitsos M-S, Koukouras A, Lampadariou N, Laxamana E, López-Fé de la Cuadra CM, Lotze HK, Martin D, Mouillot D, Oro D, Raicevich S, Rius-Barile J, Saiz-Salinas JI, San Vicente C, Somot S, Templado J, Turon X, Vafidis D, Villanueva R, Voultsiadou E (2010). The biodiversity of the Mediterranean Sea: estimates, patterns, and threats. PloS one 5(8): e11842.

Coll, M., Piroddi, C., Albouy, C., Ben Rais Lasram, F., Cheung, W.W.L., Christensen, V., Karpouzi, V.S., Guilhaumon, F., Mouillot, D., Paleczny, M., Palomares, M.L., Steenbeek, J., Trujillo, P., Watson, R. and Pauly, D. (2012). The Mediterranean Sea under siege: spatial overlap between marine biodiversity, cumulative threats and marine reserves. Global Ecology and Biogeography 21, 465–480.

Coscieme, L., Sutton, P.C., Anderson, S., Liu, Q. and Elvidge, C.D. (2017). Dark Times: nighttime satellite imagery as a detector of regional disparity and the geography of conflict. GIScience & Remote Sensing 54, 118–139.

D'Ortenzio, F. and Ribera D' Alcala, M. (2009). On the trophic regimes of the Mediterranean Sea: a satellite analysis. Biogeosciences, 6: 139–148.

DG MARE (2015). Summary of the results of the public consultation on international ocean governance. http://ec.europa.eu/dgs/maritimeaffairs_fisheries/consultations/ocean-governance/doc/ocean-governance-summ ary_en.pdf.

EC (2006). European Council Regulation (EC) No 1967/2006 of 21 December 2006 concerning management measures for the sustainable exploitation of fishery resources in the Mediterranean Sea. http://eur-lex.europa.eu/LexUriServ/LexUriServ.do?uri=OJ:L:2006:409:0011:008 5:En:PDF.

EC Directive 2001/42/EC ('Strategic Environmental Assessment' – SEA Directive).

EC Directive 2011/92/EU ('Environmental impact assessment') and its amendment Directive 2014/52/EU.

Encyclopaedia Britannica (2015). "Mediterranean Sea". Encyclopædia Britannica. Encyclopædia Britannica Online. Encyclopædia Britannica Inc., 2015. Web. 29 Oct. 2015. <http://www.britannica.com/place/Mediterran ean-Sea> http://www.britannica.com/place/Mediterranean-Sea

ENMC website. European Network of Maritime Clusters. http://enmc.eu/news.

EU (2008). Protocol on Integrated Coastal Zone Management in the Mediterranean. http://eur-lex.europa.eu/LexUriServ/LexUriServ.do?uri=OJ:L: 2009:034:0019:0028:EN:PDF

EU (2009). Towards an Integrated Maritime Policy for better governance in the Mediterranean. COM(2009) 466 final. http://eur-lex.europa.eu/legal-content/EN/TXT/PDF/?uri=CELEX:52009DC0466&from=EN

EUNETMAR (2014a). Report 1 – Analysis of Blue Growth needs and potential per country. In: Studies to support the development of sea-basin cooperation in the Mediterranean, Adriatic and Ionian, and Black Sea. CONTRACT NUMBER MARE/2012/07 – Ref. No 2.

EUNETMAR (2014b). Report 4 – Task 5: Mediterranean Sea – Identification of elements and geographical scope of maritime cooperation, September 2014. In: Study to support the development of sea-basin cooperation in the Mediterranean, Adriatic and Ionian, and Black Sea – Analysis of Blue Growth needs and potential per country. CONTRACT NUMBER MARE/2012/07 – Ref. No 2.

FT (2013) Gaza Strip gas project poised for approval. Financial Times 9 October 2013. https://www.ft.com/content/13474ef2-3027-11e3-80a4-00144feab7de

Galil BS, Boero F, Fraschetti S, Piraino S, Campbell ML, Hewitt CL, Carlton J, Cook E, Jelmert A, Macpherson E, Marchini A, Occhipinti-Ambrogli A, Mcenzie C, Minchin D, Ojaveer H, Olenin S, Ruiz G (2015). The enlargement of the Suez Canal and introduction of non-indigenous species to the Mediterranean Sea. Limnology and Oceanography Bulletin 24(2): 43–45.

Galil, B. S. (2006). The marine caravan – the Suez Canal and the Erythrean invasion. In: Bridging divides. Maritime canals as invasion corridors, Gollasch S, Galil BS, Cohen AN (eds.). Springer, Netherlands: 207–300.

GFCM (2016) – The State of Mediterranean and Black Sea Fisheries. FAO, Rome: 134 pp

Halpern BS, Walbridge S, Selkoe KA, Kappel CV, Micheli F, D'Agrosa C, Bruno JF, Casey KS, Ebert C, Fox HE, Fujita R, Heinemann D, Lenihan HS, Madin EMP, Perry MT, Selig ER, Spalding M, Steneck R, Watson R (2008). A Global Map of Human Impact on Marine ecosystems. Science 319: 948–952.

IHO 2002. Limits of Oceans and Seas. Chapter 3 Mediterranean Region and its sub-divisions. 4th Edition 2002. International Hydrographic Organization. 23 pp. https://www.iho.int/mtg_docs/com_wg/S-23WG/S-23WG_Misc/Draft_2002/Draft_2002.htm.

ISO 14001:2015 Environmental management systems — Requirements with guidance for use. http://www.iso.org/iso/catalogue_detail?csnumber=60857

ISO 31000:2009 — Risk management — Principles and guidelines, provides a set of principles, a framework and a process for managing risk. http://www.iso.org/iso/home/standards/iso31000.htm

Lopes M.H.T. (2014) – The Mediterranean and the voices transported by time. In: Goffredo S., Dubinsky Z. (eds) The Mediterranean Sea. Its history and present challenges. Springer, Dordrecht: 553–557.

MAP Phase I 1975. http://195.97.36.231/dbases/webdocs/BCP/MAPPhascI_eng.pdf

MAP Phase II 1995. http://195.97.36.231/dbases/webdocs/BCP/MAPPhaseII_eng.pdf (http://www.unepmap.org/index.php?module=content2&catid=001001002)

Maritime Forum 2010 Global ship density. https://webgate.ec.europa.eu/maritimeforum/sites/maritimeforum/files/AIS_Issue3_AllShip_densityMap_0.pdf

Micheli F, Halpern BS, Walbridge S, Ciriaco S, Ferretti F, Fraschetti S, Lewison R, Nykjaer L, Rosenberg AA (2013). Cumulative human impacts on Mediterranean and Black Sea marine ecosystems: Assessing current pressures and opportunities. PlosONE 8(12): e79889.

MRAG, et al. (2013). Costs and benefits arising from the establishment of maritime zones in the Mediterranean Sea. Final Report MARE-2010-05.

MSP Platform website. http://www.msp-platform.eu/msp-practice/countries. Last accessed 23 March 2017.

Pegaso project website. http://www.pegasoproject.eu/index.html

Portman, M.E. (2015) Marine spatial planning in the Middle East: Crossing the policy-planning divide. Marine Policy 61, 8–15. http://dx.doi.org/10.1016/j.marpol.2015.06.025.

PRC (2008). EU DG MARE study 2008 The role of Maritime Clusters to enhance the strength and development in European maritime sectors. November 2008 Commissioned by the European Commission. Policy Research Cooperation. 38pp. https://ec.europa.eu/maritimeaffairs/documentation/studies/clusters_en.

PRC (2011). Final report. Exploring the potential of Maritime Spatial Planning in the Mediterranean Sea. Policy Research Corporation. Framework contract FISH/2007/04. Specific contract No 6. https://ec.europa.eu/maritimeaffairs/sites/maritimeaffairs/files/docs/body/msp-med_final_report_en.pdf

Robinson, A. R., Leslie, W. G., Theocharis, A., and Lascaratos, A. (2001). Mediterranean Sea Circulation, Encyc. Ocean. Sci., Academic Press, 1689–1706.

Sala, E. (2004). The past and present topology and structure of Mediterranean Ssubtidal rocky-shore food webs. Ecosystems 7: 333–340.

Salomidi M, Katsanevakis A, Borja A, Braeckman U, Damalas D, Galparsoro I, Mifsud R, Mirto S, Pascual M, Pipitone C, Rabaut M, Todorova V, Vassilopoulou V, Vega Fernández T (2012). Assessment of goods and services, vulnerability, and conservation status of European seabed biotopes: A stepping stone towards eco system-based marine spatial management. Mediterranean Marine Science 13(1): 49–88.

SEAS ERA Med (2012). Strategic Research Agenda for the Mediterranean Sea Basin. http://www.seas-era.eu/np4/%7B$clientServletPath%7D/?newsId=149&fileName=SEAS_ERA_D.7.1.1_Med_SRA.pdf

Suárez de Vivero, J.L., Rodríguez Mateos, J.C., (2015). Marine Governance in the Mediterranean Sea. In Gilek, M and Kern, K (eds): Governing Europe's Marine Environment: Europeanization of Regional Seas or Regionalization of EU Policies. Ashgate Publishing, 2015, pp. 203–224.

Suez Canal Authority (2017). http://www.suezcanal.gov.eg/English/About/SuezCanal/Pages/NewSuezCanal.aspx. Website last accessed 22 March 2017.

Suez Canal vessel traffic http://www.marinetraffic.com/en/ais/details/ports/17181/Egypt_port:SUEZ%20CANAL

Taghizadeh Moghaddam, H., Sayedi, S.J., Emami Moghadam, Z., Bahreini, A., Ajilian Abbasi, M. and Saeidi, M. (2017) Refugees in the Eastern Mediterranean Region: Needs, Problems and Challenges. International Journal of Pediatrics 5, 4625–4639.

Tardif, E. (2017). Migration Crisis in the Mediterranean: Reconciling Conflicting Agendas. Human Rights Brief.

TIMP (ongoing) Technion Israel Marine Plan. Israel Institute of Technology. http://msp-israel.net.technion.ac.il/en/ http://msp-israel.net.technion.ac.il/en/about/plan/

UfM (2015). Ministerial Declaration on Blue Economy. http://ufmsecretariat.org/wp-content/uploads/2015/11/2015-11-17-declaration-on-blue-economy_en.pdf

UNEP (1995a). Action Plan for the Protection of the Marine Environment and the Sustainable Development of the Coastal Areas of the Mediterranean (MAP Phase II), replacing the MAP of 1975. http://195.97.36.231/dbases/webdocs/BCP/MAPPhaseII_eng.pdf

UNEP (1995b). Convention for the Protection of the Marine Environment and the Coastal Region of the Mediterranean (Barcelona Convention). http://195.97.36.231/dbases/webdocs/BCP/bc95_Eng_p.pdf

UNEP (2008). Protocol on Integrated Coastal Zone Management in the Mediterranean. http://www.unep.org/NairobiConvention/docs/ICZM_Protocol_Mediterranean_eng.pdf

UNEP (2016). Athens Declaration. UNEP(DEPI)/MED IG.22/28. https://we docs.unep.org/rest/bitstreams/8363/retrieve

UNEP MAP website. http://www.unepmap.org/index.php?module=content2 &catid=001001001

UNESCO website (2017). Marine Spatial Planning Programme. Israel. http://msp.ioc-unesco.org/world-applications/middle-east/israel/. Website last accessed 31 March 2017.

Urgeles, R., Camerlenghi, A. (2013). Submarine landslides of the Mediterranean Sea: Trigger mechanisms, dynamics, and frequency-magnitude distribution. Journal of Geophysical Research: Earth Surface 118: 2600–2618.

US Geological Survey Fact Sheet 2010–3014, March 12, 2010. https://pubs. usgs.gov/fs/2010/3014/pdf/FS10-3014.pdf

Vella, K. (2015). Speech – 16 November 2015 Union for the Mediterranean (UfM) Ministerial Conference on Blue Economy, in Brussels. https://ec. europa.eu/commission/2014-2019/vella/announcements/union-mediterr anean-ufm-ministerial-conference-blue-economy-brussels_en

VT Explorer website. http://www.vtexplorer.com/vessel-tracking-strait-of-gibraltar.html. Website last accessed 22 March 2017.

13

Regulation and Planning in the Sea Basins – The Caribbean Basin

Irati Legorburu* and Kate Johnson

Heriot-Watt University, Scotland
*Corresponding Author

13.1 Introduction and Geography

Located in the American continent, the Caribbean Sea is an arm of the Atlantic Ocean (Figure 13.1). Its geographical boundaries are: (i) the islands of the West Indies to the N and E; (ii) South America to the S; and (iii) the Central American Isthmus to the W. Extending between 9–22°N and 61–88°W, it is considered one of the largest seas in the world. It connects to the Gulf of Mexico through the Yucatan channel, to the Pacific through the Panama canal and to the Atlantic by multiple straits. It is divided into two main sub-regions: (i) the continental Caribbean, formed by the countries from North, Central and South America bordering its waters; and (ii) the insular Caribbean, formed by the Greater Antilles (Cuba, Hispaniola – containing Haiti and Dominican Republic-, Jamaica and Puerto Rico) and the Lesser Antilles (Islands between the southeast of Puerto Rico and the north coast of Venezuela). The economy of many of these countries is based principally in the exploitation of their natural resources, the tourism and fisheries sectors are critical activities for the economic development of the area.

The complex political structure (including sovereign countries, outermost regions and overseas countries territories) is reflected in the maritime governance of the region. Although the sea is one of the main resources for the economic, social and cultural development of most of these countries, the many initiatives and regional governance commissions (some examples are given in Table 13.1), are highly uncoordinated and fragmented, resulting often in duplication and ineffectiveness. However, three integrative ocean

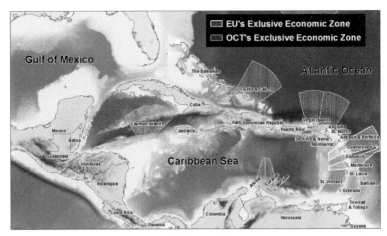

Figure 13.1 Caribbean Sea Basin. [OCT = EU Overseas Countries and Territories] [OR = Outermost Region].

Table 13.1 List of countries and organisations

Country	STATUS	ECLAC	ACS-CSC	CARIFORUM	CARICOM	OECS
Anguilla	OCT (UK)	*		‡		*
Antigua & Barbuda	Sovereign					
Aruba	OCT (NL)	*	*	‡		
Barbados	Sovereign					
Bonaire	OCT (NL)		*	‡		
British Virgin Islands	OCT (UK)	*		‡		*
Cayman Islands	OCT (UK)	*		‡		*
Curaçao	OCT (NL)	*	*	‡		
Dominica	Sovereign					
Grenada	Sovereign					
Guadeloupe	OR (FR)	*	*	‡		
Martinique	OR (FR)	*	*	‡		
Montserrat	OCT (UK)	*		‡		
Saba	OCT (NL)		*	‡		
St. Barthelemy	OCT (FR)		*			
St. Kitts & Nevis	Sovereign					
St. Lucia	Sovereign					
St. Martin	OR (FR)		*			
St. Vincent and the Grenadines	Sovereign					
St. Eustatius	OCT (NL)		*	‡		
St. Maarten	OCT (NL)	*	*	‡		
Trinidad & Tobago	Sovereign					
Turks and Caicos Islands	OCT (UK)	*		‡		*

* Associate country; ‡ Observer country.

management policy frameworks can be considered of relevance under the scope of this chapter.

1. **Caribbean Sea Initiative** – establishes the basis for a regional maritime governance framework through the following fields for action: marine pollution; coastal and marine resources management; climate change and disaster risk reduction; social and economic development; sustainable ocean governance; and human capacity development. Established by the Association of Caribbean States (ACS), the Caribbean Sea Commission (CSC) has been recognised as the body that can potentially provide policy harmonisation and coordination for the achievement of the objectives established by the Initiative (UN, 2014).

2. **Eastern Caribbean Regional Ocean Policy (ECROP)** – adopted by the Organisation of Eastern Caribbean States (OECS), the ECROP has the following policy goals (OECS, 2013): secure access to marine resources; maintain and improve ecosystem integrity; promote social and economic development; adopt multi-use ocean planning and integrated management; promote public awareness, participation and accountability; support research and capacity building; and, building resilience and managing for uncertainty. Although it is not a legally binding document, its guiding principles are based on international law.

3. **Integrated Maritime Policy** – Although ORs (Guadeloupe, Martinique and Saint Martin) are full members of the EU, given their exceptional situation (geographical, economic, etc.) the application of the European marine legislation within their EEZs differs in comparison with the remainder of European Sea basins (i.e., Atlantic, Baltic, Mediterranean). However, ORs not only have a great potential for maritime activities, but due to their geographical location they provide a global dimension to the maritime space of the EU. Being aware of this, the IMP seeks to promote and facilitate the maritime development of these regions, as well as to increase cooperation at the regional level (EC, 2007). Among the challenges for the region the following must be noted: adaptation of the IMP and its action plans to the specific characteristics of the ORs (including Blue Growth, Marine Data and Knowledge, Maritime Spatial Planning, Integrated Maritime Surveillance, and Sea Basin Strategies); increase the knowledge of the maritime affairs and marine environment; creation of maritime research networks; promotion of regional maritime governance policies; and, enforce maritime relations with neighbouring countries (EC, 2008a).

As ORs of France, Guadeloupe, Martinique and Saint Martin are full members of the EU and as such, their waters are under the EU's jurisdiction. Further, given its colonial past, the EU maintains close ties and strategies in terms of cooperation and development with the region, which are of particular relevance in the case of the Lesser Antilles. In fact, many of them are classified as Overseas Countries Territories (OCT), which gives them a special relationship status with some of the EU's Member States (Table 13.1). Considering that, this chapter will focus in the countries with which the EU has some relevant policy relationship (i.e., Lesser Antilles). Recognising their economic and geographical inequalities (remoteness, insularity, small surface, economic dependence...) the Strategy for the ORs, sets as main objectives for these regions the reduction of their accessibility deficit, the increase of their competitiveness and the strengthening of their regional integration. Given their close dependence with their coastal resources, many of the proposed measures closely link to the management of the marine resources of the region (EC, 2012a). In the case of OCTs and remaining islands, the development and the sustainable use of marine resources is generally one of the central pillars of the signed agreements. Among these, the Joint EU-Caribbean Strategy can be highlighted. Adapting the bases of the Cotonou agreement to the specific characteristics of the signatory Caribbean countries (CARIFORUM), this strategy aims to promote the economic, social and cultural development of the region (EC, 2012b). Again, the achievement of these objectives is closely linked to the management of marine resources (e.g., development of renewable energy, food security, promotion of the blue economy, protection of marine habitats, etc.).

13.2 Current and Planned Environmental Policies

Adopted in 1983 the Convention for the Protection and Development of the Marine Environment of the Wider Caribbean Region (Cartagena Convention) is the main framework for the protection of the marine environment in the Caribbean (UNEP, 2012a). The Convention with its three protocols (Annex I), provides the legal basis for the implementation of the Action Plan of the Caribbean Environment Programme (CEP), which aims to promote regional cooperation in different aspects related to the protection and development of the marine environment. This includes: land-based pollution; fisheries management; critical habitats; urbanization and coastal development; agriculture and forestry; sustainable tourism; oil spills; and, capacity-building (Parris, 2013). To carry out their actions the CEP consists of 3 sub-programmes

(Table 13.2). At the OECS level, the St. George's declaration adopts the requirements of the Cartagena convention and establishes the benchmark for environmental management (OECS, 2006). Composed of 21 principles, the declaration gives to environmental management a central role in the socio-economic development of OECS countries (Table 13.2). The declaration

Table 13.2 Summary of environmental conventions in the Caribbean and OECS

Caribbean Environment Programme (Regional)	
Sub-Programme	Actions
Assessment and Management of Environmental Pollution (AMEP)	Control, prevent and reduce marine pollution: Coordination of LBS and Oil Spill protocols (Annex I)
Specially Protected Areas and Wildlife (SPAW)	Achievement of SPAW Protocol goals and: (i) increase and support the development of protected areas; (ii) support the conservation of endangered species and promote the sustainable use of natural resources; and (iii), coordinate and develop synergies with other initiatives related to the conservation of biodiversity (e.g., Convention on Biological Biodiversity, Ramsar Convention, International Coral Reef Initiative, etc...)
Communication, Education Training and Awareness (CETA)	Increase public awareness, involvement and skill in order to provide timely and efficient responses to problems
St. George Declaration (OECS)	
Goal	Principles
Build the capacity of Member States and Regional Institutions to guide and support processes of sustainable development	**P.1**: Integrate social, economic and environmental considerations into national development policies, plans and programmes. **P.3**: Improve on legal and institutional frameworks. **P.8**: Address the causes and impacts of climate change. **P.15**: Promote cooperation in science and technology.
Incorporate the objectives, perspectives, resources and talents of all of society in environmental management	**P.4**: Ensure meaningful participation by civil society in decision making. **P.5**: Ensure meaningful participation by the private sector. **P.7**: Foster broad-based environmental education, training and awareness. **P.15**: Promote cooperation in science and technology.

(*Continued*)

Table 13.2   Continued

St. George Declaration (OECS)	
Goal	Principles
Achieve the long-term protection and sustained productivity of the region's natural resource base and the ecosystem services it provides	**P.10**: Prevent and control pollution and manage waste. **P.11**: Ensure the sustainable use of natural resources. **P.12**: Protect cultural and natural heritage. **P.13**: Protect and conserve biological diversity. **P.16**: Manage and conserve energy.
Ensure that natural resources contribute optimally and equitably to economic, social and cultural development	**P.6**: Use economic instruments for sustainable environmental management. **P.8**: Address the causes and impacts of climate change. **P.9**: Prevent and manage the causes and impact of disasters. **P.14**: Recognise relationships between trade and environment.

is implemented by signing countries through their National Environmental Management Strategies (NEMS).

Despite this framework for the protection of the environment, the Caribbean has been subjected to a continued deterioration of its natural resources (overexploitation of resources, loss of biodiversity, increased pollution, climate change...). Pushed by civil society and being more aware of the great socio-economic potential of their natural resources, the Governments of the area have begun to adopt cross-sectoral approaches, giving a higher importance to the protection of the environment and its resources. However, this change of focus has not resulted in effective environmental protection. The main reason for this failure is the structure of the economic development model itself, which prioritises sectoral economic policies over sustainable development in the region (and possibly increasing social inequalities) (UNEP, 2012b).

13.3 Regulatory Requirements

Annex I addresses the main global regulations of relevance for the maritime sectors addressed in this book. In addition, Table 13.3 shows specific policy frameworks in relation to the sector combinations considered for the Caribbean.

Table 13.3   Caribbean sector specific policies

Sector	Policy/Agreement	Implementing body
Fisheries & Aquaculture	Caribbean Community Common Fisheries Policy (evolving)	Caribbean Regional Fisheries Mechanism (CRFM)
Tourism	Common Tourism Policy	OECS
	Caribbean Regional Sustainable Tourism Development Programme	Caribbean Tourism Organisation (CTO)
Transportation & Trade	Revised Treaty of Chaguaramas (CARICOM Single Market and Economy)	CARICOM
	EU-Caribbean Economic Partnership Agreement	EU-CARIFORUM
Renewable energy	Regional Energy Policy	CARICOM
	Eastern Caribbean Energy Regulatory Authority (ECERA; evolving)	OECS

13.4 Spatial Requirements, Conflicts and Planning/Policy

As a result of the strong dependence on maritime activities, competition and conflicts for space and marine resources are a common issue in the Caribbean (especially on small islands). This is enhanced by the proximity between countries, as the transboundary nature of the uses and resources is added to the interaction between the different uses of the sea (fishing, tourism, energy, etc.) and different factors such as natural disasters or climate change. Driven by increasing maritime activities, two main types of conflicts predominate, which weaken the goods and service provision capacity of coastal zones: (i) conflicts between human uses (user–user); and (ii) conflicts between human uses and the marine environment (user-environment) (Pomeroy et al., 2014). This has led to movement towards integrated management approaches of the marine environment (e.g., ecosystem based management, integrated coastal zone management), including marine spatial planning (MSP). Table 13.4 shows some of the major initiatives in the insular region of the Caribbean, which generally focus on the management and reduction of conflicts with fisheries. Despite these initiatives, the comprehensive implementation of MSP in the region presents a series of challenges. These, relate closely to the complex geopolitical structure and lack of political will (Pomeroy et al., 2014).

- Limitations of governance mechanisms: the different governance frameworks in the region (e.g., CSC, ECROP, IMP) advocate the use of MSP approaches. For example, one of the major policy goals of ECROP is the

Table 13.4   Examples of MSP actions developed in the Caribbean

Country	MSP in Place	Activities Covered	Comments	Links
St. Kitts and Nevis	Pilot project	Fishing; Conservation; Tourism; Transportation; Recreation; Development & Planning	Initiative/partnership: USAID and The Nature Conservancy	http://www.marineplanning.org/pdf/ StKitts_Nevis_Full_Report.pdf
Grenadine Islands	Conceptual framework	Fishing; Tourism/Recreation; Transportation/Industrial; Conservation; Mariculture	Initiative/partnership: Sustainable Grenadines Inc., NOAA, CERMES, The Nature Conservancy	http://www.grenadinesmarsis.com/ uploads/Baldwin_MZP_FinalReport .pdf
Barbuda	Pilot project	Fishing; Conservation; Recreation; Offshore energy; Aquaculture; Transportation	Initiative/partnership: Blue Halo initiative (Waitt Institute and Seasketch)	http://barbuda.waittinstitute.org/
Curaçao	Pilot project	Fishing; Conservation; Recreation; Offshore energy; Aquaculture; Transportation	Initiative/partnership: Blue Halo initiative (Waitt Institute and Seasketch)	http://curacao.waittinstitute.org/
Montserrat	Pilot project	Fishing; Conservation; Recreation; Offshore energy; Aquaculture; Transportation	Initiative/partnership: Blue Halo initiative (Waitt Institute and Seasketch)	http://montserrat.waittinstitute.org/
Wider Caribbean Region	Study to support the CBD convention (Marine mammals)	Conservation (habitats & species distribution); Fishing; Shipping; Land-based non-point organic pollution	Initiative/partnership: UNEP-Spain LifeWeb Project (Broad-scale Marine Spatial Planning of Mammal Corridors and Protected Areas in Wider Caribbean and Southeast & Northeast Pacific)	file:///C:/LocalStore/il6/Downloads/ Report%20on%20the%20LifeWeb- Spain%20UNEP-CEP%20 Meeting...%20UNEP(DEPI)- CAR%20WG.36-INF.8- en%20(1).pdf

adoption of multiple-use ocean planning as a tool for the management of maritime areas. However, these governance mechanisms lack, at least so far, sufficient strength and capacity for implementation.

- Limitations on basic geographic data: collection and dynamic integration of reliable spatial data on the activities, objectives and possible changes (e.g., growth, climate change) at different spatio-temporal scales is crucial for the effective implementation of the MSP.
- Involvement of authorities and stakeholders: the joint collaboration of marine stakeholders and authorities is required in order to obtain a complete picture of the issues and conflicts that may arise between uses of the marine environment.
- Financial resources: most of the MSP initiatives are being carried out with funds from foreign projects. Although these foreign initiatives may be valuable as a way for introducing MSP practices, the long-term sustainability of the approach requires national/local interest, support and funding.

13.5 Support Schemes

13.5.1 Support Programmes

The joint European-Caribbean strategy establishes an Economic Partnership Agreement (EPA) between the EU and the countries of CARIFORUM (EC, 2008b). The European Development Fund (EFD) is the main financing mechanism in the context of this partnership, which aims to fund projects for the economic, social and human development of the region. Similarly, Regulation No. 233/2014 establishes a financing instrument for development cooperation for the period 2014–2020 (EC, 2014). It provides a priority role to issues such as the promotion of renewable energies, strategies for employment creation, the preservation of the environment or food security. Two additional funding and support mechanisms which of interest for different maritime sectors are shown below.

1. **Caribbean Investment Facility** – based on the objectives of the EU-CARIFORUM agreement the facility mobilises resources for strategic economic projects and the support of the private sector. The investment priorities are: (i) improvement of transport and energy infrastructures (interconnectivity, security, efficiency, etc.); (ii) improvement of access to ICT infrastructure; (iii) establish better infrastructures for water and

sanitation; (iv) promote infrastructure for the prevention of natural disasters; and, (v) improvement of social infrastructure. The support is provided in the form of investment grants, technical assistance, risk capitals and other risk sharing instruments.

2. **European Investment Bank** – financing from the EIB areas include: (i) strengthening of local financial sector; (ii) credit lines and financial contributions for SMEs; (iii) projects of sustainable infrastructure in the sectors of energy, transport, water and telecommunications; (iv) industrial activities (e.g., manufacturing, mining); (v) expansion of the service sector, including tourism; (vi) food security; and (vii), climate change mitigation and adaptation measures. The main investment instruments are: subordinated loans; quasi-equity funding; equity funding; guarantees; senior loans; intermediate loans; technical assistance; and, interest rate subsidies.

13.5.2 Subsidies

Agreed by CARICOM's Member States, the Caribbean Single Market Economy establishes a strategy for cooperation, integration and economic competitiveness of the signing parties. In addition to the rules for trade between Member States, it sets the conditions for subsidies (mainly oriented to import/export activities). The Agreement prohibits direct government subsidies which may involve a disadvantage for the industries from other signing countries. However and always in a justified manner, it also establishes a series of general exceptions under which government aid would be permitted. Among the exceptions potentially linked to Blue Growth, would be those related to the prevention and relief of food shortages and the conservation of natural resources and the environment. In the same way, the Regional Energy Policy (CARICOM, 2013) advocates the phasing out and rationalisation of fossil-fuel subsidies in order to enhance the competitiveness of renewables.

13.6 Key Considerations

The policy framework in the Caribbean is extremely complex. The region is characterised by a large number of small neighbouring countries, which have different sovereignty levels (ORs, OCTs, and sovereign countries) and important socio-economic inequalities. Despite the large amount of institutions focused on the cooperation between countries and their development, they have limited implementation capacity. This is due largely to the colonial past of the region, which favoured the development of an economic model based on the massive exploitation of the rich natural resources and the strong dependency on the colonial powers. As a result, the economies of the countries studied in this report are characterised by the lack of industrial fabric and dependence on natural resources, being especially important those provided by the sea (e.g., fishing, tourism).

The EU maintains close ties with the region in terms of cooperation and development, for which the sustainable management and exploitation of marine resources plays an essential role. Objectives of these agreements such as food and energy security, improvement of water and sanitation infrastructures, sustainable tourism or the eradication of poverty through employment generation, relate directly with many of the sectors studied in this book (e.g., fisheries/aquaculture, marine renewable energies, tourism, desalination systems, etc.). In addition, the probability for spatial conflicts (user-user and user-environment) is increased by the small size of the countries and the great amount of activities carried out in their maritime space. In this context, combining technologies in the same marine space decreases conflict between users, and simultaneously facilitates better management of this important resource.

The EU-CARIFORUM Economic Partnership Agreement (EPA), provides a series of investment opportunities. The partnership aims to promote trade and investment, facilitating the access to markets of both signing partners. Although the Agreement considers some specific national reserves, it addresses directly sectors such as fisheries, mining, oil & gas, renewables and services (including transportation and tourism). Thus, it provides a good starting point for both the development of BG sectors and the socio-economic development of the region. In the same vein, the Eastern Caribbean Regional Ocean Policy (OECS, 2013), highlights specifically the need for the adoption of multiple-use ocean planning approaches. Again, given the special relationship between OECS countries and the EU (either through the EPA or because of their Overseas Country Territory (OCTs) status), Caribbean Small Developing Islands appear as a suitable location for the development of Multi-use-of space combinations and Multi-use platforms.

References

CARICOM (2013). Caribbean Community Energy Policy. Caribbean Community Secretariat, Trinidad and Tobago, March 2013.

EC (2007). Communication from the Commission to the European Parliament, the Council, the European Economic and Social Committee and the Committee of the Regions. An Integrated Maritime Policy for the European Union. COM(2007) 575 final. Commission of the European Communities, Brussels, 10.10.2007. 16 pp.

EC (2008a). Communication from the Commission. The outermost regions: an asset for Europe. COM(2008) 642 final. Brussels, 17.10.2008. 8 pp.

EC (2008b). Economic Partnership Agreement between the CARIFORUM States, of the one part, and the European Community and its Member States, of the other part. Official Journal of the European Union, L 289. Brussels, 30.10.2008.

EC (2012a). Communication from the Commission. The outermost regions of the European Union: towards a partnership for smart, sustainable and inclusive growth. COM(2012) 287 final. European Commission, Brussels, 20.6.2012. 18 pp.

EC (2012b). Council Conclusions on the Joint Caribbean-EU Partnership Strategy. Council of the European Union, Brussels, 19.11.2012. 15 pp. (http://www.consilium.europa.eu/uedocs/cms_data/docs/pressdata/en/foraff/133566.pdf).

EC (2014). Regulation (EU) No. 233/2014 of the European Parliament and of the Council of 11 March 2014 establishing a financing instrument for development cooperation for the period 2014–2020. Official Journal of the European Union, L77. Brussels, 15.3.2014.

OECS (2006). St George's Declaration of Principles for Environmental Sustainability in the OECS. Organisation of Eastern Caribbean States. 37 pp.

OECS (2013). Eastern Caribbean Regional Ocean Policy. Organisation of Eastern Caribbean States Secretariat, Saint Lucia. 72 pp.

Parris, N. N., (2013). Towards an Ocean Policy for Integrated Governance of the Caribbean Sea and the Sustainable Development of the Wider Caribbean Region (WCR): What could it look like and how would it work? Division for Ocean Affairs and the Law of the Sea. Office of Legal Affairs, United Nations, New York. 188 pp.

Pomeroy, R. S., Baldwin, K., McConney, P., (2014). Marine Spatial Planning in Asia and the Caribbean: Applications and Implications for Fisheries and Marine Resource Management. Desenvolvimento e Meio Ambiente 32: 151–164.

UN (2014). Towards the sustainable development of the Caribbean Sea for present and future generations. Report of the Secretary-General. United Nations General Assembly. Report A/69/314, 20 pp.

UNEP (2012a). Convention for the Protection and Development of the Marine Environment of the Wider Caribbean Region and its Protocols. Cartagena Convention Booklet, United Nations Environment Programme, Kingston. 126 pp.

UNEP (2012b). Chapter 12: Latin America and the Caribbean. In: GEO 5, Global Environment Outlook, Environment for the future we want. United Nations Environment Programme. pp. 317–348.

Appendix 1

Master Document of Global and Basin-Specific Regulations

International Conventions	General Objectives	Relevance to Specific Marine Sectors (if any)
United Nations Convention on the Law of the Sea (UNCLOS)	Enables the creation of maritime jurisdictional zones, outlines States roles and responsibilities, general duty to protect the marine environment.	MRE: States have exclusive rights, to exploit their renewable energy resources and to construct, authorise and regulate the construction, operation and use of artificial islands and of installation and structures to exploit those resources. States can create safety zones around installations. General duty to protect the environment e.g. remove installations after use.
Convention for the Protection of the Marine Environment of the North-East Atlantic (OSPAR)	Prevention and elimination of all kinds of pollution and covers all human activities that might adversely affect the marine environment of the North-East Atlantic.	The Biological Diversity and Ecosystems Strategy includes a list of the human activities that can adversely affect the marine environment. The considered impacts are related with dredging, sand and gravel extraction, offshore wind farms, cables and pipelines and underwater noise. OSPAR assesses those activities and, if necessary, develops programmes and measures to control those activities and to restore adversely affected marine area.

		Also produced offshore wind guidance and EIA guidance.
Convention on the Protection of the marine environmental of the Baltic Sea (HELCOM)	Reduction and prevention of land-based pollution.	Covers all marine sectors with five permanent and 3 temporary working groups on specific issues e.g. MSP. Working on the development of a new HELCOM Recommendation on sustainable aquaculture, a regional action plan for underwater noise, etc. Guidelines for sustainable and environmentally friendly coastal tourism. HELCOM Recommendation 34E/1 "Safeguarding important bird habitats and migration routes in the Baltic Sea from negative effects of wind and wave energy production at sea".
Convention for the Protection of the Mediterranean Sea Against Pollution (Barcelona)	Protection of the marine environment of the Mediterranean Sea against pollution.	Supplemented by seven protocols on Dumping from ships and aircraft; Prevention and Emergency Protocol (pollution from ships and emergency situations); Land-based Sources and Activities Protocol; Specially Protected Areas and Biological Diversity Protocol; Offshore Protocol (pollution from exploration and exploitation); Hazardous Wastes Protocol; Protocol on Integrated Coastal Zone Management (ICZM). Guidelines: Dumping of Platforms and other Man Made Structures at Sea.

Convention for the Protection and Development of the Marine Environment of the Wider Caribbean Region (Cartagena)	Umbrella agreement for the protection and development of the marine environment. Convention is supported by three additional technical agreements or Protocols on Oil Spills, Specially Protected Areas and Wildlife (SPAW) and Land Based Sources of Marine Pollution (LBS).	Contains provisions aimed at preventing, reducing and controlling pollution from ships, pollution caused by dumping, pollution from sea-bed activities, airborne pollution and pollution from land-based sources and activities. Parties are required to take measures to protect and preserve rare or fragile ecosystems, habitats of depleted, threatened or endangered species; and to develop technical and other guidelines for the planning and environmental impact assessments of important development projects in order to prevent or reduce harmful impacts within the Wider Caribbean Region.
Convention on Biological Diversity	Conservation of biodiversity sustainable use of species and natural habitats	Environmental Impact Assessment. Protected Areas. Underwater noise and its impacts. Invasive species. Marine Spatial Planning (MSP).
Cartagena Biosafety Protocol (2000) and its Supplementary Protocol on Liability and Redress (2010)	Aims to ensure the safe handling, transport and use of living modified organisms (LMOs) resulting from modern biotechnology that may have adverse effects on biological diversity, taking	The Protocol seeks to protect biodiversity from the potential risks posed by living modified organisms resulting from modern biotechnology.

	also into account risks to human health	
UNESCO Convention on the Protection of Underwater Cultural Heritage	Protection of underwater cultural heritage	State Parties must use the best practicable means to prevent or mitigate any adverse effects that might arise from activities incidentally affecting underwater cultural heritage.
UN Framework Convention on Climate Change (UNFCCC)	Stabilisation of greenhouse gas concentrations	For MRE: Reduction of greenhouse gases, Clean Development Mechanism, Technology Mechanism
IMO Protocols and Resolutions	Measures to regulate marine traffic and its operation.	Ships routing, Safety of navigation around offshore installations and structures, Decommissioning of offshore structures
International Convention for the Prevention of Pollution from Ships (MARPOL)	Prevention and control of pollution from ships	MRE: Article 1 generally applicable to 'ships' servicing energy installations
The Convention on the International Regulations for Preventing Collisions at Sea (COLREGs)	Regulation of international marine traffic	Traffic separation schemes, navigation schemes etc.
Convention on International Civil Aviation	Convention establishes rules of airspace, aircraft registration and safety, and details the rights of the signatories in relation to air travel.	OW: Offshore wind turbine heights, location and lighting, implications for radar and aerial navigation

International Convention for the Safety of Life at Sea (SOLAS)	Safety at sea, security of ships, safety on board	Chapter V, Safety of Navigation
Convention on Migratory Species (CMS or Bonn Convention)	Aims to conserve terrestrial, aquatic and avian migratory species throughout their range. Strict protection of species facing extinction involving the conservation or restoration of the places where they live, mitigation of obstacles to migration and control of other factors that might endanger them	MRE: Implications of renewable energy for migratory species (CMS Draft Resolution: Renewable Energy and Migratory Species (CMS and ASCOBANS), 8th Sept. 2014) Resolution 7.5 (2002) addresses the impact of wind turbines on migratory species.
Convention on Wetlands (Ramsar)	Designation of wetlands of international importance	Guidance on how to consider wetlands in planning and operating energy infrastructure
Agreement on the Conservation of Cetaceans of the Mediterranean and the Black Sea and contiguous Atlantic Area (ACCOBAMS)	Reduction of threats to cetaceans in specified waters	Resolution 4.17 on Guidelines to address the impact of anthropogenic noise on Cetaceans in the ACCOBAMS area adopted. Man-made noise is recognised as a form of pollution.
Berne Convention on European Wildlife and Habitats	To conserve wild flora and fauna and their natural habitats, especially species and habitats whose	States must consider the conservation of wild flora and fauna in their planning and development policies, and in their measures against pollution. In the

	conservation requires the cooperation of several States.	EU the Natura 2000 network (Recommendation No. 16 (1989) of the Standing Committee to the Berne Convention/Emerald Network) implements the Berne Convention.
Agreement on the Conservation of Small Cetaceans of the Baltic, North East Atlantic, Irish and North Seas (ASCOBANS)	Reduction of threats to cetaceans in specified waters	See under Bonn Convention. The 2009 ASCOBANS Resolution No 2 of the 6th Meeting of the Parties sets up a range of recommendations applying to offshore construction activities for renewable energy production.
Agreement on the Conservation of Albatrosses and Petrels (ACAP)	Seeks to conserve albatrosses and petrels by coordinating international activity to mitigate known threats to their populations	One of the agreements under CMS / Bonn Convention: see above.
Espoo Convention on EIA	Obligation on States to notify and consult each other on all major projects under consideration that are likely to have a significant adverse environmental impact across boundaries.	All articles relevant. Guidance on the Practical Application of the Espoo Convention. Guidelines on public participation in transboundary EIA.
Stockholm Convention on Persistent Organic Pollutants (POPs)	Treaty to protect human health and the environment from chemicals that remain intact in the	Requires parties to take measures to eliminate or reduce the release of POPs into the environment. Restricts or bans the use of certain chemicals that have been used in

	environment for long periods	marine operations in the past. PCBs can be used in existing equipment until 2025.
Cooperation Agreement for the Protection of the Coasts and Waters of the North-East Atlantic against Pollution (Lisbon Agreement)	To be prepared to deal with an incident of pollution at sea such as pollution caused by hydrocarbons or other harmful substances.	Ensures cooperation between States if there is pollution of the marine environment by hydrocarbons and other harmful substances in the area covered by the Agreement. Hydrocarbons cover oil in all its forms in particular crude oil, fuel oil, muds, hydrocarbon residues and other refined products. 'Other harmful substances' means all substances other than hydrocarbons, including hazardous waste, the release of which into the marine environment may be harmful to human health, ecosystems or living resources, coasts or the related interests of the Parties.
EUROPEAN	General Provisions	Relevance to Specific Marine Sectors (if any)
Environmental Impact Assessment Directive (85/337/EEC, 2014/52/EU)	Evaluation of environmental impact of a project at site level	Relevant to all activities listed in Annex I or Annex II and requires an assessment of the environmental effects of projects that are likely to have significant effects on the environment.
Strategic Environmental Impact Assessment Directive (2001/42/EC)	Evaluation of environmental impact of a plan or programme at strategic level	SEA requires an assessment of public plans and programmes which are likely to have significant effects on the environment.

Habitats Directive (92/43/EEC)	Conservation of natural habitats and wild flora and fauna	Provides for the protection of certain habitats and species. Enables Special Areas of Conservation to be designated to protect certain habitats and species. Development in such areas may be subject to additional consenting requirements i.e. Article 6, Appropriate Assessment.
Wild Birds Directive (2009/147/EC)	Conservation of wild birds	Provides for the protection of certain bird habitats and species. Enables Special Protection Areas to be designated to protect certain species. Directive operates in conjunction with the Habitats Directive.
Water Framework Directive (2000/60/EC)	Protection of inland and coastal waters	Coastal and marine activities should not impact negatively on inland and coastal waters or those covered by a River Basin Management Plan.
Marine Strategy Framework Directive (2008/56/EC)	Achieve good environmental status by 2020	Coastal and marine activities should not impact negatively on marine waters or those covered by a Marine Strategy with specific POMs identified.
Floods Directive (2007/60/EC)	Reduce and manage the risks that floods pose to human health, the environment, cultural heritage and economic activity.	Member States required to take adequate and coordinated measures to reduce flood risk.
Renewable Energy Directive (2009/28/EC)	Promotion of energy from renewable resources	Member States required to specify their renewable energy targets and their anticipated energy mix which may include offshore wind or wave and tidal.

Directive on Animal Health Requirements (2006/88/EC)	Prevention of diseases in aquaculture species	Applies to aquaculture animals and products thereof. Governs health monitoring of finfish, shellfish and crustaceans and puts in place controls on the movement of potential vector and susceptible species. It also provides a structure for declaring the health status of Member States and areas within them.
Bathing Waters Directive (2006/7/EC)	To safeguard public health and ensure clean bathing waters	Member States manage bathing water quality in association with the WFD and may take measures to improve bathing water quality which may impact upon certain marine and coastal activities.
Waste Framework Directive (2008/98/EC)	Regulation of waste management, recycling, recovery	Directive requires that waste be managed without endangering human health and harming the environment, and in particular without risk to water, air, soil, plants or animals, without causing a nuisance through noise or odours, and without adversely affecting the countryside or places of special interest.
Urban Wastewater Directive (91/271/EEC, 98/15/EEC etc.)	Objective is to protect the environment from the adverse effects of urban waste water discharges and discharges from certain industrial sectors	Requires pre-authorisation of all discharges of urban wastewater, of discharges from the food-processing industry and of industrial discharges into urban wastewater collection systems. Covers the fish processing sector. Covers transitional (estuarine) and coastal waters.

SEVESO III Directive (2012/18/EU)	Prevention of major accidents involving dangerous substances and also limiting the consequences of such accidents for human health and the environment.	Covers establishments where dangerous substances may be present (e.g. during processing or storage) in quantities above a certain threshold. Includes petroleum products and alternative fuels.
Industrial Emissions Directive (2010/75/EU)	Commits Member States to control and reduce the impact of industrial emissions on the environment	Replaces IPPC Directive though the same requirements still apply. Applies to industrial installations and their emission to air, soil, water etc.
Public Participation Directives (2003/4/EC, 2003/35/EC etc.)	Public access to environmental information and participation in decision-making	The public must have access to environmental information and that such information is accessible to the public.
Directive on environmental liability with regard to the prevention and remedying of environmental damage (2004/35/EC)	Establishes a framework of environmental liability, based on the polluter-pays principle, to prevent and remedy environmental damage.	Covers damage to protected species and natural habitats, damage to water and damage to land. Damage caused and financial consequences will be borne by the economic operator who caused the harm (doesn't include third party right to compensation).
Directive on safety of offshore oil and gas operations (2013/30/EU)	Contains rules to help prevent accidents.	Amends the Environmental Liability Directive by extending its scope of damage to marine waters. For oil and gas, before exploration or production begins, companies must prepare a Major Hazard Report for their offshore installation. National authorities must verify safety provisions, environmental protection measures, and the emergency preparedness of rigs and platforms. 'Offshore'

		means situated in the territorial sea, the Exclusive Economic Zone or the continental shelf of a Member State within the meaning of UNCLOS.
Directive concerning common rules for the internal market in natural gas (2009/73/EC)	Establishes common rules for the transmission, distribution, supply and storage of natural gas.	Applies to natural gas and includes biogas, gas from biomass and LNG.
Directive imposing an obligation on Member States to maintain minimum stocks of crude oil and/or petroleum products (2009/119/EC)	Imposes an obligation on Member States to maintain minimum stocks of crude oil and/or petroleum products.	International marine bunkers are not included in the calculation of stock levels. Member States must be able to ensure that the total oil stocks maintained at all times within the Community for their benefit correspond, at the very least, to 90 days of average daily net imports or 61 days of average daily inland consumption, whichever of the two quantities is greater.
Directive on the conditions for granting and using authorisations for the prospection, exploration and production of hydrocarbons (94/22/EC)	Prescribes rules to help to reinforce the integration of the internal energy market, encourage greater competition within it and improve security of supply.	Aim is to prevent a single entity from having exclusive rights for an area whose prospection, exploration and production can be carried out more effectively by several entities. Procedures for granting authorisations in Member States must be introduced in a transparent manner based on objective, non-discriminatory criteria.

PART IV

Combining Uses

14

Multi Use Platforms (MUPs) and Multi Use of Space (MUS)

**Gordon Dalton[1,*], Kate Johnson[2]
and Ian Masters[3]**

[1]University College Cork, Ireland
[2]Heriot-Watt University, Scotland
[3]Swansea University, Wales
*Corresponding Author

14.1 Introduction

Our oceans are important drivers of economic growth. They provide natural resources, access to trade and transport and opportunities for leisure activities. As maritime activity increases, however, so does the competition for space as coastal areas become overcrowded. This led the European Commission to publish a call in 2014 asking researchers to prepare for the 'future innovative offshore economy' (BG5 2014). Expecting economic activities to move further offshore as competition for space increased, this call was designed to promote smarter and more sustainable use of our seas. It was in response to this call that the project "Marine investment in the blue economy" (or Maribe www.maribe.eu) was initiated with the aim of promoting growth and jobs within the blue economy. The Maribe project started in March 2015, with a duration of 18 months and a total budget of 2 million euros under the European Commission's Horizon 2020 programme, it was led by the MaREI Centre in University College Cork (www.marei.ie). A total of 11 partners contributed to the project from Ireland, United Kingdom, Belgium, Spain, Italy, Malta and the Netherlands, including FAO, who add an international extra-EU dimension to the consortium.

The primary objective of Maribe was to promote smarter and more sustainable use of the sea through the sharing of space. It investigated the potential of combining maritime sectors in the same place (Multiple-Use-of-Space (MUS)) or on a specifically built platform (Multi Use Platform (MUP)) in order to make more efficient use of space and resources. It paid particular attention to new and emerging industries featured in the other chapters of this book that could benefit greatly from the synergies created, increasing their chances of survival and enabling future growth. In order to achieve its aim, Maribe conducted:

- A study on *"socio-economic trends and EU policy in the offshore economy"*, to review each sector from a business lifecycle and socio-economic perspective. A review of the policy and planning frameworks that applied to the sectors was conducted for each of the sea basins under study: Baltic basin, Atlantic basin, Mediterranean and Black sea Basin, and the Caribbean Basin.
  WP4 (http://maribe.eu/blue-growth-deliverables/blue-growth-work-pac kages/) The results of this work package formed the basis for this book.
- A study on *"Technical and non-technical barriers facing Blue Growth sectors"*, to look at barriers by sector and also by combination and to identify the barriers that existed when two sectors shared marine space or multi-use platforms; WP5 (http://maribe.eu/download/2581/).
- An *"investment community consultation"* to assess the current investment environment, as well as best practices and key barriers for investment; WP6 (http://maribe.eu/download/2575/).
- A *"business model mapping and assessment"* to analyze and map the business models that lie behind Blue Growth/Economy industries. WP7 (http://maribe.eu/download/2569/).

Building on the above studies, Maribe then assessed the potential for each of the sectors falling within its scope to combine their activities with those of other Blue Growth or Blue Economy sectors.

14.2 A Methodology for the Selection of a Promising Combination of Blue Growth Sectors

The 4 Maribe sea basins were reviewed using an Excel based spread sheet for each basin:

1. Atlantic and North Sea
2. Baltic

Table 14.1 Maribe Blue Growth Matrix selection template showing the Atlantic basin and average of all marks for the 5 headings listed below[1]

	Wave	Tidal	Tidal Lagoon	Desalination	Offshore W	Offshore Wind	Aquaculture	Biotechnology	Seabed Mining	Offshore fixed ter	Toursim	Oil & Gas	Fisheries
				Atlantic Basin - Final Score									
Wave													
Tidal													
Tidal Lagoon													
Desalination													
Offshore Wind fixed													
Offshore Wind floating													
Aquaculture													
Biotechnology													
Seabed Mining Offshore													
Offshore fixed terminal													
Tourism													
Oil & Gas													
Fisheries													

3. Mediterranean
4. Caribbean

A two dimensional matrix was created with the 13 Blue Growth and Blue Economy sectors on both x and y axis, and is visual presented in Table 14.1

For each basin, the potential for combination of Blue Growth sectors was rated from 1–5 (5 was maximum rating = best) under the headings of:

- technical,
- environmental,
- socio-economic,
- financial and
- commercial perspective.

The top 24 potential Blue growth combinations were initially selected. Blue Growth Companies were then matched to the chosen Blue Growth combinations and shortlisted in liaison with the European Commission.

It was a difficult task to find existing Blue Growth companies to match the top ranked Blue growth combinations arising from Table 14.1. It was even more challenging to gain cooperation form these companies to participate in the Maribe case studies. Finally, Maribe succeeded in securing 9 companies related to the top ranking blue growth combinations in Table 14.1. Table 14.2 lists the 9 case studies, the companies that participated and their relevant Blue growth sectors.

This chapter will present three representative case studies and will conclude with an overall evaluation of the viability of Blue Growth combinations.

---

[1]The marks ratings are not disclosed here, but are available from the Maribe website.

Table 14.2   Case studies of Blue Growth combinations

Case Study[2] and Maribe report link	Company 1	Technology 1	Company 2	Technology 2
A1	Floating Power Plant, Denmark	Floating wind		Wave Energy
A2	ASC, Spain	Floating Wind	Cobra/Besmar, Canaries	Aquaculture
A3[3]	Grant Port Guyane	Floating shipping Terminal		Aquaculture
B1	Float Inc, USA	Floating shipping terminal		Wave Energy
B2	Wave Dragon, Denmark	Wave Energy	Seaweed Energy Solutions	Seaweed macro Algae
B4	Albatern, Scotland	Wave Energy	Aquabiotech	Aquaculture Finfish
B5[3]	JJ Cambells, Ireland	Wave Energy		Floating wind
B6[4]		Fixed Wind		Mussel farm
B8	EcoWindWater, Greece	Desalination		Floating wind

14.3 Case Study Description Methodology

Each Maribe case study was assessed under 4 sub-sections:

1. Technical brief
2. Financials
3. Business plan
4. Risk Assessment

Extensive efforts were made during the project to protect the value proposition of the companies that contributed to the case studies. There were extensive discussions around non-disclosure agreements and protection of IP. Therefore, the sections on the financial details of the proposition and the risk identified will not be discussed here. However, the technical brief and business plan of the three case studies below provides very interesting reading and shows how blue growth combinations can become viable companies.

---

[2]Maribe numbering: Project B3 and B7 were dropped from the case studies, due to lack of sufficient data. Partners in A3 and B5 declined permission to make their reports publically available.

[3]Companies did not provide permission for their reports to be made publically available on Maribe website.

[4]No companies were found to provide data for project. Maribe had enough expertise to complete the case study.

The reader should also note that dynamic start up companies such as those featured here can change significantly in a short space of time. The case studies here reflect the status of the company during the course of Maribe in approximately 2016. Each of the case studies was subject to extensive investigation by the consortium and by a panel of experts, the three projects chosen here were among those that received positive feedback and ratings from this assessment.

14.3.1 Technical Brief Methodology

The technical brief of each case study contains the following information:

- Size and scale: *Number of units, rating of each unit, for instance: wind: one unit rated at 5 MW; wave: four units rated at 2MW total wave power take-off.*
- Footprint incl. boundary: *Information on the site the deployed asset is expected to occupy. For instance: "approx. 0,25 km^2".*
- Located: *Information on the proposed or intended location for the proposed project. For instance: "24 km (15 miles) off the coast of Wales".*
- Water depth: *Range or maximum water depth at the specific location based on readily available information in the literature of company's survey. For instance: "60 m".*
- Cable to shore or power source (if applicable): *For electricity exporting projects only. Informing if grid connection is available near the project deployment site, size of cables to be used and any other relevant information. Even at pilot stage some project developers are looking to demonstrate electricity exported to the grid as this often presents as a milestone in the development stages. For instance: "Grid connection available at Galway Bay test site".*
- Moorings: *Information on what mooring technology will be used to secure the assets to the seabed if a structure is floating, as this could present a major risk if not addressed appropriately.*

14.3.2 Business Plan Using Business Model Canvas

The Business Model Canvas tool is a method that is used by companies to describe their business models using nine building blocks. It was developed by Alexander Osterwalder based on his earlier Business Model Ontology[5],[6].

---

[5]Osterwalder, A., & Pigneur, Y. (2010). *Business model generation: a handbook for visionaries, game changers, and challengers.* John Wiley & Sons.

[6]Osterwalder, A. (2004) *The Business Model Ontology – A Proposition in a Design Science Approach* PhD Thesis, University of Lausanne.

Each of these building blocks represents a key organisational structure that is required to have a functioning business. It has been utilised by companies and organisations such as NASA, Intel, Microsoft, PWC, and Ernst & Young.[7]

A business model canvas has 9 building blocks. Maribe added four extra blocks (to total 13): competition, market, management and Financing/investment. Due to space limitations, this chapter will only present the four of the most relevant building blocks. The following presents their description:

1. Competition

The competition section for each combination was drawn up to include a cross-section of direct and indirect competitors based on the companies' information and publicly-available information regarding projects with similar characteristics. Indirect competitors were included, for example conventional carbon and nuclear energy generation for wave and offshore wind energy, to reflect the reality that in order for the businesses to reach commercial viability, they must compete on price with conventional established industries. These competitors were listed, together with 'Key Differentiators' and a competitive threat rating. The competitive threat rating was based on the companies' rating of the perceived threat from their perspective.

2. Business model

Bringing together different types of business into a single location is, by definition, a new way of working that normally involves more than one company. This building block contains a description of how the company or companies will work together to create the value proposition, and get it to market. This may include business models such as special purpose vehicle, partnerships, mergers etc.

3. Value proposition

Value proposition provides a unique combination of products and services which provide value to the customer by resulting in the solution of a problem the customer is facing or providing value to the customer.

4. Market analysis

Market analysis of the sectors involved in each combination was compiled from a combination of desk studies of existing market research and reports, together with information provided by the companies involved. This was then used to calculate:

---

[7]http://www.businessmodelgeneration.com/canvas/bmc

- Total Available Market (TAM),
- Serviceable Available Market (SAM) and
- Serviceable Obtainable Market (SOM).

14.4 Case Study 1 – Floating Wind and Wave – Floating Power Plant

14.4.1 How Floating Power Plant Was Selected by Maribe

The Maribe ranking of BG combinations exercise determined that floating offshore wind with wave energy would be both technical, and economically viable, whilst also having large socio-economic and environmental benefits. The combination of the two sectors could either be sharing one space (MUS), or in a MUP. The following are the Maribe findings for both MUS and MUP possibilities:

- Technically:
    - MUS: there are no technical barriers faced by mixing wave energy with floating offshore wind parks.
    - MUP: One company has championed this technology and is at TRL8. Although technical challenges still exist, the commination looks technically favourable.
- Economic: Both sectors are economically viable by themselves providing they receive appropriate Feed in Tariffs. Higher tariffs for wave energy may prove beneficial for the floating wind part of the project. Cost savings should accrue in shared costs, and reduced materials, both in MUS and MUP.
- Environmentally: The environmental impact of these two sectors in MUS or MUP would not increase, than if they were deployed completely separately.
- Socio-economically: it was determined that public acceptance would be high, and assist in green renewable energy, increasing job prospects.
- Commercial development: floating wind is at TRL 7 of higher. Wave energy is still at TRL5.

The Atlantic was deemed the best basin, due to extremely good resources of both wave and wind.

The MUS option theoretically had the most possibilities. There are already a number of pilot prototype floating offshore wind farms in development: e.g. Hywind Park off Scotland, Windfloat project off the Mediterranean cost of France, and ASC NER 300 FLOCANS 5 project in Canaries. There would also be a number of wave energy candidate technologies that could be deploy in the midst of the floating wind farms. However, to date, there are no current projects exploring this combination. This is probably because floating offshore wind is currently emerging from the high risk prototype stage. Therefore, combining another technology with floating wind may be seen to increase the risk above investor tolerance or interest.

Fixed offshore wind, which is now considered in the mature commercial phase, is also hesitant about combining other technology types within its wind farm. Significant progress was made in 2016, when Wavestar was awarded €28M H2020 award to test pilot deployment of one of its devices within a Belgian wind farm. Unfortunately, a combination of key investor withdrawal, and difficulties in finalising a deployment location suitable for Wavestar ended in termination of the project.

The other combined option discussed earlier, is the MUP option. Currently the only technology that is successfully exploring this option is Floating Power Plant. The Business plan fitted perfectly within the Maribe ranking criteria, of an Atlantic deployment and TRL 5 or higher having been already achieved.

14.4.2 Company Description

Floating Power Plant A/S[8] is a Danish clean-tech company that develops, designs and provides a unique patented technology integrating wave energy convertors into a floating offshore wind device. The company is entering the commercialization stage, based on over 8 years of R&D, testing and business development. The company is backed by 156 private shareholders and leading industrial development partners having raised more than €15m to date. The hybrid device/technology has been developed over the last eight years, from concept to four offshore grid connected test phases totalling two years of operation with a scaled prototype. FPP is the only company in the world that has supplied power to the grid from a combined floating wind and wave device.

The FPP hybrid technology consists of five key technology elements, four of which are existing solutions from the oil and gas and offshore wind industries and one is a unique FPP solution[9]:

1. A semi-submersible floating platform
2. An offshore wind turbine (5–8 MW)

---

[8]http://www.floatingpowerplant.com/
[9]All patents and IPR are placed with the company.

3. A disconnectable and vaning turret mooring system allowing 360-degree rotation
4. Flexible subsea cables and power export system
5. A unique wave energy and PTO systems placed on a known stable structure (2–3, 6 MW in total).

The design combination of a highly efficient wave energy device on a stable structure connected to a disconnectable turret ensures that the platform passively vanes to face the primary wave direction. The disconnectable turret mooring allows the device to be completed constructed and commissioned in harbour and returned to harbour for major maintenance activities, eliminating the extra risk of performing these offshore and avoiding the use of costly specialist vessels. The platform orientation into the primary wave directly combined with the high wave absorption (50–70% of the wave energy is absorbed) results in an artificial offshore harbour effect at the aft end of the system, enabling significantly increased accessibility for routine maintenance and repairs.

14.4.3 Technical Specification of Technology

14.4.3.1 Current status

The technology is at TRL 6. FPP is the only wind/wave hybrid technology in the world that has been proven in the offshore environment. P37, the half-scale grid connected prototype, has operated for 4 offshore tests periods constituting more than 2 years of data. The commercial development plan for the technology has three key stages and is performed with an Irish end customer (project developer). The first stage is a full scale pilot demonstration platform, deployed at a Welsh or Scottish offshore site (Welsh project presented here and shown in Figure 14.1).

- Located: 24 km (15 miles) off the coast of Wales
- Water depth: 60 m

Pilot Demonstration project has a 7MW total capacity with 3 years' operational lifetime (assets to be transferred to early commercial follow up project). The pilot comprises of a single P80 first generation MUP hosting:

- Wind: one unit rated at 5 MW
- WECs: four units rated at 2MW total wave power take-off

Other details

- 77 GWh delivered to the grid (over 3 years) from a single MUP during the demonstration project

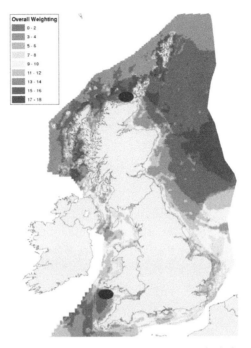

Figure 14.1   UK map, showing hotspots of combined wave and wind resource. Circles show FPP proposed deployment sites in Scotland and Wales.

- Total capex incl. grid capacity for the worlds full scale hybrid 7 MW: €64.1 million

- After 3 years operation, the next stage will see the addition of 27 P80 platforms to the same site, totalling 28 platforms.
  - *Power generated: 13600 GWh*
  - *Total capex: €889 million*

The combined array project will be commercial and will make a profit over the indented 20 year design life at the current feed-in-tariff with a CfD contract mechanism (agreed "strike" price set with the UK Government).

1. The third stage is a commercial outlook case, deploying 2nd and 3rd generation P80 platforms at high energy sites, where each generation represents a significant step up in technology improvement.

   P80 3rd Gen specs:

   - *Wind: 8 MW wind turbines*
   - *WECs: 3, 2 MW wave power take-offs*

The 3rd generation P80 commercial array will have a total power capacity of 464 MW and 20 years' operational lifetime.

Other details:

- *Footprint approx. 72km²*
- *Located 100 km of the west Scottish coast*
- *Water depth: 75 m*
- *Power generated: 31746 GWh*
- *Total capex: €1383 mill*

14.4.3.2 Advantages of floating power plant combination

1. Increased uptime through greater access for O&M, due to the harbour effect
2. Greater addressable market (exploitable sites) through improved operability
3. Greater energy density (MW/km^2) with a smoother, more predictable power output
4. Cost savings in construction, installation and operation
5. Conforms to EU directives multi-use of space and MSP directives
6. Low LCOE relative to competitors, declining further with maturation

14.4.4 Business Section

14.4.4.1 Competition

Wave device developers are not considered direct competition due to their high projected LCOE and the small number of other hybrid devices under

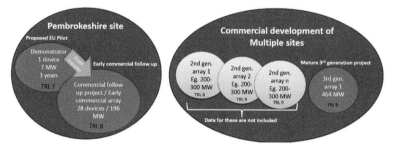

Figure 14.2 Diagram showing the project stages in FPP commercialisation process. The Pembrokeshire site contains the first two project stages and the commercial development of multiple sites is the 3rd stage. Projects in dark blue are described and costed in this document reflecting the 1st and 3rd technology generation.

development, are significantly less mature than FPP, who are the only developer to have undergone grid connected offshore testing. The direct competitors can be divided into separate sub groups as represented in Figure 14.3.

1. Windfloat project – a single demonstrator with a small array planned
2. Hywind Demo – A single turbine deployed on a spar buoy
3. Hywind Pilot – An array of 5 Hywind turbines, fully consented and under construction
4. Kincardine – An array of up to 8 floating turbines, consent application submitted

No project or technology is currently targeting the green area of the map due to the operational challenges of the high wave resource (Figure 14.3). FPP provides a unique option to exploit this vast area of resource.

14.4.4.2 Value proposition

FPP presents a simple 3 pronged value proposition in Figure 14.4, based on a versatile product which is both profitable and low cost of energy. The value proposition is sensibly linked to technology offering and technology risk mitigation plans.

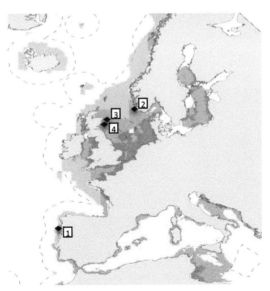

Figure 14.3 Combined Wind and wave resource map (green indicating highest combined potential where Floating Power Plant will be deploying).

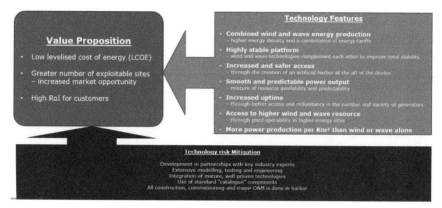

Figure 14.4   Floating Power Plant value proposition.

14.4.4.3 Business model

One key difference is that the goal of FPP is not to be producer of the technology, but the designer and manager thereof. The production and assembly will be handled by preferred partners and in some cases sub contracts. This is also the reason for FPP's partnering model, the value chain is built up alongside the technology development. This strategy reduces the capital burden and increase global flexibility. As the company develops, a strategic partner/value chain investor will be taken in to the required balance sheet to provide cheaper finance and trustworthy warranties/guaranties.

14.4.4.4 Market capture

Figure 14.5 presents FFP TAM, SAM and SOM: predicting a SOM (Serviceable Obtainable Market) of 20GW of installed product by 2050, 10% of the SAM (Serviceable Available Market).

IEA	7200 GW of new global power 2040	2,8t$ pr. year
IEA	4300 GW renewables by 2040	1,7t$ pr. year
EWEA/OEE	400 GW floating wind and wave power demand by 2050 (TAM)	
Consultancy analysis	200 GW within FPP's market segment (SAM) by 2050	Wave resource and bathymetry restrictions
FPP analysis	20 GW is an obtainable market for FPP (SOM) by 2050	Value chain, Consenting, etc.

Figure 14.5   Floating Power Plant market capture.

14.5 Case Study 2 – Floating Wind and Aquaculture – Besmar and Cobra/ACS

14.5.1 How Besmar Cobra/ACS Was Selected by Maribe

The Maribe ranking of BG combinations exercise determined that floating offshore wind with aquaculture would be both technically and economically viable, whilst also having large socio-economic and environmental benefits.

There are no MUP floating offshore wind and aquaculture enterprises. Therefore, Maribe explored whether floating offshore wind parks could host aquaculture farm in its midst. The following were its findings:

- Technically: there are no technical barriers faced by mixing aquaculture with floating offshore wind parks.
- Economic: Both sectors are economically viable by themselves, and together might prove to have increased economic viability.
- Environmentally: The environmental impact of these two sectors in MUS would not increase, than if they were separate.
- Socio-economically: it was determined that public acceptance would be high, and assist in moving aquaculture offshore, increasing job prospects.
- Commercial development: both are at TRL 6 of higher.

The Atlantic was deemed the best basin, due to extremely good resources of both wind and aquaculture production.

There are only a few floating offshore wind parks currently in deployment: e.g. Hywind Park off Scotland, Windfloat project off the Mediterranean cost of France, and ASC NER 300 FLOCANS 5 project in Canaries. Hywind will be exploring combining offshore aquaculture with their wind farm in the future once the Scottish pilot has passed a certain stage of testing. Thus they were not willing to cooperate with Maribe at this stage. WindFloat are in a similar phase. Fortunately ACS were in a position to consider combining with an aquaculture partner, which was Cobra Besmar.

14.5.2 Company Description

Grupo **COBRA**[10] is a subsidiary of ACS, a Spanish multinational company with long experience in the construction and operation of fixed wind farms. **BESMAR**[11] Aquaculture Company was established in 2004 as a specialist

---

[10]http://www.grupocobra.com/content/page/group-companies/

[11]http://www.besmaraquaculture.com/

offshore aquaculture company to consult and develop commercial projects that are unique and front runners at both the commercial and technical level.

14.5.3 Technical Specification of Technology

COBRA/ACS floating wind technology is at TRL5/6 and has successfully completed the testing of their floating wind turbine at 1:40 scale, 78 Kg in weight in 2014 in a laboratory test channel in the Canaries (more information at http://www.cehipar.es/). COBRA has been awarded €34 million (FLOCAN 5) from the €1 bn NER 300 Program in 2014 to deliver the FLOCAN 5 project (i.e. Pilot TRL 8 pre-commercial farm Gran Canaria) which is expected to be operational in 2017–18.

BESMAR aquaculture technology is at TRL9 with a 1st commercial aquaculture farm deployed in Gran Canaria. The BESMAR organic aquaculture plant has been operating since 2012 and has the latest technology cages constructed from heavy duty Polyethylene (PE). Measuring 25m diameter and 5000 m^3 volume the main frame of the cage is composed of three 400 mm diameter rings with heavy walled pipe to resist impact and kinking, it also has a 5 tonnes weight "froya ring" to tension the net pen.

The Cobra Besmar project is planning 2 phases of development:

1. Phase 1: TRL 7/8 per-commercial pilot
2. Phase 2: TRL9 Commercial project

14.5.3.1 Phase 1: TRL 7/8 pre-commercial pilot

The pilot project will consist of:

- COBRA: 5 floating wind turbines rated at 5 MW each total capacity, total 25 MW
- BESMAR: 6 fusion type offshore aquaculture cages with 40 tons capacity (organic sea bass production)

The location for the pilot project will be South-East coast of Gran Canaria 5.2 Km from shore, at water depth range 40 to 200 m for wind turbines, and 40m for aquaculture (Figure 14.6).

Other relevant technical details:

- Cable to shore or power source: submarine cable 2×(5MW/13.2kV) linking the wind farm to an offshore floating substation
- Array connection or autonomous power: 33 kV

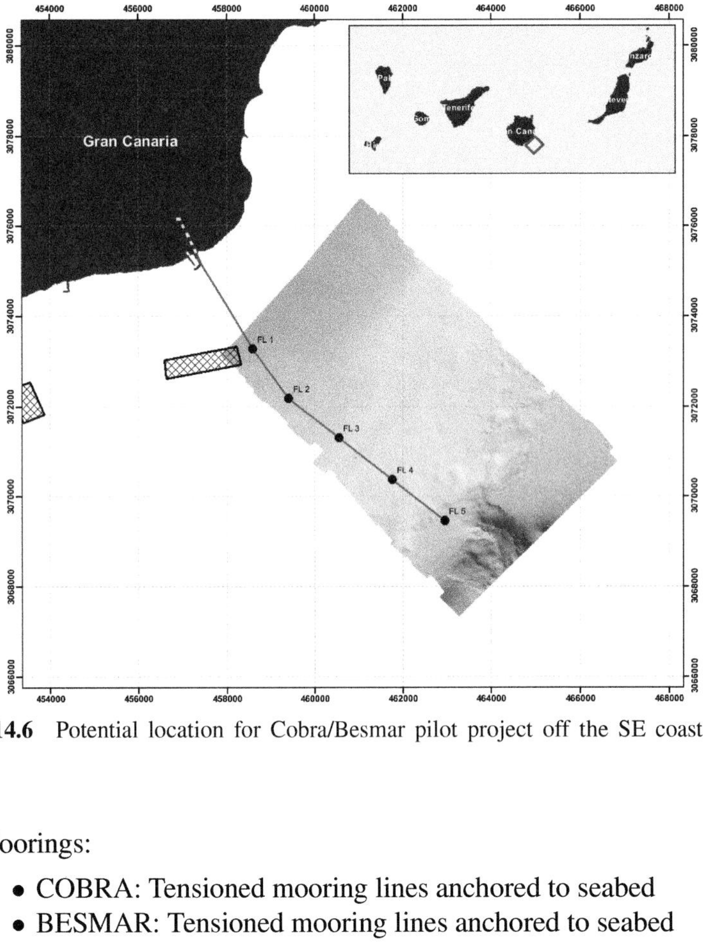

Figure 14.6 Potential location for Cobra/Besmar pilot project off the SE coast of Gran Canaria.

- Moorings:
  - COBRA: Tensioned mooring lines anchored to seabed
  - BESMAR: Tensioned mooring lines anchored to seabed
- Operations and Maintenance: COBRA: Wind turbines O&M will be supported by the port and shipyard of "Puerto Las Palmas". BESMAR: supported by port "Puerto de Taliarte"

This pilot project has secured part of the funding required from NER 300 (project FLOCAN).

14.5.3.2 Phase 2: TRL 9 commercial

Expansion to a full commercial farm will be the next phase with additional wind units and aquaculture cages installed at the same site. Market entry will be completed with a 2nd commercial farm, followed by a 3rd commercial project. The commercial projects will deployed in a new site at PLOCAN testing site in South-East coast of Gran Canaria, Plataforma Oceanográfica de Canarias (PLOCAN) a Research Institute co-funded by the Economy and Competitiveness Ministry of the Spanish government and the Canary Islands government.

The commercial project will consist of a total of 125 MW and 1300 tons/ year organic sea bass

- *COBRA: 25 floating wind turbines rated 5MW each*
- *BESMAR: 24 fusion type offshore aquaculture cages with 40 tons sea bass production capacity each*

Other relevant technical details:

- *Footprint combined approx. 23 Km^2*
- *Water depth: 600m*

14.5.3.3 Advantage of floating wind and aquaculture combination

14.5.3.3.1 *General for both sectors*

- Cost savings on O&M due to shared vessels, using multi-purposes vessel should have all the equipment and facilities to operate for both activities.

14.5.3.3.2 *Aquaculture farm*

- Cost savings on energy due to energy supplied by wind farm.
- Wind farm provides protected calmer waters for aquaculture cages, increasing cage longevity and also increasing performance at earlier stages by reducing fish losses due to broken nets.
- Healthier product, the fish will have less stress and clean water, increasing animal welfare and the final quality of the product.
- Less environmental pollution due to better dispersion by currents due to distance from coast.
- Security camera and radar systems can be installed at the turbine to protect finfish farm from robbery.
- Automatic feeding systems could be installed (not included in this project).

14.5.3.3.3 *Offshore platform wind farm*

- Good public perception, allowing the companies to advertise their products as environmentally friendly produced.
- Tax exemption: is considered in the Spanish law for those companies providing renewable energy.

14.5.4 Business Section

14.5.4.1 Competition

Key Competitors

Table 14.3 presents the 3 main competitors to the Cobra Besmar cooperative venture. Competition from thermo-electric electricity company is perceived to be the main threat, mainly due to the the fact that thermo-electric power will be cheaper than the offshore offering, at least initially, and enjoys a monopoly on the island at present.

There will be competition from other aquaculture producers, but only one so far is targeting organic produce; Kefalonia.

14.5.4.2 Value proposition

A floating platform enables the device to generate electricity in areas that typically have more powerful wind resources. The floating wind platform can also partially protect juvenile fish cages and security systems for the fish farm. The power generated can also aid installation of other equipment such as automatic fish feeder, underwater CCTV, etc.

"Green energy" integration with ecological fish production, maximizes the use of the space, and increase consumer perception of ecological fish production.

Table 14.3   Key competitors

Competitor	Key Differentiators	Competitive Threat Rating (1–5)*
Unelco-Endesa	The main Spanish electrical producer company. Electricity produced by ENDESA is mainly generated by thermo-electrical power plant in the Canary Islands. http://www.endesa.com/es/home	5
Kefalonia	Greek Sea bass and Sea bream producers. Their production is focusing on organic fish production. http://kefish.gr/mobile/organic/en_organic.html	3
Nireus	One of the biggest aquaculture companies in Greece. Their produce standard Seabass and Seabream. No reference on organic production of these fish. http://www.nireus.com/1_2/Home	1

*Competitive threat based on companies' appraisal of perceived threat with 5 being severe competitive threat.

14.5.4.3 Business model

The long term business model proposed is that COBRA & BESMAR cooperate through an SPV or similar to sell combined wind and aquaculture installations. BESMAR will subsequently operate the aquaculture elements.

14.5.4.4 Market capture

The target in the Canary Islands is to have at least 60% of the total electricity produced coming from renewable energy by 2020 through the PECAN (Canary strategic energy plan). The 25 MW produced in the pre-commercial farm will increase the renewable electricity generation up to 17% for Gran Canaria Island. Thus the offshore wind platform electricity generated by COBRA will have a secure market in Gran Canaria for the 25 MW produced. Long term the model is based on selling installations in Europe, North and South America and Japan, and plans to capture 10% of the Serviceable Available Market (SAM) (Figure 14.7). Currently BESMAR produce 240

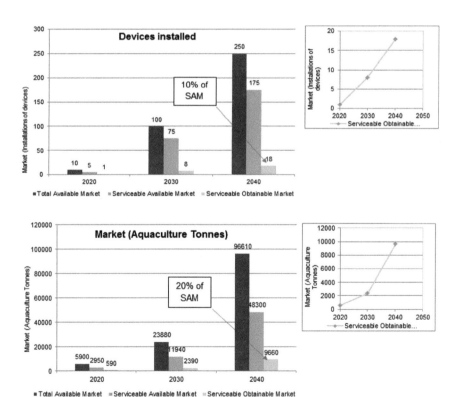

Figure 14.7   Market analysis: Floating wind devices and aquaculture production.

tons of organic Sea bass annually, and through their commercial branch, Naturally Atlántico, sell their products in Canada, US, France and Spain. Only 5% of sales are in the Canary Islands, therefore the focus of increased production will be on exports. The company plan to expand to capture 20% of SAM by 2050 (Figure 14.7).

14.6 Case Study 3 – Mussel Aquaculture in Borssele Offshore Wind Parks

14.6.1 How Mussel and Offshore Wind Farm Case Study Was Selected by Maribe

The Maribe ranking of BG combinations exercise determined that fixed offshore wind with an aquaculture farm would be both technical, and economically viable, whilst also having large socio-economic and environmental benefits. MUS option was the only combination considered.

The following are the Maribe findings for the MUS:

- Technically: there are no technical barriers faced by mixing aquaculture with fixed offshore wind parks.
- Economic: Both sectors are economically viable by themselves, and together might prove to have increased economic viability.
- Environmentally: The environmental impact of these two sectors in MUS would not increase, compared to the situation where they are separate. In some cases, such as seaweed, the environmental impact would improve.
- Socio-economically: it was determined that public acceptance would be high, and assist in moving aquaculture offshore, increasing job prospects.
- Commercial development: both are at TRL 8 of higher.

The Atlantic was deemed the best basin, due to extremely good resources of wind, as well as the best basin for most types of aquaculture.

The 3 types of aquaculture were considered for the case study:

- Finfish: The Mermaid project explored this combination. Unfortunately, Maribe were unable to secure the cooperation of the project as a case study in Maribe.
- Seaweed: Seaweed production is mostly in the North Sea and Baltic, and near shore. There were no fixed offshore wind farms currently exploring this combination.

- Mussels: There are a number of wind farms currently being constructed in Belgium and Netherlands. The waters of these coast are perfect for mussel production. Maribe selected this final combination as the basis for its case study, due to expertise within the consortium.

14.6.2 Project Background and Description

Maribe project did not succeed in obtaining candidate companies in fixed wind and mussel farms to cooperate in the case study. Never the less, Maribe consortium decided to undertake the study as there was sufficient expertise within the consortium to complete the relevant sections, and commercial interest in the case study is increasing. Since the Maribe project completion, an actual project has commenced in Belgium, and described in the following link: https://www.offshorewind.biz/2017/06/02/belgians-start-growing-mussels-on-offshore-wind-farms/. Further development of offshore wind in the North Sea is expected and now that Belgian and Dutch governments have established support schemes, various new wind parks are being proposed off the coast of Belgium and the Netherlands.

The project case study presented is based on a previous FP7 research project, MERMAID, which explored multiple use concepts for four European basins, either multiple use of space or multiple-use platforms. One of the most promising designs emerging from the project was wind farms with bottom-fixed offshore wind turbines and mussel aquaculture for the North Sea area. The case study uses the planned phased development of the Borssele windpark (Figure 14.8). This wind park consists of 4 plots. The total site area equals 344 km^2. Due to existing pipelines and cables that cross the site, the plot has been subdivided into 4 parcels (25.2 km^2, 17.8 km^2, 2.1 km^2, and 4.1 km^2). Plot II does not have cables or pipelines crossing this site; it consists of one parcel with an effective area of 63.5 km^2. It is located 12 nautical miles (22–39 km) offshore. Grid connection is scheduled to be ready 31 August 2019. The tender document provides detailed information on routing of the underwater cables and land connection. Delivering this is the responsibility of the Dutch offshore grid operator, Tennet, and not part of the tender. The permitted foundations are monopile, tripod, jacket, gravity based and suction bucket for turbines in the range of 4 to 10 MW. The government support scheme is a contract-for-difference (CFD).

In this combination, the wind park includes space for the production of mussels. The relevant system in this case study has a culture grown on simple structures such as ropes and frames on subsea lines as shown schematically in Figure 14.9. These lines are connected to the sea bottom through a mooring

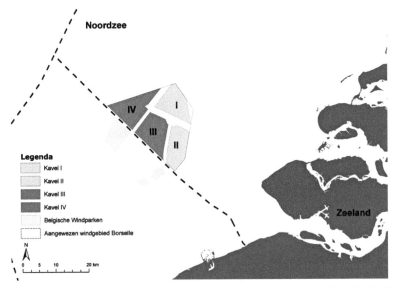

Figure 14.8 Potential location of Maribe case study project for mussel farm in fixed wind farm.

system. Installation time for this option is less than a week. The Dutch Mussel industry and NGO's have agreed that best practice for the collection of mussels will be the use of long-lines. These long lines are mainly used in the Wadden Sea. It is assumed that the mussels are not restocked during growth (i.e. taken of the longline and put back with greater distance between them). Instead, the system is thinned out. The resultant mussel spat and half-growns are transported to the Eastern Scheldt to grow further.

14.6.3 Technical Specification of Technology

14.6.3.1 Phase 1: TRL 7/8 per-commercial pilot

The pilot project will be located in Wind park Amalia using 60 Vestas V80, 2 MW turbines.

The footprint of the wind park 49.5 km², located: 23 km off the coast of Netherlands, water depth: 19–24 m.

Aquaculture will have a target output of 0.5 million kg mussel seed in 21.4 hectares. The aquaculture long line system will be fabricated at Machinefabriek Bakker or comparable.

CAPEX is likely to be in the order of €12 million so grant funding required in the order of €5 million (No detailed estimates were carried out for this project case study).

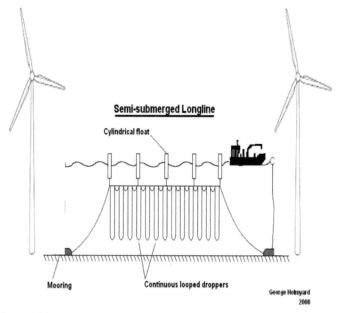

Figure 14.9 Artist impression of mussel string in between wind turbines.

14.6.3.2 Phase 2: TRL 9 commercial

The next phase plans to be situated in a larger Wind park: 380 MW installed capacity using 4 to 10 MW turbines, with a footprint of 49.5 km^2, located 22–38 km off the coast of Netherlands and Belgium, in 15 to 35 m of water.

The aquaculture target output is planned to be 5.5 million kg of mussel seed over the project period, covering 235 ha.

Based on the literature, the estimated annual production per unit is as follows, in two years:

- 14,064 kg of mussel seed, harvested in autumn
- 14,064 kg of half-growns, harvested in early spring
- 9,376 kg of consumption size mussels, harvested in autumn

14.6.3.3 Advantage of floating wind and aquaculture combination

14.6.3.3.1 *Aquaculture farm*

- The wind park provides the mussel companies with an area not accessible for large other vessels, reducing risk that the mussel facilities are negatively affected by these vessels.

14.6.3.3.2 *Offshore platform wind farm*

- Mussel aquaculture makes areas less accessible for other vessels, reducing risk of collisions with unfamiliar vessels. Mussel aquaculture can have a wave dampening effect, reducing fatigue and resultant O&M for wind farm structures. Dampened seas will also enable access for O&M for longer periods increasing wind farm availability.

14.6.4 Business Section

14.6.4.1 Competition

Competitor	Key Differentiators	Competitive Threat Rating (1–5)*
Conventional Mussel Farming	One of the advantages of this combination is the multiple use of space. This gives the combination an advantage over conventional mussel farming, especially in countries which have increasing demands on limited space.	4
Floating offshore wind	Fixed offshore wind is the most cost-effective technology given the low water depths in this area. Other foundations are not eligible under the prevailing subsidy scheme.	1
Wave and/or tidal energy	Both wave and tidal energy are not cost-competitive as they are much more expensive the fixed offshore wind. They are also not eligible under the prevailing subsidy scheme.	1
Wave energy/aquaculture concept (e.g. Albatern WaveNET)	The Albatern WaveNET devices may not be suitable for the conditions found at the mussel farms. It would also take up space that could be more suitable for fixed offshore wind. Wave energy is not eligible under the prevailing subsidy scheme.	1

14.6.4.2 Value proposition

- Company A – reliable, less expensive renewable electricity due to combination, more sustainable image, possibility of easier consenting if government policy advocated more efficient use of space. Concept is easily transferrable to other sites once concept is proven.
- Company B – cheaper mussels due to combination, more sustainable image, mussels with less toxins, increase in the areas available to the industry to utilise.

14.6.4.3 Business model

Two independent companies will operate each sector (fixed offshore wind and aquaculture) separately. The offshore wind energy company (Company A) will be an offshore wind project developer who source, install and operate offshore wind farms. Examples of such companies include DONG Energy, Vattenfall, etc. The aquaculture company (Company B) will be a company who has experience in operating aquaculture farms. Examples of such a company include Prins & Dingemans, Delta Mosselen, Roem van Yerseke, etc. While both companies will remain separate and not form a joint venture or special purpose vehicle (SPV), both companies will have a legal agreement to install and operate wind farms with integrated aquaculture installations (most likely mussel farms in this case). Company A will sell electricity with revenue from CFD supplied by government for 15 years, and Company B will sell mussels with revenue from mussel markets.

14.6.4.4 Market capture

The market for this combination is twofold: Electricity wholesale market and mussel wholesale market. In the Netherlands, electricity production capacity equals 31.5 GW, out of which 20.1 GW consists of centralised production (i.e. powerplants) and 11.4 GW is decentralised production. In 2014, total installed wind capacity in the Netherlands equalled 2.7 GW. This capacity was used to produce 5.627 million kWh of electricity from wind, making it the second-largest source of renewable electricity.

The aquaculture sector of the Netherlands can be divided into two different sectors, namely shellfish and finfish. The shellfish sector is an older and more established sector, and consists of 50 companies growing blue mussels which result in between 50,000 to 60,000 tonnes of mussels per year. Shellfish culture takes place in the estuarine waters in the southwest Netherlands and in the shallow Wadden Sea in the North of the country. The mussel wholesale market is based in Yerseke where mussels are auctioned. By 2020, the case study analysis estimates that the project will attain 20% of SAM (Figure 14.10).

14.7 Conclusion

There has been much scepticism of the value of combining Blue Growth sectors together, or Blue Growth with mature Blue Economy sectors. Market forces today has favoured the more established single technology BG sectors and enterprises (e.g. fixed offshore wind) which are continuing to thrive

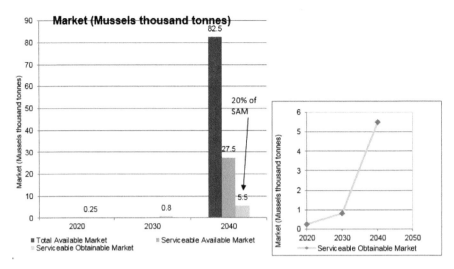

Figure 14.10   Market assessment for mussel production.

having proven they can achieve EC and member state targets for lowering the cost of energy. The EC have tried to stimulate Blue Growth, by funding a number of projects exploring multi-use of space and multi-purpose platforms. The most notable funded activities were the four Oceans of Tomorrow projects (OoT): Tropos, Mermaid, H2Ocean, Marina. The projects were not successful in developing strong IP from the projects that was able to progress to commercialisation. One observation of the results developed from these projects was that the focus was on the technology, creating a situation of "technology push" rather than "market pull". To put this another way, there may have been great technology developments but did not find a market willing to purchase these products. Unfortunately, these unsuccessful OoT projects have reduced EC confidence in MUS/MUP.

The Maribe case studies demonstrated that combining Blue Growth sectors with more established or mature Blue Economy sectors can make their overall value proposition more attractive. The advantages of combining are substantial, benefiting the newer technology tremendously, reducing the risk for the newer technology, and enabling learning. More importantly, MUS/MUP combinations conforms to EU Maritime Spatial Planning directive, thus should attract continued EC funding.

Case study of projects combining aquaculture with fixed or floating wind energy presented attractive business cases, and projects were highly rated by independent experts organised by Maribe. The mature sector of each project assisted in de-risking the less mature sector by ensuring good financial returns for the combined projects. Aquaculture benefited from using green powered electricity thereby increasing its public image. Incorporating wind energy is also relatively easy for aquaculture, and increases its position within the MSP directives, leasing/licencing etc.

Combining floating wind and wave also received very high independent review scores in Maribe assessment. The business case presented was very thorough and well researched. It highlighted the importance of a holistic approach, project consortiums that start off with well-developed business plans have a great chance of success, in comparison to those that rely for success solely on their technology.

Maribe cautioned that combining a new technology sector with a more mature technology (either MUP or MUS) will never fully compare financially with the mature sector operating by itself. For example, offshore wind parks operated by itself will always be a large competitor to MUS/MUP combination projects. Thus, MUS/MUP combination projects will consistently require EC support in the medium term.

In summary, the Maribe cases studies identified a range of MUS and MUP combinations that have the potential to become attractive business cases by their third phase of commercial deployment. The results should give confidence to the EC to pursue policy to promote appropriate MUS and MUP combinations both in the Strategic Energy Technology Plan (SET-Plan) and continued funding for MUS and MUP in H2020. The Commission's drive to promote multiple-use of space (MUS) (an important part of the Marine Spatial Planning directive) and multi-use platforms (MUP) has been justified by the positive outcomes from the Maribe evaluation.

Endwords

Kate Johnson[1,*], Gordon Dalton[2] and Ian Masters[3]

[1] Heriot-Watt University, Scotland
[2] University College Cork, Ireland
[3] Swansea University, Wales
*Corresponding Author

The preceding chapters have described in detail each of the nine industries which comprise most of the ocean economy and its anticipated future. The four traditional sectors (Chapters 6–9) are undergoing change, with Fisheries and Oil & Gas looking to consolidate while shipping and tourism will grow; and the five new Blue Growth sectors (Chapters 1–5) which are new innovative sectors on which the hope of new jobs and growth are founded. The new industries are coming into their own as markets emerge and technology is proven; these two aspects combine to create workable businesses. Overall, when traditional and emerging sectors are considered, there is an expectation of an expanding future maritime economy creating jobs and growth. In addition, there are plans for multi-use marine platforms (Chapter 14) where combinations of industries can support each other in common resources with reduced demand on space and consequent environmental harm.

Technology is moving very fast and there are several promising prospects which are not covered in this book. For example, energy storage will be transformative, whether it uses any one of a number of known, and possibly unknown, forms. Other marine renewable energy sources, such as OTEC and salinity gradients were also not covered in this publication, either due to their early stage of development or reasons of intermittence or lack of scale in European markets. Similarly, local distribution, storage and use of energy may enable industries in areas previously thought to be off limits. Consequently, the availability of reliable, sustainable, power in offshore locations enables offshore aquaculture or even large offshore platforms combining accommodation and several blue economy/growth sectors to become more realistic options.

Coincident work on maritime policy and regulation (Chapters 10–13) seeks to modernise and streamline ocean governance – easing the transition of the oceans from a status quo where any type of vessel has a right to open access commons to a future planned and controlled space where industries can flourish in harmony with each other and the environment. High minded ideals which are fraught with difficulties in practice. As UNESCO emphasise in their definition of marine spatial planning (MSP), it is a political process.

While this book has focused on the markets and industries in European sea basins, the ocean is also important globally. The global gross value added (GVA) of the ocean economy is estimated by the OECD to grow to more than US$3 trillion (at 2010 prices) by 2030, about 2.5% of total global GVA[1]. In 2010 it was recorded at about US$1.5 trillion when it was dominated by the established industries of offshore oil and gas (34%) and maritime tourism (26%). Ship transport and shipbuilding accounted for 9% with a further 13% attributed to port activities and 11% for marine equipment. Catch fisheries featured at only a 1% share of the ocean economy with a further 5% in fish processing. The new Blue Growth industries hardly featured at all with only aquaculture and offshore wind noticeable at less than 1% each. Reference to employment statistics paints a different picture. Catch fisheries dominated in 2010 with about 11 million people employed or over 40 million if artisanal fisheries are included. Maritime tourism employed about 7 million but all other activities employed 2 million each or less across the world. In forecasting the future to 2030 the OECD considered only a 'business as usual' scenario which recognises growth (or decline) of what exists but does not allow for new technologies. In fact, all existing sectors are forecast to grow in financial value and employment and on this scenario the most significant changes are in shares of the whole. Most prominently offshore oil and gas is forecast to fall to 21% by value and offshore wind is forecast to rise to 8%. The other sector changes are quite small with maritime tourism remaining at 26% making it the largest maritime sector by value (US$800 bn) in 2030 and only just second to catch fisheries in employment (8.5 million).

Summarising in turn each of the sectors described in this book, the availability of proven technology is perhaps the most significant factor among several others which constrain progress towards an ocean economy even larger than the one predicted in the OECD 'business as usual' scenario for 2030. The resources exist in large quantities in most of the sectors described. Policy will have a big part to play in overall growth, all the sectors are

---

[1] OECD (2016), The Ocean Economy in 2030, OECD Publishing, Paris.

regulated in some way, some have market incentives (green electricity tariffs), some require licences for use of space and planning decisions in the coastal zone is often contentious.

1. **Aquaculture.** A rapidly expanding sector meeting demand for food fish but constrained by the price which the market will bear in competition with wild fish and other sources of food protein in the form of meat and vegetables. New technologies are aimed at reducing feed and operations costs while meeting concerns about pollution, disease and escapes, particularly in a diminishing number of suitable sites close to shore. Techniques and equipment suitable to extend operations into the offshore will open up new sites for development, however, this could be limited by the marine spatial planning strategy in place in a particular location. Having achieved 50% of the food fish market in 2014, aquaculture is set to be a key factor in world food security. Commercial methods of farming an increased number of species are also a target.

2. **Blue Biotechnology.** The marine realm represents 70% of the biosphere and representatives of 34 of 36 known phyla are found there. The oceans are a reservoir of potentially useful molecules which may yield important pharmaceuticals, nutrients and other compounds. Identification of molecules and access for harvesting have constrained the sector so far but interest and practice is accelerating. Global Industry Analysts estimate a marine Blue Biotechnology sector industry value approaching US$5 billion by 2020, which sounds impressive but is under 5% of the whole biotechnology industry. At this value it is about 0.3% of the ocean economy. Blue biotechnology figures strongly in public policy ideas for the ocean economy but the industry has yet to respond with significant investment, probably due to the lack of speculative investment capital following the financial crisis. However, the potential appears to be promising.

3. **Seabed Mining.** The existence of valuable minerals in high demand lying on the seabed has been known and surveyed for exploitation for several decades. However, prospective recovery technologies have not so far kept pace with the vagaries of prices on the commodities markets. On several occasions the recovery of seabed minerals has looked to be approaching a commercial enterprise, only to be knocked back by a collapse in demand and the availability of cheaper terrestrial sources. This seems bound to change at some point, possibly in the near future

as demand increases and terrestrial sources are exhausted. It is likely that a particular "niche" mineral will be the first to be exploited, while large scale mining may be a long-term market. A further constraint has been one of ownership and the return on risk capital in developing the recovery technologies. Most of the resource lies in the UN defined 'Area' under the High Seas. They are beyond the limits of national jurisdiction and require international agreement. They are designated as *"...the common heritage of mankind"*[2].

4. **Wave and Tidal Energy.** Wave and tidal technologies are still at an early stage of development and a fully commercial enterprise is some years away. The size and apparently ubiquitous nature of the resource is a key driver in research and a number of technologies have achieved full-size prototypes generating power to the grid. Tidal stream devices are more advanced with the first arrays consented but their ultimate capacity is relatively small compared to wind and wave (but still worth pursuing) with the majority of potential concentrated into a few suitable locations. Wave has failed so far to achieve a potentially commercial power take off (PTO) technology and the development effort has divided into two. First, is consideration of bulk generation of power direct to grid; and second are smaller devices generating to local networks or activities (e.g. aquaculture, desalination, island communities) or, as yet undetermined, methods of energy storage.

5. **Offshore Wind Energy.** Offshore wind energy is the most rapidly expanding sector in the ocean economy. The technology has advanced rapidly over ten years from small 0.5MW turbines to new designs of turbines reaching up to 10MW capacity or more generating electricity direct to grid. The levelised cost of electricity (LCOE) from offshore wind has tumbled from over €300 per MWh to some recent (2017) strike prices below €75 per MWh. It is approaching a fully competitive level with other forms of electricity generation including fossil and nuclear. Constraints on expansion of the industry are suitable sites for fixed turbine towers and the large areas of marine space needed in possible conflict with other activities. Visual intrusion on seascapes has also been a significant factor in objections to some proposals. The innovation of floating turbines is progressing which, if successful and commercial, will release the industry from many of the site constraints.

---

[2]UNCLOS (1982), United Nations Convention on the Law of the Sea, United Nations, New York.

6. **Catch Fisheries.** Commercial catch fisheries have faced a storm of regulation and controls in response to the effects of overfishing and climate change. In addition, they face increased competition from farmed fish. Total annual catch has stalled at a little over 80 million tonnes for several decades and the World Bank forecasts a similar level to at least 2030. Fleet sizes have been drastically reduced but the power (KW) of fishing vessels has increased both by vessel and in aggregate. The catch fishery is a small component of ocean economy by value at less than 1%. However, in employment terms it is large at over 11 million people in the commercial fishery and on FAO figures perhaps more than 40 million if the artisanal fishery is added in. The fisheries are therefore relatively insignificant in global economic terms but when the economic wellbeing of remote coastal communities is considered, they are important and even essential in social and subsistence terms. They are under severe pressure for sustainability in catching and methods but will continue to be a cornerstone of the ocean economy for some time to come.

7. **Offshore Oil and Gas.** Offshore oil and gas exploration and operation pioneered the move of heavy industry out to sea in the form of fixed platforms. It has been the highest value ocean economy sector by far for some decades. People have lived, and heavy equipment has worked, for years on end on artificial structures in deep water often hundreds of kilometres from land. The development of technology and maritime skills in the building, operating and servicing these platforms has been one of the great achievements (or disasters depending on your point of view) of the latter half of the 20th century. However, the offshore oil sector is in decline either because reservoirs are exhausted or due to the additional costs of marine operations compared to terrestrial sources in an oversupplied market for oil. Alternative sources of energy and the need to control emissions have left their mark. The decommissioning of platforms is gathering pace although the industry will exist for several decades to come. Geopolitical forces for energy security and foreign exchange often have as much influence as economics in the energy sectors. The platform technologies and operating skills are transferable to the rising Blue Growth sectors; however, these do not have the same turnover and profit margins and so significant cost cutting is needed to realistically transfer the technology.

8. **Shipping and Shipbuilding.** The shipping and shipbuilding industries are highly cyclical and dependent on global economic activity but, measured by total freight moved around the world, have doubled in capacity

in the last twenty years and are forecast by the OECD to double again in the next twenty. The construction of cruise ships is a significant part of the sector in Europe. The time taken to build large vessels can lead to a glut of certain types of vessel when they eventually come to market. For example, a very high demand for dry bulk carriers at the height of the commodities boom at the beginning of the 21st century created too much capacity which was only ready when the market had already crashed. The sector is expected to continue to grow overall but with large fluctuations in the quantum of freight and its type. The challenge for Europe is to retain the value of this growth in a globalised sector.

9. **Tourism and Recreation.** The draw of the sea and coast for purposes of tourism and recreation appears to be insatiable. The sector combines high value, second only to oil and gas, and high employment, second only to catch fisheries. OECD forecasts anticipate the sector to retain its 26% share of the ocean economy through to 2030 and to increase employment to over eight million people. The sector has engaged with direct sales using internet technologies and this will drive future market changes, as it allows a multitude of businesses to exist at all levels of the supply chain, however, the major multinationals are fully operating in the digital space to retain market share. Constraints are experienced from time to time with the strength of national economies and the disposable income of populations but experience shows the high priority in spending patterns attached by people to their annual holidays. Other constraints relate to environmental pollution and disturbance concerns to nature and other industries. However, it looks set to be the main sector by value and employment in the maritime economy for some decades to come.

The traditional ocean economy sectors are expected to maintain their dominance in the next decades led by tourism, expected to grow, and offshore oil, expected to decline but to remain very large. Shipping will fluctuate with trade volumes but in an era of globalisation is also expected to grow. Catch fisheries remain largely static and relatively insignificant in value but enormously important in terms of employment and subsistence. Of the new Blue Growth sectors, only offshore wind is, so far, set on a strong path of expansion and set to be a major contributor to the ocean economy by 2030. The basic technologies are set and are currently being refined to ever greater cost effectiveness and productive capacity to meet high market demand. The other Blue Growth sectors offer exciting potential but are dependent on transformational technology development or changes in market perspectives. It is likely that niche markets will be the first profitable businesses, for example, wave energy displacing diesel generators on remote islands. Aquaculture is well established and growing but technology developments for offshore aquaculture could rapidly advance the significance of the industry.

The other critical factors in development of the ocean economy are related to ocean governance, planning and management. The evolution of the oceans from being owned by no-one (*res nullius*) to being owned by everyone (*res communis*) to accommodating areas of private property rights is a hard road to follow. Changes to national and individual rights are, we know from experience, bound to be controversial and political, a matter for negotiation. Any restriction to de-facto navigation routes or fishing grounds is often robustly resisted. The offshore wind industry is a good precursor to the introduction of change and the future needs of the ocean economy. With its high demand for marine space in areas relatively close to shore, it focuses attention on the key questions of jurisdiction, ownership, existing rights, distribution of benefits, consenting and marine planning. A world movement towards marine spatial planning (MSP), designed to balance competing claims of industries and the ecosystem, has developed in the early years of the 21st century. It is apparent from the sea basin studies (Chapters 10–13) that the practical implementation of MSP has been led by the development of the offshore wind industry in the North Sea and the Celtic/Irish Seas (included here with the North East Atlantic). It is in these busy areas of activity that MSP is most in evidence in combination with a modernisation of maritime governance and streamlined procedures for the consenting of developments. Without these kinds of development pressures there is little incentive to change the practice of centuries and to implement the right policies that understand the slight difference between MSP and marine protected areas (MPAs).

The final word must go to the concept of multi-use platforms (Chapter 14). The European Union in particular has promoted the investigation of designs and business plans for multi-use platforms (MUPs) at sea. It is thought that the various ocean economy sectors may benefit from the sharing of infrastructure and facilities or might, in the case of renewable energy, enable activities in areas which were otherwise impractical. In addition, the focusing of activities into common platform areas will save the use of marine space and reduce environmental harm. Here again, it is offshore wind which is providing the practical lead in the concept. Active research and trials are underway to combine the wind farm areas with catch fisheries and aquaculture. This is largely in response to the need to offset opposition created by displacing established fisheries from such large areas of sea. It sets the standard. Offshore wind is also setting the standard in community benefit payments to the coastal communities affected by their operations. From the findings of the Maribe project, it was found that many types of combination simply add complexity and hence reduce profitability. However, when the examples described in this book are considered, the right combination, operating in the right niche market, enables the development of technologies that would be less viable when developed alone. Therefore, we conclude that a carefully structured Blue Growth multi-use business can create a profitable operation and provide the jobs and value that Blue Growth aims to provide.

This book has completed what it set out to do. We have provided a detailed analysis of the sectors and we hope that the reader can draw lessons from the "old" sectors that will help to shape the "new" sectors. We have shown that policy and its implementation through market support and marine spatial planning has considerable influence on the Blue Economy. Finally, we hope that the reader has been inspired to search out the right business opportunity to make the next successful steps into Blue Growth.

Index

About the Editors

Dr. Kate Johnson, MSc, PhD, CEng, MICE

Kate graduated as a civil engineer in 1967 and worked for over thirty years in the international construction business holding senior management positions including those of overseas director and managing director. She joined Heriot-Watt University (HWU) in 2009 as a Research Fellow before leading the HWU role in an EU FP7 funded multi-partner project to design and test a generic model for marine spatial planning across Europe. She was made Assistant Professor in 2013 and now conducts research into the emerging ocean economy and teaches at postgraduate level leading MSc programmes in marine planning and development. Projects include the EU H2020 funded MARIBE (Marine Investment for the Blue Economy) project completed in 2016. Kate was awarded her PhD in 2004 by Heriot-Watt University with a thesis entitled 'Perspectives on inshore fisheries - case studies in conflict, participation and cooperation'. She retains a strong international interest with recent work in Ecuador, Vanuatu and Philippines as well as across Europe.

Dr. Gordon Dalton, PhD

Gordon is an Ocean Renewable Energy Economics Engineer, a staff member of MaREI Research centre, UCC, Cork Ireland. His specialty is techno-economics, socio-economics and business plans.

Gordon was the leader of the BG5 MARIBE H2020 project (www.maribe.eu). MARIBE was a Co-ordination and Support Action, spanning 1.5 years, € 2M funds, and 10 international partners, including FAO from the United Nations and BVG associates. This book is motivated by the findings of the project. Gordon is chair of the International Consortium of Research Staff Associations (ICoRSA), consisting of 14 member associations, and a research network of 0.5 million researchers globally. Gordon is the chair of the Irish Research Staff Association (IrishRSA http://www.irishrsa.ie/) and the Vice Chair of the Cork Branch in UCC (www.ucc.ie/en/rsa). As a mature student, he completed a degree in electronic engineering in Trinity College Dublin 2002, and a PhD in Australia in renewable energy economics in 2007. Prior to the pursuit of an engineering career, Gordon had a career in dentistry spanning 12 years.

Professor Ian Masters, BSc, PhD, CMath, MIMA, CEng, MIMechE, SFHEA

Ian is Professor in Mechanical Engineering at Swansea University. He founded the Marine Energy Research Group in 2001 and is author of many academic papers and patents. Ian's research areas include: Tidal current turbines, tidal range technology, wave energy and multiple use of space. In particular, he specialises in the development of computational models to improve the performance of tidal turbines. He also has an interest in the link between design and cost of energy in order to develop cost effective ocean energy devices. Ian is a Senior Fellow of the Higher Education Academy and teaches engineers about sustainable design and energy policy.